博士后文库

中国博士后科学基金资助出版

Unstable Hydraulic Fracture and Stress Disturbance in Heterogeneous Rock

（非均质岩体非稳定水力裂缝与应力扰动）

Wang Yongliang（王永亮）　著

Science Press
Beijing

Responsible Editor: Wang Yun

ISBN 978-7-03-074837-9
Copyright© 2023 by Science Press
Published by Science Press
16 Donghuangchenggen North Street
Beijing 100717, P. R. China
Printed in Beijing

Introduction of the author

Dr. Wang Yongliang is currently a researcher in the Department of Engineering Mechanics, School of Mechanics and Civil Engineering, State Key Laboratory of Coal Resources and Safe Mining, at China University of Mining and Technology (Beijing), and the head of computational mechanics group. He obtained his Ph.D. degree from the Department of Civil Engineering at Tsinghua University in 2014. In 2015, 2016, 2017, and 2019, he successively visited the Zienkiewicz Centre for Computational Engineering at Swansea University, UK, the Applied and Computational Mechanics Center at Cardiff University, UK, and the Rockfield Software Ltd., UK, to carry out cooperative research. In 2022 and 2023, he visited University of California, Berkeley and San Diego, USA, as visiting scholar.

His research interests include high-performance adaptive finite element method, computation and analysis of rock damage and fracture, and structural vibration and stability. He has also taught computational mechanics at the undergraduate level and the basic theory of the finite element method, computational solid mechanics, and rock fracture mechanics, and frontier and progress in mechanics at the graduate level. He is a board member of the Soft Rock Branch of the Chinese Society for Rock Mechanics and Engineering and a member of the Chinese Society of Theoretical and Applied Mechanics, China Civil Engineering Society, and China Coal Society; moreover, he serves as a project expert of the National Natural Science Foundation of China, and project expert of the Degree Center of Ministry of Education of China.

As the person in charge, he presided over eighteen research projects, including the National Natural Science Foundation of China, Beijing Natural Science Foundation, Teaching Reform and Research Projects of Undergraduate Education, the China Postdoctoral Science Foundation, Fundamental Research Funds for the Central Universities, Ministry of Education of China, Key Laboratory Open Project Foundation of Soft Soil Characteristics and Engineering Environment, and Yue Qi Young Scholar Project Foundation, CUMTB. He participated in twelve research projects, including the Major Scientific Research Instrument Development of the National Natural Science Foundation and the National Key Research and Development

Program of China. He published 4 books in English and more than 100 academic papers; furthermore, he obtained more than 50 software copyrights. He has received the Rock Mechanics Education Award, Chinese Society for Rock Mechanics and Engineering (2021), the Excellent Supervisor Award, CUMTB (2019, 2020), the Science and Technology Award, China Coal Industry Association (2019), the Yue Qi Young Scholar Award, CUMTB (2019), the Emerald Literate Highly Commended Paper Award (2018), and the Frontrunner 5000 (F5000) Top Articles in Outstanding S&T Journals of China (2016, 2023).

"博士后文库"编委会

主　任　李静海

副主任　侯建国　李培林　夏文峰

秘书长　邱春雷

编　委（按姓氏笔画排序）

王明政　王复明　王恩东　池　建

吴　军　何基报　何雅玲　沈大立

沈建忠　张　学　张建云　邵　峰

罗文光　房建成　袁亚湘　聂建国

高会军　龚旗煌　谢建新　魏后凯

"博士后文库"序言

　　1985 年，在李政道先生的倡议和邓小平同志的亲自关怀下，我国建立了博士后制度，同时设立了博士后科学基金。30 多年来，在党和国家的高度重视下，在社会各方面的关心和支持下，博士后制度为我国培养了一大批青年高层次创新人才。在这一过程中，博士后科学基金发挥了不可替代的独特作用。

　　博士后科学基金是中国特色博士后制度的重要组成部分，专门用于资助博士后研究人员开展创新探索。博士后科学基金的资助，对正处于独立科研生涯起步阶段的博士后研究人员来说，适逢其时，有利于培养他们独立的科研人格、在选题方面的竞争意识以及负责的精神，是他们独立从事科研工作的"第一桶金"。尽管博士后科学基金资助金额不大，但对博士后青年创新人才的培养和激励作用不可估量。四两拨千斤，博士后科学基金有效地推动了博士后研究人员迅速成长为高水平的研究人才，"小基金发挥了大作用"。

　　在博士后科学基金的资助下，博士后研究人员的优秀学术成果不断涌现。2013 年，为提高博士后科学基金的资助效益，中国博士后科学基金会联合科学出版社开展了博士后优秀学术专著出版资助工作，通过专家评审遴选出优秀的博士后学术著作，收入"博士后文库"，由博士后科学基金资助、科学出版社出版。我们希望，借此打造专属于博士后学术创新的旗舰图书品牌，激励博士后研究人员潜心科研，扎实治学，提升博士后优秀学术成果的社会影响力。

　　2015 年，国务院办公厅印发了《关于改革完善博士后制度的意见》（国办发〔2015〕87 号），将"实施自然科学、人文社会科学优秀博士后论著出版支持计划"作为"十三五"期间博士后工作的重要内容和提升博士后研究人员培养质量的重要手段，这更加凸显了出版资助工作的意义。我相信，我们提供的这个出版资助平台将对博士后研究人员激发创新智慧、凝聚创新力量发挥独特的作用，促使博士后研究人员的创新成果更好地服务于创新驱动发展战略和创新型国家的建设。

　　祝愿广大博士后研究人员在博士后科学基金的资助下早日成长为栋梁之才，为实现中华民族伟大复兴的中国梦做出更大的贡献。

中国博士后科学基金会理事长

Preface

The unstable dynamic propagation of multistage hydrofracturing fractures in heterogeneous rock (e.g. rock embedded with the multi-layers, granules, and pre-existing natural fractures) leads to complex fracture network, and research on the mechanism controlling this phenomenon indicates that the stress disturbance around the fractures is the main mechanism causing this behaviour. Further studies and simulations of the stress disturbance are necessary to understand the controlling mechanism and evaluate the fracturing effect. The stress disturbance, continuum-discontinuum methods and models for stress-dependent unstable dynamic propagation of multiple hydraulic fractures are well summarized and analysed in this book. It can provide a reference for those engaged in the research of unstable dynamic propagation of multiple hydraulic fractures, and have a comprehensive grasp of the research in this field.

The book consists of: (1) dual bilinear cohesive zone model for fluid-driven propagation of multiscale tensile and shear fractures; (2) multi-thread parallel computation method for dynamic propagation of hydraulic fracture networks; (3) heterogeneous continuum-discontinuum computation method for dynamic diversion and penetration of hydraulic fractures contacting multi-layers and granules; (4) dynamic propagation and intersection of hydraulic fractures and pre-existing natural fractures involving the sensitivity factors: orientation, spacing, length, and persistence; (5) unstable propagation of multiple hydraulic fractures and stress shadow effects in multilayered reservoirs; (6) unstable propagation of multiple hydraulic fractures and shear stress disturbance in multi-well hydrofracturing; (7) unstable propagation of multiple three-dimensional hydraulic fractures and shear stress disturbance in heterogeneous reservoirs; (8) unstable propagation of multiple three-dimensional hydraulic fractures and shear stress disturbance considering thermal diffusion.

The author gratefully acknowledges the financial supports from the research projects led by the author, i.e. the National Natural Science Foundation of China (Grant Nos. 41877275 and 51608301), the Beijing Natural Science Foundation (Grant L212016), the China Postdoctoral Science Foundation (Grant Nos. 2018T110158, 2016M601170, and 2015M571030), the Key Laboratory Open Project Foundation of Soft Soil Characteristics and Engineering Environment (Grant No. 2017SCEEKL003), the Fundamental Research Funds for the Central Universities, Ministry of Education of China (Grant No. 2019QL02), the Teaching Reform and Research Projects of Undergraduate Education, CUMTB (Grant Nos. J210613, J200709, and J190701), the Innovation Training Projects for Undergraduates, CUMTB (Grant Nos. 202106001, 202106030, C202006976, and C201906327), and the Yue Qi Young Scholar Project Foundation, CUMTB (Grant No. 190618).

The author gratefully acknowledges the guidance and advice from the respectable tutors during the master, Ph.D., and postdoctorate stages, Prof. Yuan Si and Prof. Zhuang Zhuo of Tsinghua University, and Prof. Wu Jianxun and Prof. Ju Yang of the China University of Mining and Technology (Beijing). During the postdoctorate stage, the author visited several research centres for computational mechanics in famous foreign universities as visiting scholar; the author gratefully acknowledges the advice and comments from the collaborators, Prof. Li Chenfeng, Prof. Feng Yuntian, and Prof. D. Roger J. Owen of the Zienkiewicz Centre for Computational Engineering at Swansea University in the UK, Prof. David Kennedy and Prof. Frederic W. Williams of the Applied and Computational Mechanics Center at Cardiff University in the UK, John Cain, Melanie Armstrong, and Fen Paw at Rockfield Software Ltd. in the UK, and Prof. Robert L. Taylor of the Department of Civil and Environmental Engineering at the University of California, Berkeley, in the USA. The author gratefully acknowledges the participation and work from the graduate students, including Nana Liu, Xuguang Liu, Yifeng Duan, Xin Zhang of the China University of Mining and Technology (Beijing). The author also gratefully acknowledges the editor, Mr Yu Pei, at the Science Press in China, for providing many suggestions and much assistance on formatting modifications and typesetting adjustments for improving this manuscript.

The key contents of this book, such as the multiphysical fields coupling, high-performance continuum-discontinuum methods and models, are crucial in analysing the unstable hydraulic fractures and stress disturbance, which are challenging the researchers' best knowledge. Further work on these fields is needed for both theoretical and algorithm advancements. Because this book is restricted by the limited knowledge of the author, a few errors are unavoidable. The author hopes that all experts, scholars, and other readers of this book will provide helpful suggestions for the book's improvement.

Dr. Wang Yongliang

Department of Engineering Mechanics

School of Mechanics and Civil Engineering

State Key Laboratory of Coal Resources and Safe Mining

China University of Mining and Technology (Beijing)

D11 Xueyuan Road, Beijing, 100083, China

Homepage: www.wangyongliang.net

Email: wangyl@cumtb.edu.cn

January, 2023

Contents

Chapter 1   Introduction

1.1   Research background and significances

Multistage hydrofracturing of horizontal wells is an important technical measure for the development of unconventional reservoirs such as shale gas and tight gas. Therefore, it is necessary to study fracture propagation law in order to obtain effective fracturing effect (Zhang *et al.*, 2016; Ju *et al.*, 2018; Wang *et al.*, 2021a, 2021b). However, the characterisation of perforation cluster spacing and fracturing scenarios of horizontal wells involved in multistage hydrofracturing, which significantly affects the evolution of the stress field and three-dimensional (3D) morphology of the fracture network, is still not well understood and has become the tough challenge (Cai *et al.*, 2017; Lu *et al.*, 2015; Shi *et al.*, 2017; Tang *et al.*, 2016). Figure 1.1 shows a schematic of the unstable dynamic propagation and deflection of multiple hydraulic fractures in a tight reservoir. The cluster spacing between the initial five perforations is identical, that is, equal to an assigned variable a, and the well spacing between the Well 1 and Well 2 is equal to an assigned variable b. The labels of the five perforation clusters to be formed into hydraulic fractures in Figure 1.1 from left to right are 1–5, respectively. For different fracturing scenarios, the fracturing fluid can be implemented for the perforation clusters being injected using the sequence of the perforation clusters. The sequential, alternate, and simultaneous fracturing are implemented by the sequential order 1→2→3→4→5, alternate order 1→3→2→5→4, and simultaneous order 1→2→3→4→5, respectively. The blue area in Figure 1.1 is the fracture surface, and the deflection of fractures in the sequential fracturing gradually increases from perforation clusters 1 to 5. It should be noted that a and b in Figure 1.1 are variables that can significantly affect the propagation behaviours of the fracturing fractures. The fracture propagation and stress field changes during the sequential fracturing are shown in Figure 1.2. The blue plane represents the fracture plane, and the section *xoz* of the fractures is intercepted to reflect the deflection of the fractures. The black line represents the intersection line between the fracture plane and section *xoz*, and the deflection degree of the fracture increases from perforation clusters 1 to 5. The unstable dynamic propagation behaviours of multistage hydrofracturing fracture networks are mainly due to the stress shadow effects, known as induced stress, generated by the fractures formed during multistage hydrofracturing of horizontal wells (Manríquez, 2018; Wang, 2021; Wang *et al.*, 2019). The stress field of section *xoz* is shown in Figure 1.2(b), where the area surrounded by red dotted line represents the variation area of *in-situ* stress caused by fracture propagation and the area surrounded by blue dotted line is the interference area of the stress field between the adjacent fractures. With the propagation of the fractures, the superposition of the stress field occurs around the fractures, resulting in the stress shadow effect, which further affects the unstable dynamic growth of the fracture. As the superposition of the stress field and stress shadow between the fractures gradually increases, the deflection of the fractures also gradually increases.

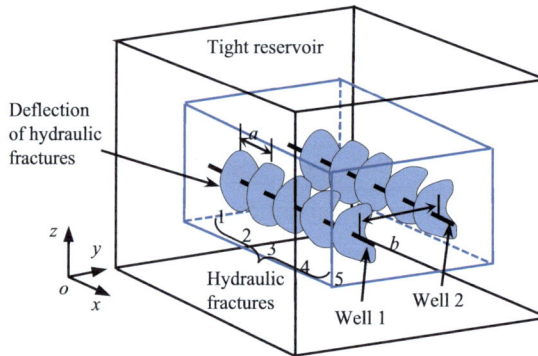

Figure 1.1. Schematic of unstable dynamic propagation and deflection of multiple hydraulic fractures in a tight reservoir.

Under a certain spacing between fractures, the stress shadow effects lead to an increase in the subsequent perforation initiation pressure, increase the difficulty of actual construction, change the original *in-situ* stress field, and change the direction of fracture propagation. The existence of stress interference between fractures was proven for the first time through the analysis of microseismic monitoring data of Barnett shale and from this, the concept of the aforementioned "stress shadow" was proposed (Bunger *et al.*, 2011; Olson and Dahi, 2009). Some scholars have studied the fracture steering mechanism during multistage hydrofracturing and proposed that stress interference between fractures can be used to induce complex fracture systems in the formation (Manchanda and Sharma, 2014; Nagel *et al.*, 2013). Stress shadow effects impose additional stress on the surrounding rock and adjacent fractures, which can reduce the fracture width, change the direction of fracture propagation, and affect the geometry of the fracture network and the placement of proppant in the fracture. Subsequently, some models of multistage hydrofracturing of horizontal wells are developed to study the interference between stress shadows and fractures (Olson and Dahi, 2009).

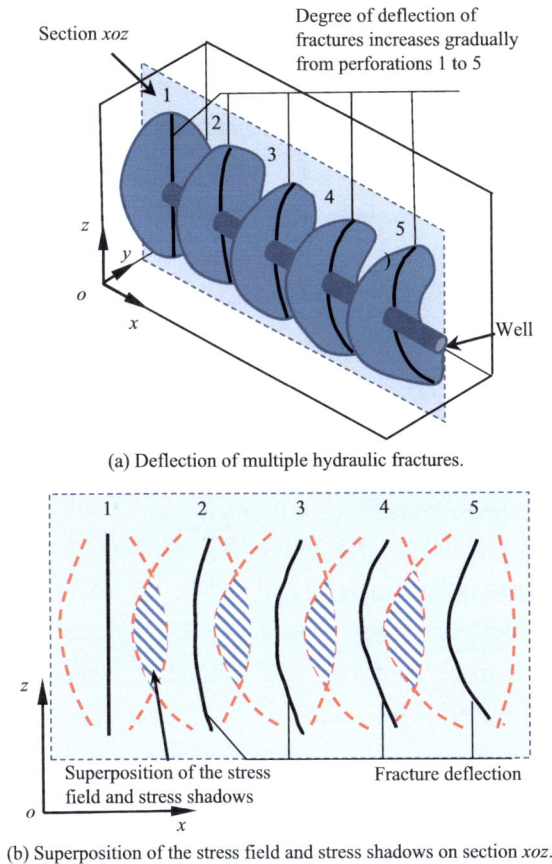

(a) Deflection of multiple hydraulic fractures.

(b) Superposition of the stress field and stress shadows on section *xoz*.

Figure 1.2. Schematic diagram of multiple fracture propagation with sequential fracturing.

At present, some scholars have implemented extensive researches on the simulation of stress-dependent unstable dynamic propagation of fractures and stress shadow effects. However, there are few relevant review articles. Table 1.1 is the list of hydraulic fracturing review articles. Chen *et al.* (2021) reviewed the history and current status of hydraulic fracturing numerical models, highlighting the advantages and disadvantages of different numerical methods. He *et al.* (2016) reviewed two possible methods to generate directional controlled hydraulic fractures in caving mining, pointed out that directional hydraulic fracturing and stress shadow effect could make the propagation trajectory of hydraulic fractures contrary to the direction predicted by theory, and suggested that proppant should be introduced into cave mining to enhance the stress shadow effect. Maulianda *et al.* (2020) summarized the numerical modeling of hydraulic fracturing and elaborated the basic theory of hydraulic fracturing simulation. Sheng *et al.* (2015) reviewed the existing mathematical and computational models for modeling rock formation and hydrodynamic behaviour, as well as some key studies and their basic results. But more research is needed to fully understand the mechanics of hydraulic fracturing. Qian *et al.* (2020) reviewed the research status of laboratory hydraulic fracturing experiments. Some reviews of hydraulic fracturing have pointed to methods of hydraulic fracturing modeling and the addition of proppant enhanced stress shadowing effects (Chen *et al.*, 2021; He *et al.*, 2016; Maulianda *et al.*, 2020; Sheng *et al.*, 2015).

The evaluation behaviours, control and optimization techniques of fracture network of hydraulic fracturing have always been important issues in the tight unconventional oil and gas production of deep reservoirs. In recent years, some scholars have summarized the research problems in this field, which provides a good reference for learners, researchers and engineering practitioners, as shown in Table 1.1:

(1) British scholars (from University of Leeds, UK) (Sheng *et al.*, 2015) review the existing mathematical and computational models for modeling rock formation and hydrodynamic behaviours in multiscale and multiphase hydraulic fracturing process. But more research is needed to fully understand the mechanics of hydraulic fracturing. This summary describes the hydraulic fracturing mechanism and provides an overview of past developments of the research performed towards better understandings of the hydraulic fracturing and its potential impacts, with particular emphasis on the development of modelling techniques and their implementation on the hydraulic fracturing.

(2) Australian scholars (from School of Mining Engineering, Faculty of Engineering, UNSW Australia, Australia) (He *et al.*, 2016) reviewed two possible methods to generate directional controlled hydraulic fractures in caving mining, pointed out that directional hydraulic fracturing and stress shadow effect could make the propagation trajectory of hydraulic fractures contrary to the direction predicted by theory, and suggested that proppant should be introduced into cave mining to enhance the stress shadow effect induced by multiple hydraulic fractures.

(3) Some scholars (from Universiti Teknologi Petronas, Malaysia; University of

Calgary, Canada; Kansas University, USA) (Maulianda *et al.*, 2020) summarized the numerical modeling of hydraulic fracturing and elaborated the basic theory of hydraulic fracturing simulation, and the effort is made to cover the analytical and numerical modelling, while focusing on a typical numerical method for continuum-discontinuum analysis: Extended Finite Element Modelling (XFEM).

(4) Compared with the previous reviews of numerical models and methods, some scholars specially study the propagation behaviour of hydraulic fracturing fractures from the perspective of experimental testing. Chinese scholars (from Hefei University of Technology, China; Hunan University of Science and Technology, China; Central South University, China; Beijing Institute of Technology, China) (Qian *et al.*, 2020) reviewed the research status of laboratory hydraulic fracturing experiments. This chapter presents a review of the state of the art of laboratory-scale hydraulic fracturing experiments, focusing on the scaling analysis, experimental setup, fracturing fluids, and sample preparation. Some discussions of the directions for future researches are also provided with the intention of stimulating the development of the experimental techniques for investigating the fracture propagation behaviours in hydraulic fracturing process.

(5) Computational mechanics experts working with software developers and industry engineers (from Swansea University, UK; Rockfield Global, UK; FracMan Technology Group, UK; Halliburton, USA) (Chen *et al.*, 2021) reviewed the history and current status of numerical models in hydraulic fracturing, highlighting the advantages and disadvantages of different numerical methods for detecting the propagation behaviours of hydraulic fractures.

The existing literature reviews summarize the modeling methods of fracture propagation behaviour in hydrofracturing, and the deflection behaviour and interaction effects of hydraulic fractures are detected. With the deepening of research, researchers become aware of that this behaviour is an important factor affecting the control and optimization of hydraulic fracture network. However, there is no detailed introduction and review of stress shadow effects and induced fracture deflection behaviours. In order to further clarify the mechanism of unsteady propagation of hydraulic fracturing fracture network and the influence of stress shadow effect on fracture propagation, it is necessary to sort out and forecast fracture deflection phenomenon, stress shadow effect and related numerical simulation methods.

Table 1.1. List of review articles for behaviours of hydraulic fractures in hydrofracturing process

Published year	Topic of the review	References
2015	Recent developments in multiscale and multiphase modelling of the hydraulic fracturing process.	(Sheng et al., 2015)
2016	Review of hydraulic fracturing for preconditioning in cave mining.	(He et al., 2016)
2020	Recent comprehensive review for extended finite element method (XFEM) based on hydraulic fracturing models for unconventional hydrocarbon reservoirs.	(Maulianda et al., 2020)
2020	Review of advances in laboratory-scale hydraulic fracturing experiments.	(Qian et al., 2020)
2021	Review of hydraulic fracturing simulation for the advantages and disadvantages of different numerical methods for detecting the propagation behaviours of hydraulic fractures.	(Chen et al., 2021)
2023	Review of stress shadow effects and continuum-discontinuum methods for unstable dynamic propagation of multiple hydraulic fractures.	This chapter

This chapter reviews the research progress on unstable dynamic propagation behaviours of multistage hydrofracturing fracture networks. In this process, there are both continuous and discontinuous stress fields. Therefore, in order to study the stress-dependent unstable dynamic propagation of multistage fracture networks, a series of continuum-discontinuum numerical methods and models are reviewed in this chapter, including the extended finite element method (XFEM) (Lecampion, 2009; Sobhaniaragh et al., 2018), displacement discontinuity method (DDM) (Wu et al., 2012; Yamamoto et al., 2004), boundary element method (BEM) (Ooi et al., 2012; Kumar and Ghassemi, 2016), and finite element (FE)-discrete element (DE) method (FE-DE method or FDEM) (Munjiza et al., 1995; Wang et al., 2021c; Zhang et al., 2015). The main contents of this study are shown in Figure 1.3 and the remainder of the article is presented as follows. Section 2 introduces the unstable dynamic propagation behaviours of multistage hydrofracturing fracture networks, including fracture deflection under initiation sequence and perforation cluster spacing and well spacing, and presents the disturbance of the stress field in the propagation process of multiple fractures. Section 3 presents the mechanisms of induced deflection of fluid-driven fractures (i.e. stress shadow effects between multiple fractures, and controlling factors of stress shadows). Section 4 presents a numerical analysis of the continuous stress field and discontinuous fracture (i.e. continuum-discontinuum numerical methods and models, and simulation of stress-dependent unstable dynamic propagation of fracture under different stress fields). Finally, Section 5 summarises the main conclusions of this study.

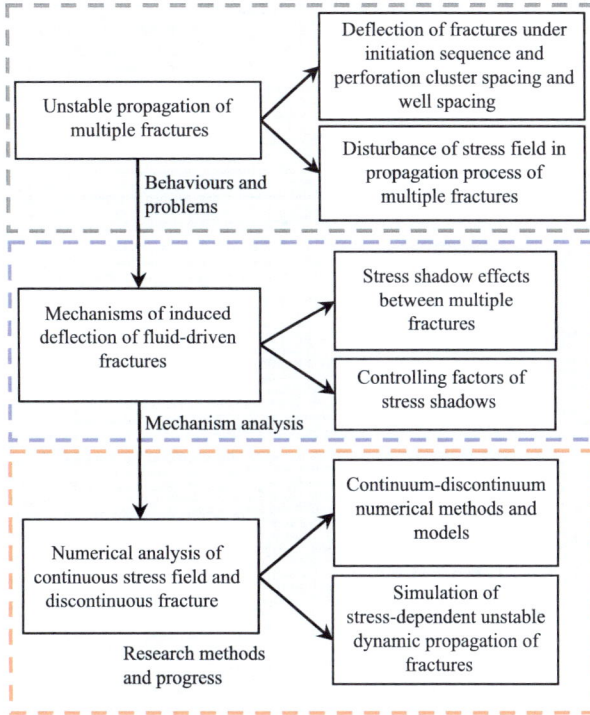

Figure 1.3. Unstable dynamic propagation behaviours of hydrofracturing fracture networks: behaviours and problems, mechanism analysis, and continuum-discontinuum numerical methods and models.

1.2   Unstable dynamic propagation behaviours of the multistage hydrofracturing fracture network

This section introduces the unstable propagation behaviours of the multistage hydrofracturing fracture network. The unstable dynamic propagation behaviours of multistage hydrofracturing fracture networks are illustrated in Figure 1.4, including unstable dynamic propagation of fractures with different initiation sequences, perforation cluster spacing, and well spacing. First, unstable fracture propagation behaviours under different initiation sequences and fracture spacing were analysed, including the deflection phenomenon and behaviours of unstable propagation of multiple fractures. Second, the optimal distances under different perforation cluster spacing and well spacing was introduced. The results show that the deflection of fracture propagation is different with the change in these three factors. Third, the stress interference between fractures in the process of multiple fracture propagation was introduced.

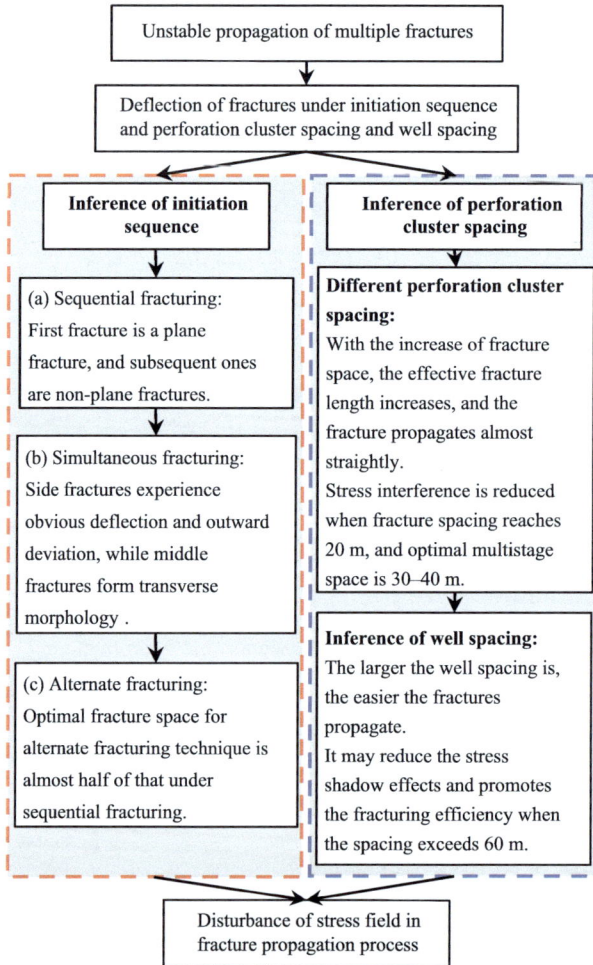

Figure 1.4. Unstable dynamic propagation behaviours of the multistage hydrofracturing fracture network: different initiation sequences and spacings.

1.2.1   Deflection of fractures under the initiation sequence and perforation cluster spacing and well spacing

The unstable propagation of fractures varies with the different initiation sequence and perforation clusters spacing. Therefore, considering the deflection of fractures is one of the factors for choosing a fracturing scenario that is more favourable for increases of the connectivity and permeability of fracture network in reservoir. The phenomenon of unstable dynamic propagation of hydraulic fractures in different fracturing scenarios is summarized in Table 1.2.

Table 1.2. Unstable dynamic propagation of hydraulic fractures in different fracturing scenarios.

Different fracturing scenarios	Unstable dynamic propagation of hydraulic fractures	References
Initiation sequence	(a) **Sequential fracturing:** first fracture is a plane fracture, and subsequent ones are non-plane fractures.	(Cheng *et al.*, 2017a; Damjanac *et al.*, 2018; Sesetty and Ghassemi, 2013)
	(b) **Simultaneous fracturing:** side fractures experience obvious deflection and outward deviation, while middle fractures form transverse morphology.	(Haddad and Sepehmoori, 2016; Li *et al.*, 2019b; Tang *et al.*, 2019a)
	(c) **Alternate fracturing:** optimal fracture space for alternate fracturing technique is almost half of that under sequential fracturing.	(Li *et al.*, 2019b; Wang *et al.*, 2016; Wang and Liu, 2021)
Perforation cluster spacing	(a) With the increase of fracture spacing, the effective fracture length increases, and the fracture propagates almost straightly. (b) The comparing the fracture width of the injection point is always greater than the fracture width of the deflection point with an increase in cluster spacing, and the fracture interaction phenomenon is particularly serious.	(Escobar *et al.*, 2019; Li *et al.*, 2019b; Liu *et al.*, 2020; Tang *et al.*, 2019a)
Well spacing	(a) Some fractures originating from the two wells connect to one major fracture along the direction of the major principal stress. (b) The larger the well spacing is, the easier the fractures propagate, when the hydrofracturing is completed.	(Duan *et al.*, 2021; He *et al.*, 2020)

1.2.1.1   Deflection of fractures under initiation sequence

The experimental results of sequential fracturing are shown in Figure 1.5. Longer and shorter fractures emerge alternately from the bottom to the top of perforations.

Fractures 1 and 3 propagate mainly along a straight plane, while fractures 2 and 4 deviate from the expected direction because of the interference by fractures 1 and 3 (Bunger *et al.*, 2011). By numerical simulation, the final morphology of the 3D fracturing fracture network for sequential fracturing is shown in Figure 1.6 (Wang *et al.*, 2021c). Since the initial stage of fracture propagation, deflections begin to appear in fracture 2, and the deformation of fracture 5 is larger and induces spatial propagation morphology. Numerical analysis of hydraulic fracture propagation under sequential fracturing of a horizontal well was carried out such that, with variations in the space and boundary conditions of the early fractures, the subsequent fractures will bend in or out of the previous fractures. The fracture geometries of subsequent fractures are dependent on the boundary conditions of earlier fractures (i.e. pressurised fracture or propped fracture). The first fracture is a plane fracture, and the subsequent fracture is a non-plane fracture, which is 'repulsed' by the first fracture due to stress interference (Cheng *et al.*, 2017a; Damjanac *et al.*, 2018; Sesetty and Ghassemi, 2013).

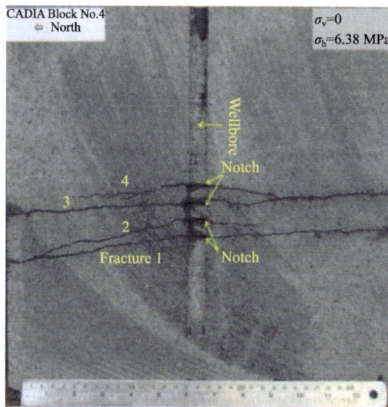

Figure 1.5. Experimental results of fractures propagation under sequential fracturing (Bunger *et al.*, 2011).

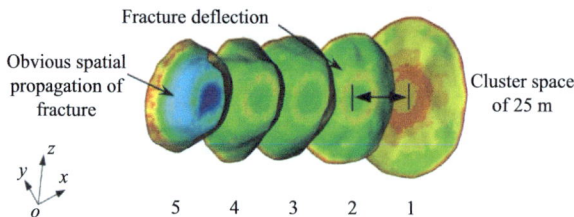

Figure 1.6. Fracture deflection under sequential fracturing (Wang *et al.*, 2021c).

By numerical simulation, the final morphology of the 3D fracturing fracture network for simultaneous fracturing is shown in Figure 1.7 (Wang *et al.*, 2021c). Symmetrically notable deflections of multiple fractures (i.e. fractures 1 and 5, fractures 2 and 4) occur on both the symmetric external positions of the perforation clusters, while the rest of fracture 3 still propagates along the plane. The simulation and analysis of fracture propagation under simultaneous fracturing were carried out, and in the process of simultaneous fracture expansion, the middle fracture can break through and then rapidly move away from the influenced region of the side fractures. Therefore, the middle fracture is much longer than the side fractures. The side fractures are significantly suppressed by the middle fracture and tend to form fractures of a more curved nature, which are shorter and wider. Conversely, as shown in Figure 1.8, the excessive opening of side fractures will result in excessive closure of the middle fracture along the propagation paths of the side fractures. Using simultaneous fracturing in close spacing, the side fractures experience obvious coalescence and outward deflection, while the middle fracture forms a transverse fracture. Simultaneous fracturing leads to overclosure of the middle fracture, which in turn leads to a high proppant injection (Haddad and Sepehrnoori, 2016; Li *et al.*, 2019a; Tang *et al.*, 2019a, b).

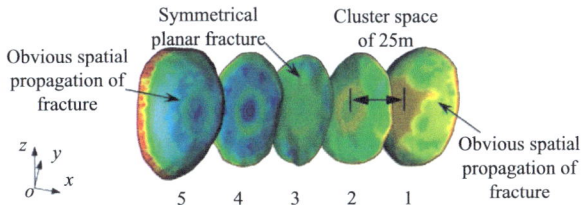

Figure 1.7. Fracture deflection under simultaneous fracturing (Wang *et al.*, 2021c).

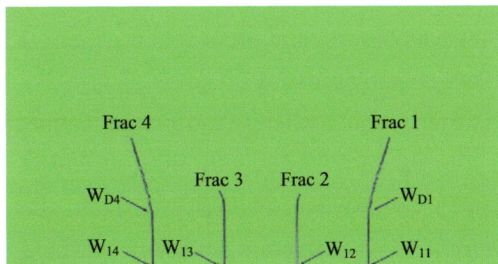

Figure 1.8. Fracture morphology of simultaneous fracturing for four fractures (Tang *et al.*, 2019).

By numerical simulation, the final morphology of the 3D fracturing fracture network for alternate fracturing is shown in Figure 1.9 (Wang *et al.*, 2021c). Fractures 1, 3, and 5 begin to deflect increasingly and produce an agglomerated large deformation, while fractures 2 and 4 propagate in an almost planar manner under the influence of squeeze and compression among fractures 1, 3, and 5. Alternate fracturing techniques reduce the perforation cluster spacing because the middle fracture propagates in the stress-reversal region between the first two fractures. Therefore, the optimal cluster spacing for alternate fracturing is almost half that of sequential fracturing. Alternate fracturing reduces the interference between fractures and leads to more fractures than sequential fracturing (Li *et al.*, 2019b; Wang *et al.*, 2016; Wang and Liu, 2021).

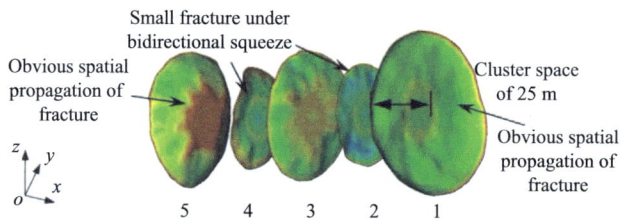

Figure 1.9. Fracture deflection under alternate fracturing (Wang *et al.*, 2021c).

1.2.1.2   Deflection of fractures under perforation cluster spacing and well spacing

The spacing of continuously activated perforation cluster varies with the initiation sequence of clusters. Therefore, the difference in fracture deflection due to the sequence of fracturing is actually caused by the difference in the perforation cluster spacing. Reducing the spacing between clusters is important when designing multistage hydrofracturing to increase production (Cipolla *et al.*, 2009; Wang, 2016). However, there may be too much interaction between dense clusters of fractures, which may lead to an undesirable fracture propagation pattern (Kumar and Ghassemi, 2016). From another research perspective, Pearson *et al.* (1992) studied the problem of minimum perforation cluster spacing when microfracture connectivity was ensured in an inclined borehole. Two micro-fractures with close distance propagate along two routes. If the perforation cluster spacing is sufficiently small, all micro-fractures on the perforation wall will form a 'zigzag' fracture network connected with each other. Yew *et al.* (1989) had also discussed the maximum perforation cluster spacing when adjacent fractures were connected end-to-end, but in that research, it was difficult to compute the result specifically.

A simulation diagram of the fracture morphologies with different cluster spacings under simultaneous fracturing of three hydraulic fractures is shown in Figure 1.10 (Liu *et al.*, 2020). The initial length of the middle fracture is 3.2 m, under a cluster spacing of *a*=2 m. As the cluster spacing increases to 5 m, the length increases to 9.3 m. The propagation of the middle fracture is restrained by the adjacent ones on both sides. Meanwhile, the fractures on left and right sides deviate from each other. The deflection angle decreases when the cluster spacing increases. Fractures show a strong interference, as the transverse fracture becomes shorter and deflects outward; the middle fractures are longer and straighter, which occurs when the perforation cluster spacing is relatively large. With a decrease in the perforation cluster space, the fracture was more prone to deflection, and the degree of deflection was greater. In the case of larger spacing between perforation clusters, the three fractures propagate almost in a straight plane (Li *et al.*, 2019b; Wang and Liu, 2021; Escobar *et al.*, 2019; Liu *et al.*, 2020). Reducing cluster spacing can increase hydraulic fracture complexity and improve horizontal well productivity (Luo *et al.*, 2021).

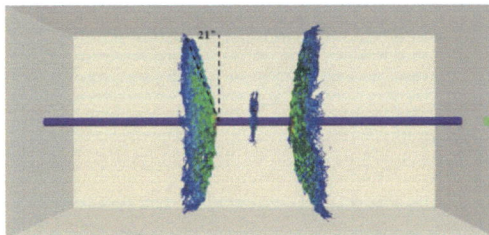

(a) Fracture morphology under a cluster spacing of *a* = 2 m.

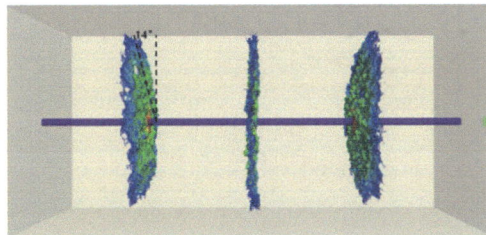

(b) Fracture morphology under a cluster spacing of *a* = 5 m.

Figure 1.10. Fracture morphology of different cluster spacing (Liu *et al.*, 2020).

To detect the fracture width with varying cluster spacing of simultaneous fracturing, the half-length and width of four fractures with different cluster spacings are simulated and shown in Figure 1.11 (Tang *et al.*, 2019a). The cluster spacing has a significant impact on the half-length of fractures 2 and 3. The growth rate of the half-length for fracture 2 was similar to that of fracture 3. In addition to the fracture morphology, comparing the fracture width of the injection point (CWIP) with the fracture width of the deflection point (CWDP) is also an important criterion for optimising the cluster spacing and the number of fractures. The CWDP is always greater than the CWIP with an increase in cluster spacing, and the fracture interaction phenomenon is particularly serious.

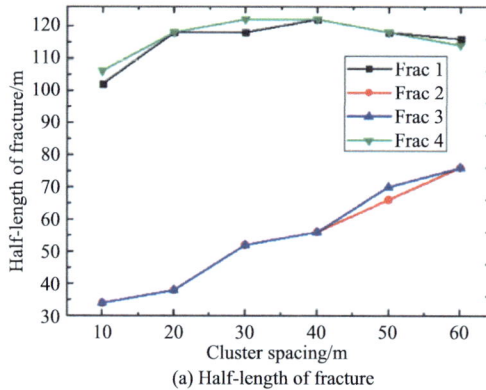

(a) Half-length of fracture

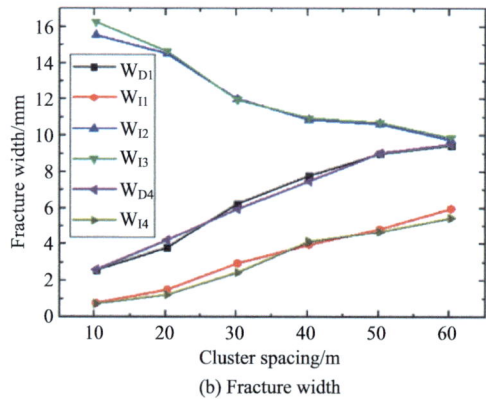

(b) Fracture width

Figure 1.11. Fracture width with varying cluster spacing of simultaneous fracturing for four fractures (Tang *et al.*, 2019a).

To detect the unstable propagation of fractures variation with different well spacing, the 2D and 3D fracture geometries are simulated and shown in Figure 1.12 and Figure 1.13, respectively. Some fractures originating from the two wells connect to one major fracture along the direction of the major principal stress. These dominant fractures hinder the development of other fractures. The larger the well spacing is, the easier the fractures propagate, when the hydrofracturing is completed.

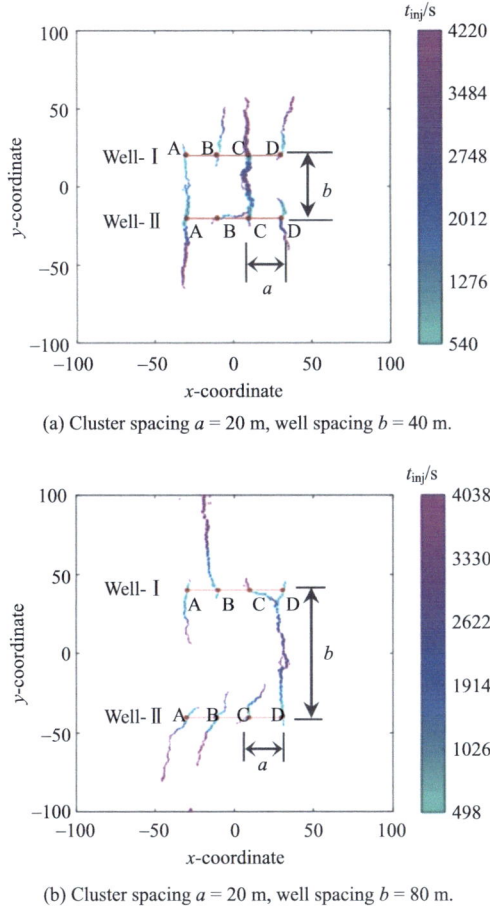

(a) Cluster spacing $a = 20$ m, well spacing $b = 40$ m.

(b) Cluster spacing $a = 20$ m, well spacing $b = 80$ m.

Figure 1.12. Simulation of unstable dynamic propagation of 2D fractures under different well spacings (Duan *et al.*, 2021).

1.2.2   Disturbance of the stress field in the propagation process of multiple fractures

In stress-dependent fracture propagation, the stress field is disturbed, and the stress field between the fractures is superimposed, resulting in mutual interference and deflection between the fractures. The stimulation principle of multistage hydrofracturing mainly consists of the following viewpoints (Liu *et al.*, 2019):

(1) Stress interference occurs between sequential fractures, and early fractures

generate induced stress on the surface of subsequent fractures. The induced stress interference is beneficial for increasing the reconstructed volume of the horizontal shale wells.

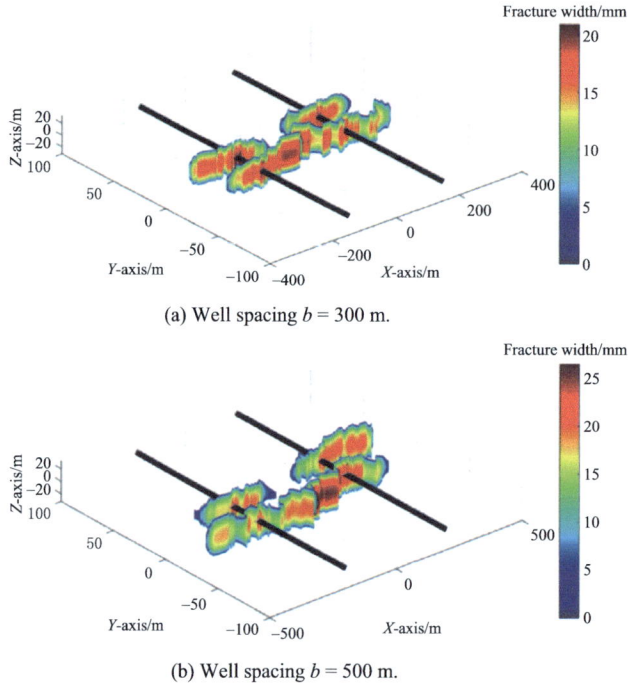

(a) Well spacing $b = 300$ m.

(b) Well spacing $b = 500$ m.

Figure 1.13. 3D fracture geometries with different well spacings (He *et al.*, 2020).

(2) Stress interference can increase the compressive stress on the surfaces of subsequent fractures, thus preventing the formation of parallel fractures closer to the initial fracture and increasing the spacings between subsequent fractures and the initial fractures.

(3) Stress interference causes the fractures to bend along the propagation direction of the length, forming a fracture network of natural and artificial fractures.

1.3   Mechanisms of induced deflection of fluid-driven fractures

Owing to the superposition of the stress field in the process of fracture propagation, the stress shadow effects become a controlling factor for fracture deflection. As shown in Figure 1.14, this section mainly introduces the mechanisms of induced deflection of fluid-driven fractures and involved stress shadow effects. The concept of stress shadow effects, the computation of induced stress, and the controlling factors of the stress shadow effect are summarised. The main factors controlling the stress shadow effects discussed here are the fracture initiation sequence, perforation cluster spacing, and well spacing.

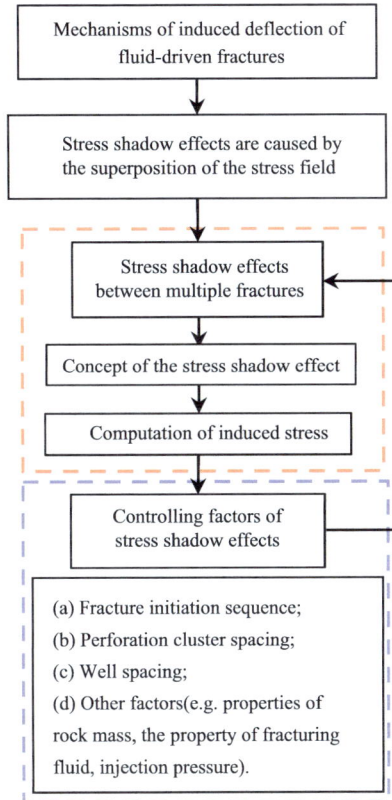

Figure 1.14. The mechanisms of induced deflection of fluid-driven fractures and involved stress shadow effects.

1.3.1   Stress shadow effects between multiple fractures

The stress shadow effects may lead to an increase in subsequent perforation initiation pressure, increase the difficulty of actual construction, change the original *in-situ* stress field, and change the direction of fracture propagation. In multistage hydrofracturing stimulation, stress shadows can affect the fracture geometries: length, height, direction, and aperture (Wu *et al.*, 2017).

Fracture propagation is accompanied with a change in the stress field, and some numerical studies have shown that different fracture initiation sequences produce different stress interferences. Taking the shear stress field as an example, the change in the stress field during fracture propagation under sequential fracturing is shown in Figure 1.15(a). The shear stress areas of each fracture cause similar superposition and reduction, and the fractures begin to propagate at the left side of the larger stress areas, showing notable fracture deflection, because the fracture in the later initiation sequence more easily forms larger stress accumulations and serious deflections. When the perforation cluster propagates at the same time, the external fractures are deflected by the influence of the intermediate fractures. As shown in Figure 1.15(b), the five perforation clusters propagate simultaneously, and the induced shcar stress region between fractures is superimposed and reduced. Fractures 2, 3, and 4 propagate in plane form and form short fractures owing to stress shadow effects. The perforation clusters near the left and right ends of the horizontal wells are disturbed by other internal fractures, which skew the clusters outward and formed a larger fracture area. In alternate fractures, the first two fractures are well developed, and the fractures between the two are affected by them and develop shorter. As shown in Figure 1.15(c), fracture 2 begins to propagate, inducing areas of shear stress variations that are affected by the superposition of stress fields from fractures 1 and 3 on the right and left sides. Owing to the interaction of fractures 1 and 3, the induced shear stress variation region of fracture 2 decreased. The fracture then propagates in a planar form and forms a short fracture zone due to compression.

In order to reflect the variation of stress field, the stress near the fracture under sequential fracturing is defined as a stress contrast coefficient G_n and shown in Figure 1.16. All the subsequent hydraulic fractures are curved except the first hydraulic fracture, because the subsequent hydraulic fracture is 'repelled' by the previous hydraulic fracture because of the stress interaction. Three areas were observed with respect to the stress contrast coefficient G_n, as shown in Figure 1.16. In the area far from the hydraulic fractures, G_n goes to $(G_n)_0$; in the area along the wellbore and among the hydraulic fractures, G_n is less than $(G_n)_0$; in the area around the fracture tips, G_n is greater than $(G_n)_0$. G_n indicates that the stress interaction focuses in the adjacent area of the multiple fractures and vanishes in the far field.

τ_{xy}/Pa

(a) Sequential fracturing.

τ_{xy}/Pa

(b) Simultaneous fracturing.

τ_{xy}/Pa

(c) Alternate fracturing.

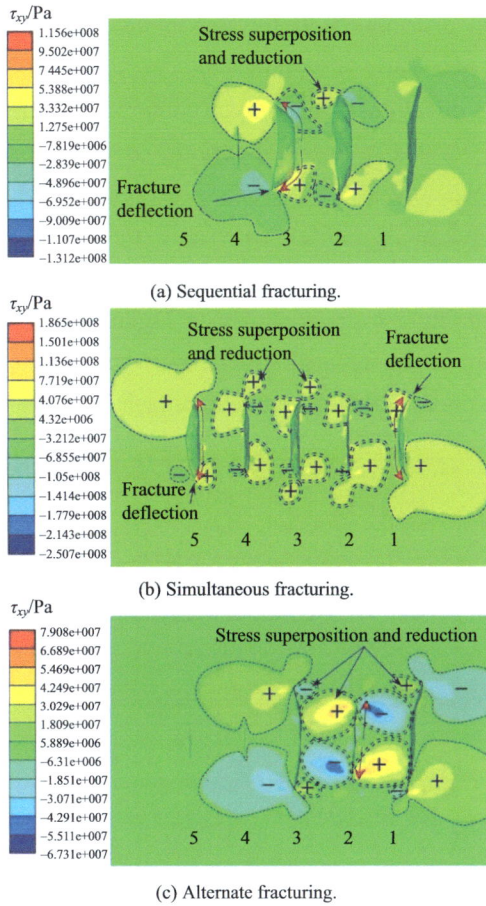

Figure 1.15. The evolution of shear stress in different initiation sequences (note "+" represents positive shear stress and "–" represents negative shear stress) (Wang *et al.*, 2021c).

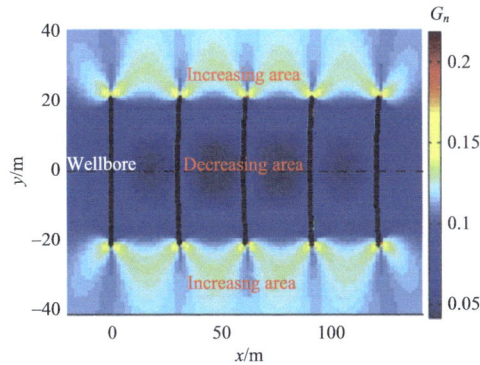

Figure 1.16. Stress contrast coefficient G_n and fracture geometry of sequential fracturing (Cheng *et al.*, 2017a).

The direction and distribution of the maximum and minimum stress between multiple fractures are also derived to reflect the stress shadows. Figure 1.17 shows gradual change of the direction of maximum compressive effective principal stress (σ_3', negative) (Yu et al., 2017). In initial periods, σ_3' is just along the direction of original maximum in-situ stress. The diversion becomes more obvious with further fracture propagation, and an asymmetric pattern of stress reversal region is formed in the reservoir (denoted with blue ellipses in Figure 1.17(b)). The minimum horizontal stress distribution in multiple wells is shown in Figure 1.18 (Kumar and Ghassemi, 2016). Under the action of the induced stress field near the fracture tip, the fractures propagating from each horizontal well tend to approach each other.

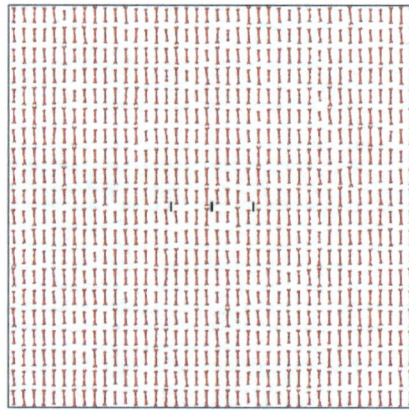

(a) Initial period.

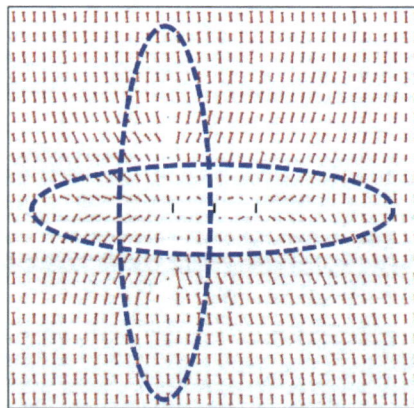

(b) Fracture propagation.

Figure 1.17. The gradual change in the direction of maximum compressive effective principal stress (σ_3', negative) during the injection process (red arrow denotes σ_3' the direction and the length of each arrow is proportional to the local size of σ_3') (Yu et al., 2017).

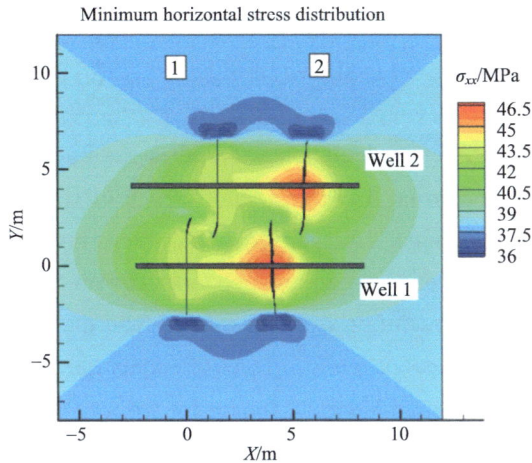

Figure 1.18. Minimum horizontal stress distribution in multiple wells (Kumar and Ghassemi, 2016).

1.3.2   Controlling factors of stress shadows

Investigation on the controlling factors of stress shadow effects is useful for studying the mechanisms of stress interference. The stress shadow effects are affected by many factors, among which initiation sequence, perforation cluster spacing, and well spacing are the main influencing factors (Wang *et al.*, 2015). The study showed that stress interference (or redirection) increases with the number of fractures and depends on the initiation sequence (Roussel and Sharma, 2011a). Spacing and the relative timing of fracture initiation control whether the fractures compete against each other to form a divergent pattern or coalesce into a single primary fracture (Wang, 2016). The cluster spacing that satisfies the requirement of preventing sand plugs and creating a large field of fracture network is optimal. It should be noted that the optimal spacing is not prominently sensitive to varying physical parameters of the rock matrix, such as the Young's modulus (Liu *et al.*, 2016); therefore, these physical factors are not discussed in this analysis.

To detect the controlling factor of fracture initiation sequence, the pore pressure and maximum principal stress distribution around the fracture during different initiation sequences under different cluster spacings are computed and shown in Figure 1.19. Under simultaneous fracturing, as shown in Figure 1.19(a), only the middle fracture follows a direction perpendicular to the minimum horizontal principal stress. The side fractures tend to deviate because of stress interference induced by the middle fracture. The side fractures are suppressed to be short and wide compared with those of the middle fracture. The distributions of stress field for sequential and alternate fracturing at fracture spacing 50 m shown in Figures 1.19(b) and 1.19(c) indicate that, only the second fracture deviated from its desired orthogonal path under alternate fracturing; the extent of fracture deviation under alternate fracturing was smaller than that under sequential fracturing. The stress interference of simultaneous fracturing is greater than that of conventional sublevel sequential fracturing, which is beneficial for improving the complexity of the fracture network. However, the 3D fracture area produced by simultaneous fracturing is much smaller than that produced by sequential fracturing, and the fracture propagation effects of sequential fracturing is better than that of simultaneous fracturing. Alternate fracturing reduces the stress shadow effects by adjusting the order of perforation clusters activated by the injected fracturing fluid.

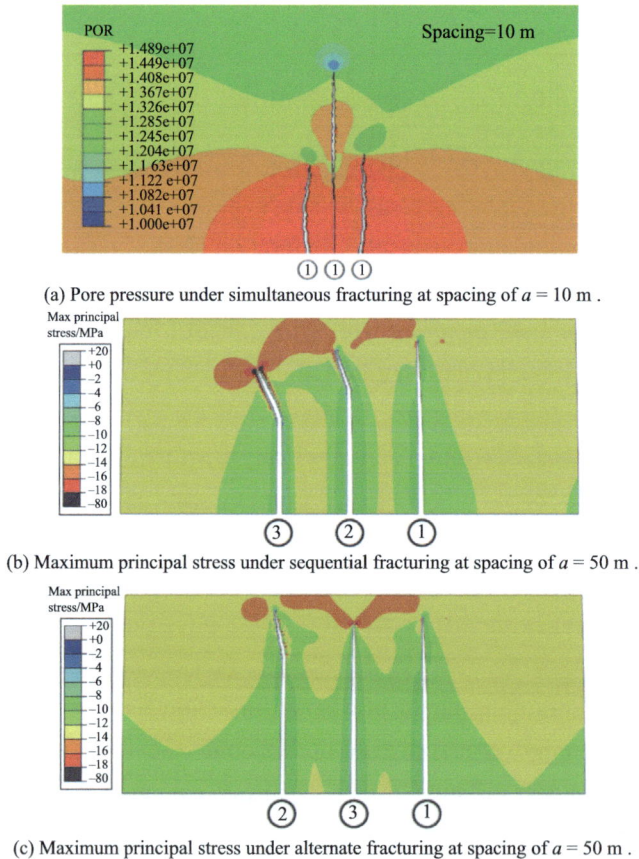

(a) Pore pressure under simultaneous fracturing at spacing of $a = 10$ m .

(b) Maximum principal stress under sequential fracturing at spacing of $a = 50$ m .

(c) Maximum principal stress under alternate fracturing at spacing of $a = 50$ m .

Figure 1.19. Pore pressure and maximum principal stress distribution around the fracture during different initiation sequences under different cluster spacings (Li *et al.*, 2019b; Wang *et al.*, 2016).

Alternate and sequential fracturing have potential advantages in avoiding fracture deflection. Compared to sequential and simultaneous fracturing, alternate fracturing can yield more fracturing fracture areas and improve the fracturing effects under narrow perforation cluster spacing (Roussel and Sharma, 2011a; Kumar and Ghassemi, 2016; Wang *et al.*, 2021c; Wang *et al.*, 2016; Xia and Zeng, 2018).

To detect the controlling factor of perforation cluster spacing, the different perforation cluster spacings cause different degrees of stress interference to fracture propagation, which lead to different degrees of fracture deflection. The effective fracture length increases with the increase in cluster spacing, which shows that even when the cluster spacing is very close, the cluster spacing is still a key factor controlling the fracture morphology. When the cluster spacing is larger, the stress shadow effects decrease, and the fracture propagates almost in a straight line. Thus, the smaller the cluster spacing is, the stronger the stress shadow effects are, which significantly reduces the perforation efficiency (Escobar *et al.*, 2019; Xia and Zeng, 2018; Xie *et al.*, 2019). The optimal cluster spacing of simultaneous fracturing for two fractures is 20–30 m, while the optimal cluster spacing of simultaneous fracturing for three fractures is 30–40 m (Tang *et al.*, 2019a).

To detect the controlling factor of well spacing, the induced shear stress distribution in multiple wells is shown in Figure 1.20 (Kumar and Ghassemi, 2016). It can be observed that the induced relatively high shear stresses in the overlap regions of the two fractures from parallel wells will prevent the fractures from intersecting at their tips. In quantitative analysis for influence of well spacing b, the spacing between wells enhances the efficiency only when the b reaches 40 m. Similarly, enlarging the spacing between injection points does not affect the distribution uniformity index when the spacing between wells is 40 m. However, when b is more than 60 m, the stress shadow effects are effectively alleviated and the fracturing efficiency is improved (Duan *et al.*, 2021).

In addition to the initiation sequence, perforation cluster spacing, and well spacing, the properties of the rock mass and fracturing fluid also control the stress shadow effects. Physical constraints such as subsurface temperatures, pressure, and geological formation fluid complexity may also affect the performance of chemical additives and hydraulic fracturing fluids (Liu *et al.*, 2021; Harry *et al.*, 2020; Reynolds, 2020).

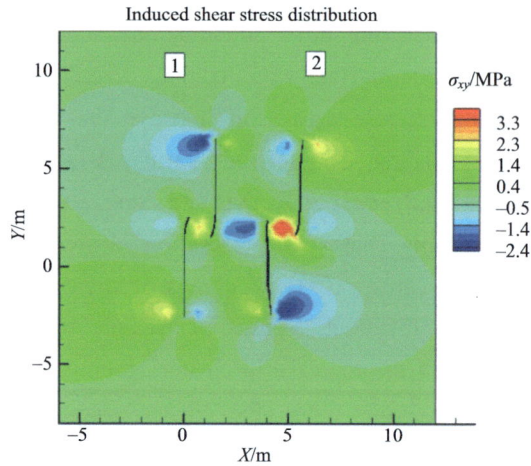

Figure 1.20. Induced shear stress distribution in multiple wells (Kumar and Ghassemi, 2016).

The analysis of influencing factors shows that the Young's modulus and horizontal principal stress difference of rock are important factors affecting the stress shadow of reservoir geological parameters, while the Poisson's ratio of rock has little effect. The size of the stress shadow depends on the shortest distance from the fracture. When the fracture length is greater than the fracture height, the stress shadow is 1.5 times greater than the fracture height. The numerical evaluation results of the stress shadow effects of multiple hydraulic fractures are functions of cluster spacing, mechanical properties of shale, and *in-situ* stress ratio (Cai *et al.*, 2017; Nagel and Sanchez-Nagel, 2011; Zeng *et al.*, 2015). In terms of the interaction between fractures, the spacing between fracture surfaces, net fluid pressure, the *in-situ* stress difference, fracture length, net pressure, and other factors have a certain influence on the range and intensity of stress interference (Manríquez, 2018; Wang *et al.*, 2016; Gai *et al.*, 2020; Rui and Yang, 2020; Sesetty and Ghassemi, 2015a, 2015b; Tarasovs and Ghassemi, 2014; Wong *et al.*, 2013). For the existing rectangular hydraulic fracture, the size of the stress reversal region depends on its short size. Proppant dosage also has a significant effect on the width of propped fracture and the accompanying stress shadows (Roussel and Sharma, 2011a; Olson, 2008; Roussel *et al.*, 2012; Roussel and Sharma, 2011b; Roussel and Sharma, 2012). The ratio of fracture spacing to fracture height and net pressure are the key parameters for determining the deflection of fractures during multistage hydrofracturing. The deflection degree of subsequent hydraulic fractures is expected to increase with the decreasing ratio of fracture spacing to fracture height and increasing net pressure in created hydraulic fractures (Shi *et al.*, 2017; Wang *et al.*, 2016; Gai *et al.*, 2020; Rui and Yang, 2020; Tian *et al.*, 2019). The stress difference has a significant effect on the range of stress disturbance, and the *in-situ* stress difference can counteract the effect of stress disturbance on fracture deflection, reduce the degree of fracture deflection, and affect the length and width of the fracture. The disturbance of the fracture also increased with an increase in the fracture length. In addition, when multiple fractures propagate simultaneously, the stress shadow effects produce additional seepage resistance, resulting in an uneven distribution of seepage velocity. Therefore, the modelling of fracture propagation must deal with the coupling of fluid flow and rock deformation (Wu *et al.*, 2017; Cheng *et al.*, 2017b; Hossain and Rahman, 2008; Wu and Olson, 2016). Each fracture is induced by uneven zoning of the flow, which is dependent on the friction resistance from the wellbore flow, perforation friction, and fracture elongation. To balance the additional resistance caused by the fracture effect, the negative effects of the stress shadow effects can be alleviated by adjusting the number or diameter of perforations and adjusting the perforation friction (Wu *et al.*, 2017; Wu and Olson, 2016).

1.4   Numerical analysis of continuous stress field and discontinuous fracture

To compute and simulate the stress shadows and discontinuous fracture propagation and deflection, some numerical methods are introduced. This section mainly includes continuum-discontinuum numerical methods and models, simulation of stress-dependent unstable propagation of fractures, and fracture deflection analysis determined by the stress field. A logical diagram of the contents is presented in Figure 1.21.

The representative continuum-discontinuum numerical methods mainly include the XFEM, DDM, and BEM and FE-DE method. In this study, the principal stress, shear stress, and *in-situ* stress are derived by these methods to simulate and analyse the stress-dependent unstable dynamic propagation of fractures.

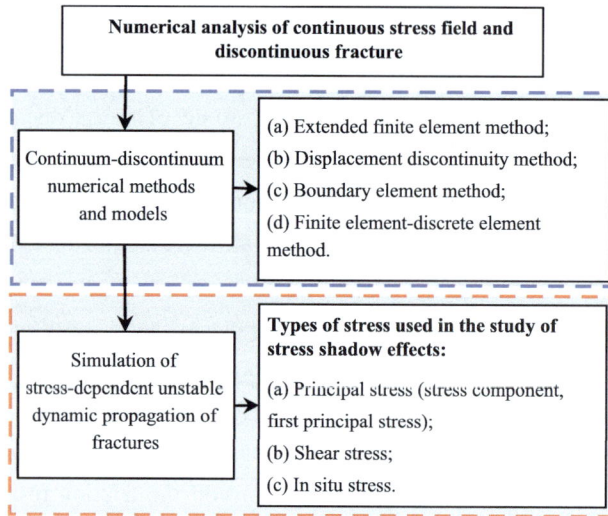

Figure 1.21. Continuum-discontinuum numerical methods and models for computation of the stress field and simulation of discontinuous fracture propagation.

1.4.1   Continuum-discontinuum numerical methods and models

By laboratory experiments, the computed tomography (CT) imaging and nuclear magnetic resonance techniques are used to analyse the behaviour of hydraulic fractures and the dynamic evolution of fractures. In addition, laboratory scale acoustic emission techniques are used to identify microseismic events generated by fracturing to track fracture propagation paths and explain the internal mechanisms of stress shadow effects (Chitrala *et al.*, 2013; Li *et al.*, 2019a; Lu *et al.*, 2020). However, neither field monitoring nor laboratory experiments can accurately obtain the evolution law of the stress field in the process of multiple well hydrofracturing in deep reservoirs, nor can they accurately reveal the internal mechanism of the influence of stress shadow effects on nucleation, propagation, and stress evolution (Vogler *et al.*, 2018).

By theoretical methods, some scholars have proposed relevant analytical models to solve continuum-discontinuum problems, such as the two-dimensional (2D) numerical fracture models, and planar 3D or pseudo-3D models (Sesetty and Ghassemi, 2015a, 2015b; Bunger and Peirce, 2014; Kresse *et al.*, 2013). From the theoretical viewpoints, the resultant stress state near the fracture plane is the superposition of the fracture-induced stress state and the *in-situ* stress state. To consider and compute the stress shadow effects, Sneddon (1946) computed the internal stress distribution around the fracture based on a theoretical plane strain model as shown in Figure 1.22. Furthermore, the induced stresses at a point (ξ, η, ζ) near the fracture is derived as (Yu *et al.*, 2017; Warpinski and Branagan, 1989)

$$\sigma_{\eta\eta,\text{induced}} = P\frac{L}{a}\left(\frac{a^2}{L_1 L_2}\right)^{\frac{3}{2}}\sin\theta\sin\left[\frac{3}{2}(\theta_1+\theta_2)\right] + P\left[\frac{L}{(L_1 L_2)^{\frac{1}{2}}}\cos\left(\theta-\frac{1}{2}\theta_1-\theta_2\right)-1\right],$$

(1.1)

$$\sigma_{\xi\xi,\text{induced}} = -P\frac{L}{a}\left(\frac{a^2}{L_1 L_2}\right)^{\frac{3}{2}}\sin\theta\sin\left[\frac{3}{2}(\theta_1+\theta_2)\right] + P\left[\frac{L}{(L_1 L_2)^{\frac{1}{2}}}\cos\left(\theta-\frac{1}{2}\theta_1-\theta_2\right)-1\right],$$

(1.2)

$$\tau_{\xi\eta} = P\frac{L}{a}\left(\frac{a^2}{L_1 L_2}\right)^{\frac{3}{2}}\sin\theta\sin\left[\frac{3}{2}(\theta_1+\theta_2)\right],$$

(1.3)

$$\sigma_{\zeta\zeta,\text{induced}} = v\left(\sigma_{\eta\eta,\text{induced}} + \sigma_{\xi\xi,\text{induced}}\right),$$

(1.4)

where (ξ, η, ζ) is the Cartesian coordinates, L and θ are linear and angular distances, P is the fluid pressure in the fracture, a is the half length of fracture, and v is Poisson's ratio.

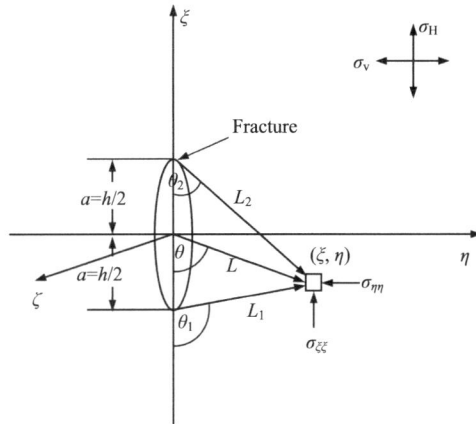

Figure 1.22. Theoretical model of the stress distribution around the fracture (Warpinski and Branagan, 1989).

However, the physical experiments are challenging to observe the evolution process of the stress field and dynamic propagation of fractures. The theoretical solutions are defective in accuracy because the models simplify the complex conditions in actual engineering. Therefore, some numerical methods and models have been proposed as powerful alternatives to do continuum-discontinuum analysis. The advantages and limitations of representative numerical methods for simulating the hydraulic fractures is shown in Table 1.3.

Table 1.3. Advantages and limitations of representative numerical methods for simulating the hydraulic fractures.

Representative methods	Advantages	Limitations	References
Extended finite element method (XFEM)	(a) Enriched functions were introduced to simulate the discontinues displacement around fractures. (b) Simulate fractures of any shape avoiding the global mesh updating.	(a) It is challenging to obtain the non-plane 3D hydraulic fracture. (b) It is challenging for handling multiphysical fields coupling.	(Lu *et al.*, 2015; Wang *et al.*, 2015; Castonguay *et al.*, 2013; Gao *et al.*, 2018; Gupta and Duarte, 2014; Wang *et al.*, 2021d)
Displacement discontinuity method (DDM)	(a) The behaviours (such as stress distribution and variation, fracture geometry and width) can be simulated. (b) Low computation cost by reducing the computational dimension.	(a) It is challenging to solve nonlinear problems. (b) It is challenging for handling multiphysical fields coupling.	(Lu *et al.*, 2015; Tang *et al.*, 2016; Tang *et al.*, 2019b; Zeng *et al.*, 2019)
Boundary element method (BEM)	(a) It can effectively simulate the singular stress field near the fracture. (b) Low computation cost by reducing the computational dimension.	(a) It is challenging to describe the elastoplastic or large deformation during fracture. (b) It is challenging to solve for large scale and number of degrees of freedom.	(Shi *et al.*, 2017; Ooi *et al.*, 2012; Kumar and Ghassemi, 2016; Crouch and Starfield, 1983)
Finite element-discrete element method (FE-DE method)	(a) Describe continuous and discontinuous fields. (b) Overcome the defects of the traditional finite element method in the simulation of fracture propagation.	(a) Large computation cost. (b) Fractures can only propagate along the edge of elements.	(Wang, 2021; Wang *et al.*, 2019; Wang *et al.*, 2021c; Ju *et al.*, 2020; Wang, 2020b)

1.4.1.1   Extended finite element method

The cohesive zone model (CZM) based on the finite element method (FEM) has been used to simulate fracture propagation, in which the inelastic behaviour before the fracture tip is described (Sobhaniaragh *et al.*, 2018). Furthermore, fractures are typically modelled as predefined interfaces in fixed directions between adjacent finite elements, and the stress fields and interactions between multiple hydraulic fractures are analysed (Salimzadeh *et al.*, 2017; Zhang, 2010). Simple plane fracture models are used to simulate multiple fractures in multiple wells using this proposed method. However, simulating the spatial propagation behaviours of 3D fractures using such simple models is difficult (Dverstorp and Andersson, 1989; Warpinski, 2000).

By introducing an extended shape function with discontinuous properties, the XFEM is developed from conventional FEM to simulate the 3D propagation of multiple fractures. Based on the method and model of Rungamornrat *et al.* (2005) it was found that the expansion of weakly singular symmetric Galerkin boundary elements can simultaneously realise the propagation of multiple 3D fractures in a single horizontal well. The propagation process of multiple hydraulic fractures was simulated using a 3D finite element model (Castonguay *et al.*, 2013; Gao *et al.*, 2018; Gupta and Duarte, 2014). In addition, the shale gas reservoir in the fracture propagation is a moving boundary problem, which makes the application of traditional finite element simulation more complicated and tedious (Guo *et al.*, 2015). The development of the XFEM solved this problem satisfactorily. In the XFEM simulation, the discontinuous displacement field is described by expanding the displacement term to make the fracture exist independently of the mesh, such that fractures of any shape can be simulated. The fracture path is independent of mesh generation; thus tedious updates of the global mesh are not required. At each time step, the geometry of the fracture can be recomputed while the global mesh remains unchanged (Belytschko and Black, 1999; Moës *et al.*, 1999).

Since the introduction of the XFEM, many authors have used the XFEM to simulate non-plane hydraulic fracture propagation (Lecampion, 2009; Dahi-Taleghani and Olson, 2011; Gordeliy and Peirce, 2013; Leonhart and Meschke, 2011). For example, an optimised perforation cluster spacing in multiple fracture model considering stress interference is established based on the XFEM, and a 2D fully coupled pore pressure-stress model based on the XFEM combined with the CZM was used to simulate multistage hydrofracturing of horizontal wells. Based on the XFEM, the function computation domain related to the stress field is introduced, and the boundary value problem of the function is solved to obtain all the envelopments along the propagation direction. The possible surface can be selected as the development direction of the fracture, which can reduce the complexity of fracture propagation trajectory planning and simulate multiple fractures (Lu *et al.*, 2015; Wang *et al.*, 2015; Wang *et al.*, 2021d).

1.4.1.2 Displacement discontinuity method

The fracture propagation is discontinuous. Therefore, many scholars have studied the fracture-induced stress field using the DDM and established the corresponding numerical model when solving the fracture propagation problem (Wu *et al.*, 2012; Yamamoto *et al.*, 2004). The solutions of stresses in the DDM can be derived as

$$\sigma_x = 2GD_x[2f_{,xy} + yf_{,xy}] + 2GD_y[f_{,yy} + yf_{,yyy}], \tag{1.5}$$

$$\sigma_y = 2GD_x[-yf_{,xyy}] + 2GD_y[f_{,yy} - yf_{,yyy}], \tag{1.6}$$

$$\tau_{xy} = 2GD_x[2f_{,yy} + yf_{,yyy}] + 2GD_y[-yf_{,xyy}], \tag{1.7}$$

where σ_x and σ_y are the normal stresses in the x- and y- directions, respectively, τ_{xy} is the shear stress, the shear modulus is denoted as G, f is the pre-defined analytic function, and D is the displacement discontinuity variable. The normal stress and shear stress of element i on fracture can be obtained using the displacement discontinuous quantity of element j ($=1, 2, \cdots, N$) from the following equations (Cheng, 2009; Crouch, 1976):

$$\sigma_x^i = \sum_{j=1}^{N} A_{xx}^{i,j} D_x^j + \sum_{j=1}^{N} A_{xy}^{i,j} D_y^j, \tag{1.8}$$

$$\sigma_y^i = \sum_{j=1}^{N} A_{yx}^{i,j} D_x^j + \sum_{j=1}^{N} A_{yy}^{i,j} D_y^j, \tag{1.9}$$

$$\tau_{xy}^i = \sum_{j=1}^{N} A_{yx}^{i,j} D_x^j + \sum_{j=1}^{N} A_{xy}^{i,j} D_y^j, \tag{1.10}$$

where A is the boundary influence coefficient for the stresses. The stress field derived by equations (1.5)-(1.10) using the DDM is shown in Figure 1.23. As can be seen, the effect of induced stress on the maximum horizontal stress considering two fractures is mainly confined in the vicinity of the fractures, with a relatively small magnitude.

Based on the DDM, a mathematical model of multi-cluster fracture propagation can be established, and the stress interference model of multi-fracture simulating non-equidistant half-length, unequal cluster spacing, and the angle between the fracture and wellbore can be established. The behaviours, such as the stress distribution and variation, fracture geometry and width, can be simulated using the boundary element

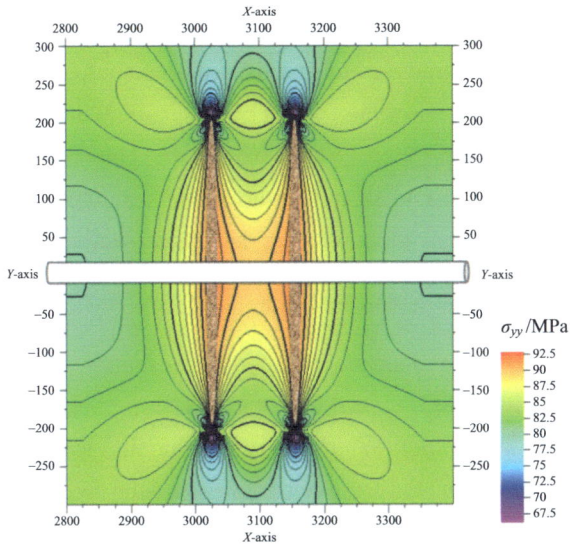

Figure 1.23. Maximum horizontal stress distribution around the wellbore after fracturing (Lu *et al.*, 2015).

model based on the DDM. The height of the finite fracture can be corrected by the enhanced 2D DDM, and the results of this model can be closer to the 3D simulation results. However, as the research on fractures is gradually transformed from 2D to 3D, the 2D DDM has been unable to meet the research needs of the fracture-induced stress field. Therefore, a 3D DDM was developed, and a multi-fracture propagation model was established based on a planar 3D DDM. To better analyse and understand the stress interference between multiple hydraulic fractures, a full 3D model based on the DDM was developed (Lu *et al.*, 2015; Tang *et al.*, 2016, 2019b; Zeng *et al.*, 2019). Optimisation strategies for the 3D model are also proposed, including adaptive mesh growth, uncrowded Gaussian points, and distance-dependent integrals. The stress redistribution induced by fracture interference in 3D spacing computed by the local method provides a more complex shape of the potential stress redirection region than the 2D results.

1.4.1.3   Boundary element method

A numerical model of fracture propagation can be established by combining the BEM and FEM models and linear elastic fracture mechanics (Ooi *et al.*, 2012; Kumar and Ghassemi, 2016; Crouch and Starfield, 1983). Based on the scale boundary finite element method (SBFEM), each polygon is considered as a subdomain, which can effectively simulate the singular stress field near the fracture. A fully coupled 3D BEM can simulate any number of fractures during the simultaneous or continuous expansion of a single or multiple horizontal wells. The model can simulate the hydraulic fractures of an elastic reservoir and a pore elastic reservoir. The BEM of induced stress interaction can be used to establish numerical models of induced stress analysis in multistage fracture propagation and multistage horizontal wells (Shi *et al.*, 2017).

1.4.1.4   Finite element-discrete element method

In general, the study of stress-dependent fracture propagation involves the simulation and computation of continuous stress fields and discontinuous fractures. In this case, both the continuous stress field and the discontinuous fracture field should be considered, and the FE-DE method can better combine the two fields (Munjiza *et al.*, 1995; Zhang *et al.*, 2015).

In the hydraulic fracture propagation model established by this method, the influences of the horizontal stress difference, number of perforation clusters, and spacing between fractures can be considered simultaneously. In order to ensure the precision of solution and the reliability of the fracture propagation paths and computational efficiency, high-performance adaptive refinement method becomes a good alternative (Wang, 2020a; Wang *et al.*, 2018); an adaptive FE-DE method and model have been proposed (Wang, 2020b, 2021; Wang *et al.*, 2019, 2021c; Ju *et al.*, 2020), used to study the stress shadows of multistage hydrofracturing effect and fracture behaviour of interaction. This method uses the local mesh refinement and coarsening strategy, to overcome the traditional FEM on the shortcomings in the 3D fracture propagation modelling. From this, the adaptive FE-DE method considering the fluid-structure coupling and seepage effect is extended to the 3D continuous simultaneous hydrofracturing analysis of multiple vertical wells in tight reservoirs, and a 3D numerical model is established, which can be used to study the influence of initiation sequence and well spacing on the stress shadow effects (Wang *et al.*, 2021c; Zhang *et al.*, 2015; Ju *et al.*, 2020).

1.4.2   Simulation of stress-dependent unstable dynamic propagation of fractures

The stress shadow effects can promote fracture propagation and form more complex fracture networks. In contrast, the stress shadow effects can also inhibit the growth of local fractures, or even prevent the formation of fractures (Manríquez, 2018; Sobhaniaragh *et al.*, 2018; Warpinski, 2000; Bai *et al.*, 2000). In the study of the influence of the stress shadow effects on the unstable propagation of fractures, the stress disturbance in the process of fracture propagation can be reflected by different types of stress fields. The typical simulation results of stress fields in unstable dynamic propagation of hydraulic fractures are shown in Table 1.4.

Table 1.4. Typical simulation results of stress fields in unstable dynamic propagation of hydraulic fractures.

Type of stress field	Typical results	References
Principle stress	(a) The first fracture is not affected by other fractures and the disturbance stress field generated around it makes the second fracture deflect. The second fracture gradually increases and eventually changes to the direction of maximum compressive stress. (b) In sequential fracturing, the stress interference increases with the increase of fracture number. Under same conditions, the deflection degree of alternately fractured fractures is smaller than that of sequentially fractured fractures, indicating that stress interference also depends on the initiation sequence.	(Sesetty and Ghassemi, 2015a; Wang *et al.*, 2016)
Shear stress	(a) Under the same stress conditions, the positive shear stress region to the left of the upper tip of the second fracture negatively interferes with the region of negative shear stress to the right of the upper tip of the first fracture. (b) The superposition of the positive and negative shear stress zones on the left and right sides of the fracture tip leads to the unbalance of the stress field on the left and right sides of the fracture tip, which leads to the fracture deflection.	(Wang *et al.*, 2021c; Hudson, 2017)
In situ stress	(a) If a fracture is initiated just outside the stress-reversal region, it will propagate away from the previous fracture, following the direction of the maximum horizontal stress. (b) As the fracture expands, the local *in-situ* stress around it changes, and the direction of the minimum horizontal stress reverses as it approaches the fracture surface. The hydraulic fracture extends into a "stress change zone" caused by another fracture, which may change direction to extend in a new direction perpendicular to the minimum horizontal stress. (c) Both minimum horizontal stress and maximum horizontal stress are affected by induced stress, and the maximum horizontal stress is much less disturbed by induced stress than that of minimum horizontal stress. The induced stress produces the maximum compressive stress at both sides of the fracture and the maximum tensile stress at the tip of the fracture due to the stress concentration. (d) As the spacing between fractures increases, the intensity of the stress shadow region decreases, which results in less curving of the fracture surfaces.	(Lu *et al.*, 2015; Manríquez, 2018; Bunger *et al.*, 2011; Kumar and Ghassemi, 2016; Wang *et al.*, 2016; Roussel and Sharma, 2011b; Bunger *et al.*, 2012; He *et al.*, 2017; Kear *et al.*, 2013)

1.4.2.1   Disturbance of principal stress

The improved maximum circumferential tensile stress criterion controls the initiation Angle and deflection of hydraulic fractures. According to the maximum principal stress of each element around the fracture tip, the elements near the fracture tip are affected by asymmetric tensile stress caused by reservoir anisotropy and perforation direction, which causes the hydraulic fracture to deflect (Zou *et al.*, 2018). Soliman *et al.* (2008) presents the results of Sneddon (1946) in solving the principal stress state around semi-infinite fracture. The ratio of principal stress of the two fractures indicates that the optimum spacing is 100 or 150 m. The simulation results of maximum principal compressive stress distribution show that the first fracture is not affected by other fractures under sequential fracturing, and the induced stress region around the first fracture changes the distribution of *in-situ* stress. The second fracture is deflected by the disturbed stress field generated by the first fracture. The second fracture gradually increases and eventually changes in the direction of maximum compressive stress (Sesetty and Ghassemi, 2015a). By simulating the distribution of the maximum principal stress, the comparison of the distribution of the maximum principal stress in different *in-situ* stress differences shows that the deviation between the fracture and the expected orthogonal path becomes more obvious with the decrease of *in-situ* stress differences. In sequential fracturing, the displacement of the third fracture is more obvious than that of the second fracture, indicating that the stress interference increases with the increase of the number of fractures. In addition, under the condition of equal *in-situ* stress difference and equal spacing, the migration degree of alternately fractured fractures is smaller than that of sequentially fractured fractures, indicating that stress interference also depends on the initiation sequence (Wang *et al.*, 2016).

1.4.2.2   Disturbance of shear stress

In order to analyse the superposition behaviour between stress fields, the positive and negative regions of shear stress around the fracture tip are used. According to the numerical analysis in the study of Soliman *et al.* (2008) the variation in the shear stress field has also been simulated. As shown in the negative and positive shear stress figure, the shear stress is larger around the tips of the fracture, and taking the upper tip as an example, the negative shear stress area is largely confined to the area propagating towards the right. Conversely, a positive shear stress propagation area forms to the left. Under the same stress conditions, the positive shear stress area to the left of the upper tip of the second fracture negatively interferes with the area of negative shear stress to the right of the upper tip of the first fracture (Hudson, 2017).

By simulating the fracture propagation and shear stress evolution in different fracturing scenarios and deriving the stress field of multistage fracturing models, it is shown that the fracture morphology and stress field are completely different in each stage of fracture propagation, especially in the later stage of fracture propagation. According to the fracture stress results of the first stage of sequential fracturing, the left side of the upper fracture tip is positive shear stress, and the right side is negative shear stress. The left side of the lower fracture is negative shear stress and the right side is negative shear stress. The superposition of these positive and negative areas of shear stress leads to the imbalance of stress field on the left and right sides of the fracture tip, resulting in behaviours of fracture deflection (Wang *et al.*, 2021c).

1.4.2.3   Disturbance of *in-situ* stress

Based on the physical property parameters of the actual tight sandstone reservoir, a multi-fracture stress interference model was established, and the variation law of *in-situ* stress was analysed. It was found that the minimum horizontal stress was greatly affected by induced stress. The maximum horizontal stress is much less affected by the induced stress than the minimum horizontal stress. Stress changes at the fracture tip indicate local elastic stress concentration. Due to stress concentration, induced stress produces maximum compressive stress at both sides of the fracture and maximum tensile stress at the tip of the fracture (Lu *et al.*, 2015). The distributions of changes in the minimum horizontal stresses in the *x-y* plane are shown in Figure 1.24 for two cases with 3 m and 6 m cluster spacing. It can be observed from Figures 1.24(a) and 1.24(b) that as the spacing between fractures increases, the intensity of the stress shadow region decreases, which results in less curving of the fracture surfaces (Kumar and Ghassemi, 2016).

The minimum cluster spacing and interaction between fractures can be studied by simulating the direction change of the minimum horizontal stress, maximum horizontal stress and angle of stress reorientation around the fracture. The minimum fracture spacing can be defined as the distance between the fracture and the end of the stress-reversal region, also known as the isotropic point. The results show that the direction of the minimum horizontal stress is initially along the x-direction; however, as the fracture propagates, it alters the local *in-situ* stresses around it, and the direction of the minimum horizontal stress can be reversed when close to the fracture surface. If a fracture is initiated just outside the stress-reversal region, it will propagate away from the previous fracture, following the direction of the maximum horizontal stress. Once a hydraulic fracture propagates into a 'stress change zone' caused by another fracture, this fracture may be redirected to propagate in a new direction perpendicular to the minimum horizontal stress. Hence, the tensile stress around the tip of the newly formed fracture may cause the original fracture to reopen and enhance the stress shadow effects of the original fracture, thus promoting the hydraulic fracture reorientation. Local stress reorientation can lead to subsequent fracture deflection and restrain

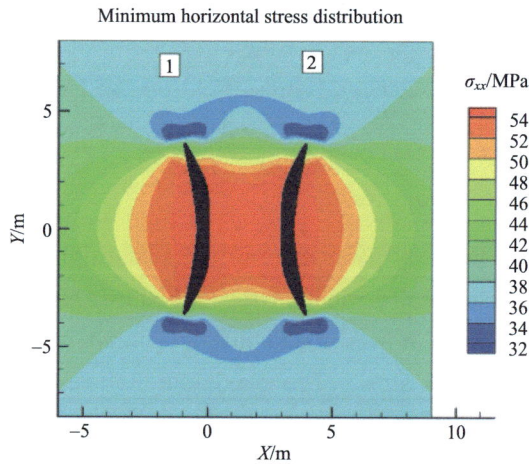

(a) Perforation cluster spacing, $a = 3$ m.

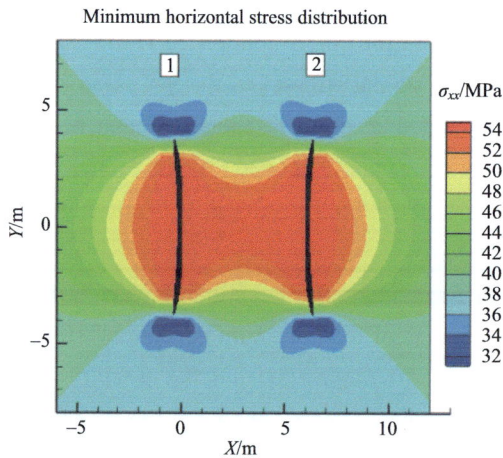

(b) Perforation cluster spacing, $a = 6$ m.

Figure 1.24. Distributions of fracture opening, fracture geometries, and minimum horizontal stresses (Kumar and Ghassemi, 2016).

straight growth of fracture. Conversely, the fracture reorientation decreased as the cluster spacing increased (Manríquez, 2018; Bunger *et al.*, 2011; Wang *et al.*, 2016; Roussel and Sharma, 2011b; Bunger *et al.*, 2012; He *et al.*, 2017; Kear *et al.*, 2013). In analogous research, the XFEM was implemented in a multi-physics framework called GeMA (Escobar *et al.*, 2019). For that, the distribution of horizontal *in-situ* stress under two types of perforation cluster spacing is used to show the stress shadow effects between fractures, so as to infer the suitable perforation cluster spacing.

1.5   Challenges and perspectives

(1) Based on the fracture propagation under different initiation sequences and perforation cluster spacing, the phenomenon of fracture propagation under the influence of different factors and the optimal perforation cluster spacing and well spacing is summarised. Under simultaneous fracturing, the excessive opening of the side fractures results in excessive closure of the middle fracture along the propagation paths of the side fractures. The alternate fracturing technique can reduce the perforation cluster spacing and reduce stress shadow interference, resulting in more lateral fractures than sequential fracturing. Compared to simultaneous fracturing, alternate and sequential fracturing have potential advantages in improving the performance. Fracturing scenario has become an important factor affecting the deflection behaviours of multiple hydraulic fractures.

(2) The stress shadow effects affecting fracture deflection is introduced, and the different factors controlling the stress shadows are summarised. Formed hydraulic fractures can produce stress shadow effects during the fracturing of horizontal wells. Under certain conditions and within a certain spacing between fractures, the stress shadow effects lead to an increase in the subsequent perforation initiation pressure, increases the difficulty of actual construction, changes the original *in-situ* stress field, and changes the direction of fracture propagation. Stress shadow effects are related to the fracture initiation sequence, perforation cluster spacing, well spacing, and other factors (e.g. the properties of rock mass, the property of fracturing fluid, injection pressure).

(3) The types of stress fields and the related models and methods used in the study of the stress shadow effects are summarised and analysed. The main types of stress fields used to study the mutual disturbance of stress and stress shadow effects during fracture propagation are: principle stress, shear stress, and *in-situ* stress. Most of the

methods and models used to simulate stress-dependent unstable dynamic propagation of fractures are numerical methods. In the study of the influence of the stress shadow effects on fracture propagation morphology, the most commonly used simulation methods are the XFEM, BEM, and DDM and FE-DE method. Especially, the FE-DE method has the potential to effectively simulate the continuous stress evolution and discontinuous fracturing fracture propagation. The developed XFEM simulates multiple hydraulic fractures and improves fracture propagation through elements, but the accuracy of the computed results still depends on the quality of mesh generation. The mesh can be refined adaptively and can effectively describe continuous and discontinuous fields, which overcomes the defects of traditional FEM in simulating propagation of multiple hydraulic fractures.

(4) Regarding the prospect, the most important aspect of studying the stress-dependent unstable dynamic propagation behaviour of multistage hydrofracturing fracture networks depends on accurately evaluating stress field and propagation of multiple hydraulic fractures. In future studies, high-performance continuum-discontinuum numerical methods and models should be developed to provide accurate fracture deflection behaviours and stress field evolution results that are consistent with those in practical engineering. Furthermore, the aims of evaluating the unstable dynamic propagation and optimizing the morphology of hydrofracturing fracture networks may be reliably achieved. Optimization of multiple wells hydrofracturing is the key to develop complex fracture network and promote the migration of unconventional oil and gas resources in tight reservoirs. Stress shadow effect between multiple wells has an important influence on the development of fracture network (Kear *et al.*, 2013). The previous study shows that the stress shadow effect between wells and the mutual interference between fractures are different under different well spacing and fracturing sequence between wells (Duan *et al.*, 2021; He *et al.*, 2020). It is a challenging issue to simulate and optimize the unstable propagation of fracture network in a well factory composed of multiple horizontal wells, but it is very helpful for practical fracturing engineering.

1.6　Conclusions

In this study, the stimulation principle of multistage hydrofracturing, fracture propagation under different fracturing scenarios, and different perforation cluster spacing are introduced, and the stress shadow effects generated in the fracturing process are analysed. By summarising the relevant literature, the research status of the controlling factors of the stress shadow effects is presented, and the relevant literatures on stress interference between hydraulic fractures and fracture morphology analysis are introduced. In order to study the stress-dependent unstable dynamic propagation of multistage fracture networks, a series of continuum-discontinuum numerical methods and models are reviewed, and the computed results of the dynamic distribution of stress-dependent unstable dynamic propagation of fractures under different stress fields are summarised. This chapter can provide a reference for those engaged in the research of unstable dynamic propagation of multiple hydraulic fractures, and have a comprehensive grasp of the research in this field.

References

Bai, T., Pollard, D.D., Gao, H. (2000), "Explanation for fracture spacing in layered materials", Nature, Vol. 403 No. 6771, pp. 753–756.

Belytschko, T., Black, T. (1999), "Elastic crack growth in finite elements with minimal remeshing", International Journal for Numerical Methods in Engineering, Vol. 45 No. 5, pp. 601–620.

Bunger, A.P., Peirce, A.P. (2014), "Numerical simulation of simultaneous growth of multiple interacting hydraulic fractures from horizontal wells", Shale Energy Engineering 2014: Technical Challenges, Environmental Issues, and Public Policy, pp. 201–210.

Bunger, A.P., Jeffrey, R.G., Kear, J., Zhang, X., Morgan, M. (2011), "Experimental investigation of the interaction among closely spaced hydraulic fractures", 45th US Rock Mechanics/ Geomechanics Symposium, American Rock Mechanics Association, ARMA-11-318.

Bunger, A.P., Zhang, X., Jeffery, R.G. (2012), "Parameters affecting the interaction among closely spaced hydraulic fractures", SPE Journal, Vol. 17 No. 1, pp. 292–306.

Cai, B., He, C., Ding, Y., Gao, Y., Jiang, W., Duan, G. (2017), "Stress shadow analysis on multi-stage fracturing stimulation of horizontal wells", 4th ISRM Young Scholars Symposium on Rock Mechanics, ISRM-YSS-2017-083.

Castonguay, S.T., Mear, M.E., Dean, R.H. (2013), "Predictions of the growth of multiple interaction hydraulic fractures in three-dimensions", SPE Annual Technical Conference and Exhibition, SPE-166259-MS.

Chen, B., Barboza, B.R., Sun, Y., Bai, J., Thomas, H.R., Dutko, M., Mark Cottrell, M., Li, C. (2021), "A review of hydraulic fracturing simulation", Archives of Computational Methods in Engineering, pp. 1–58.

Cheng, Y. (2009), "Boundary element analysis of the stress distribution around multiple fractures: implications for the spacing of perforation clusters of hydraulically fractured horizontal wells", SPE Eastern Regional Meeting, SPE-125769-MS.

Cheng, W., Gao, H., Jin, Y., Chen, M., Jiang, G. (2017a), "A study to assess the stress interaction of propped hydraulic fracture on the geometry of sequential fractures in a horizontal well", Journal of Natural Gas Science and Engineering, Vol. 37, pp. 69–84.

Cheng, W., Jiang, G., Jin, Y. (2017b), "Numerical simulation of fracture path and nonlinear closure for simultaneous and sequential fracturing in a horizontal well", Computers and Geotechnics, Vol. 88, pp. 242–255.

Chitrala, Y., Moreno, C., Sondergeld, C., Rai, C. (2013), "An experimental investigation into hydraulic fracture propagation under different applied stresses in tight sands using acoustic emissions", Journal of Petroleum Science and Engineering, Vol. 108, pp. 151–161.

Cipolla, C.L., Lolon, E.P., Mayerhofer, M.J., Warpinski, N.R. (2009), "Fracture design considerations in horizontal wells drilled in unconventional gas reservoirs", SPE Hydraulic Fracturing Technology Conference, SPE-119366-MS.

Crouch, S.L. (1976), "Solution of plane elasticity problems by the displacement discontinuity method", International Journal for Numerical Methods in Engineering, Vol. 10 No. 2, pp. 301–343.

Crouch, S.L., Starfield, A.M. (1983), "Boundary Element Methods in Solid Mechanics", London: George Allen & Unwin.

Damjanac, B., Maxwell, S., Pirayehgar, A., Torres, M. (2018), "Numerical study of stress shadowing effect on fracture initiation and interaction between perforation clusters", SPE/AAPG/SEG Unconventional Resources Technology Conference, URTEC-2901800-MS.

Dahi-Taleghani, A., Olson, J.E. (2011), "Numerical modeling of multistranded-hydraulic-fracture propagation: accounting for the interaction between induced and natural fractures", SPE Journal, Vol. 16 No. 3, pp. 575–581.

Duan, K., Li, Y., Yang, W. (2021), "Discrete element method simulation of the growth and efficiency of multiple hydraulic fractures simultaneously-induced from two horizontal wells", Geomechanics and Geophysics for Geo-Energy and Geo-Resources, Vol. 7 No. 1, pp. 1–20.

Dverstorp, B., Andersson, J. (1989), "Application of the discrete fracture network concept with field data: Possibilities of model calibration and validation", Water Resources Research, Vol. 15 No. 3, pp. 540–550.

Escobar, R.G., Sanchez, E.C.M., Roehl, D., Romanel, C. (2019), "XFEM modeling of stress shadowing in multiple hydraulic fractures in multi-layered formations", Journal of Natural Gas Science and Engineering, Vol. 70, 102950.

Gai, S., Nie, Z., Yi, X., Zou, Y., Zhang, Z. (2020), "Study on the interference law of staged fracturing crack propagation in horizontal wells of tight reservoirs", ACS Omega, Vol. 5 No. 18, pp. 10327–10338.

Gao, Q., Cheng, Y., Yan, C. (2018), "A 3D numerical model for investigation of hydraulic fracture configuration in multilayered tight sandstone gas reservoirs", Journal of Petroleum Exploration and Production Technology, Vol. 8 No. 4, pp. 1413–1424.

Gordeliy, E., Peirce, A. (2013), "Implicit level set schemes for modeling hydraulic fractures using the XFEM", Computer Methods in Applied Mechanics and Engineering, Vol. 266, pp. 125–143.

Gupta, P., Duarte, C.A. (2014), "Simulation of non-planar three-dimensional hydraulic fracture propagation", International Journal for Numerical and Analytical Methods in Geomechanics,

Vol. 38 No. 13, pp. 1397–1430.

Guo, J., Zhao, X., Zhu, H., Zhang, X., Pan, R. (2015), "Numerical simulation of interaction of hydraulic fracture and natural fracture based on the cohesive zone finite element method", Journal of Natural Gas Science and Engineering, Vol. 25, pp. 180–188.

Haddad, M., Sepehrnoori, K. (2016), "XFEM-based CZM for the simulation of 3D multiple-cluster hydraulic fracturing in quasi-brittle shale formations", Rock Mechanics and Rock Engineering, Vol. 49 No. 12, pp. 4731–4748.

Harry, D., Horton, D., Durham, D., Constable, D.J., Gaffney, S., Moore, J., Todd, B., Martinez, I. (2020), "Grand challenges and opportunities for greener chemical alternatives in hydraulic fracturing: A perspective from the ACS Green Chemistry Institute Oilfield Chemistry Roundtable", Energy & Fuels, Vol. 34 No. 7, pp. 7837–7846.

He, Q., Suorineni, F.T., Oh, J. (2016), "Review of hydraulic fracturing for preconditioning in cave mining", Rock Mechanics and Rock Engineering, Vol. 49 No. 12, pp. 49(12), 4893–4910.

He, Q., Suorineni, F.T., Ma, T., Oh, J. (2017), "Effect of discontinuity stress shadows on hydraulic fracture re-orientation". International Journal of Rock Mechanics and Mining Sciences, Vol. 91, pp. 179–194.

He, Y., Yang, Z., Li, X., Song, R. (2020), "Numerical simulation study on three-dimensional fracture propagation of synchronous fracturing", Energy Science & Engineering, Vol. 8 No. 4, pp. 944–958.

Hossain, M.M., Rahman, M.K. (2008), "Numerical simulation of complex fracture growth during tight reservoir stimulation by hydraulic fracturing", Journal of Petroleum Science and Engineering, Vol. 60 No. 2, pp. 86–104.

Hudson, M.R. (2017), "Numerical simulation of hydraulic fracturing in tight gas shale reservoirs", PhD Thesis, University of Leeds.

Ju, Y., Chen, J., Wang, Y., Gao, F., Xie, H. (2018), "Numerical analysis of hydrofracturing behaviors and mechanisms of heterogeneous reservoir glutenite, using the continuum-based discrete element method while considering hydromechanical coupling and leak-off effects", Journal of Geophysical Research: Solid Earth, Vol. 123 No. 5, pp. 3621–3644.

Ju, Y., Li, Y., Wang, Y., Yang, Y. (2020), "Stress shadow effects and microseismic events during hydrofracturing of multiple vertical wells in tight reservoirs: a three-dimensional numerical model", Journal of Natural Gas Science and Engineering, Vol. 84, 103684.

Kear, J., White, J. Bunger, A.P. Jeffrey, R., Hessami, M.A. (2013), "Three dimensional forms of closely-spaced hydraulic fractures", ISRM International Conference for Effective and Sustainable Hydraulic Fracturing, ISRM-ICHF-2013-024.

Kresse, O., Weng, X., Gu, H., Wu, R. (2013), "Numerical modeling of hydraulic fractures interaction in complex naturally fractured formations", Rock Mechanics and Rock Engineering, Vol. 46 No. 3, pp. 555–568.

Kumar, D., Ghassemi, A. (2016), "A three-dimensional analysis of simultaneous and sequential fracturing of horizontal wells", Journal of Petroleum Science and Engineering, Vol. 146, pp. 1006–1025.

Lecampion, B. (2009), "An extended finite element method for hydraulic fracture problems", Communications in Numerical Methods in Engineering, Vol. 25 No. 2, pp. 121–133.

Leonhart, D., Meschke, G. (2011), "Extended Finite Element Method for hygro-mechanical analysis of crack propagation in porous materials", Proceedings of the Institution of Mechanical Engineers, Vol. 11, pp. 161–162.

Li, J., Xiao, W., Hao, G., Dong, S., Hua, W., Li, X. (2019b), "Comparison of different hydraulic fracturing scenarios in horizontal wells using XFEM based on the cohesive zone method", Energies, Vol. 12 No. 7, pp. 1232–1250.

Li, S., Liu, L., Chai, P., Li, X., He, J., Zhang, Z., Wei, L. (2019a), "Imaging hydraulic fractures of shale cores using combined positron emission tomography and computed tomography (PET-CT) imaging technique", Journal of Petroleum Science and Engineering, Vol. 182, 106283.

Liu, C., Wang, X., Deng, D., Zhang, Y., Zhang, Y., Wu, H., Liu, H. (2016), "Optimal spacing of sequential and simultaneous fracturing in horizontal well", Journal of Natural Gas Science and Engineering, Vol. 29, pp. 329–336.

Liu, X., Rasouli, V., Guo, T., Qu, Z., Sun, Y., Damjanac, B. (2020), "Numerical simulation of stress shadow in multiple cluster hydraulic fracturing in horizontal wells based on lattice modelling", Engineering Fracture Mechanics, Vol. 238, 107278.

Liu, Y., Gao, D., Li, Q., Wan, Y., Duan, W., Zeng, X., Li, M., Su, Y., Fan, Y., Li, S., Lu, X., Zhou, D., Chen, W., Fu, Y., Jiang, C., Hou, S., Pan, L., Wei, X., Hu, Z., Duan, X., Gao, S., Shen, R., Chang, J., Li, X., Liu, Z., Wei, Y., Zheng, Z. (2019), "Mechanical frontiers in shale-gas development", Advances in Mechanical Engineering, Vol. 49 No. 1, 201901.

Liu, Z., Bai, B., Tang, J., Xiang, Z., Zeng, S., Qu, H. (2021), "Investigation of slickwater effect on permeability of gas shale from longmaxi formation", Energy & Fuels, Vol. 35 No. 4, pp. 3104–3111.

Lu, C., Guo, J., Liu, Y., Yin, J., Deng, Y., Lu, Q., Zhao, X. (2015), "Perforation spacing optimization for multi-stage hydraulic fracturing in Xujiahe formation: A tight sandstone formation in Sichuan Basin of China", Environmental Earth Sciences, Vol. 73 No. 10, pp. 5843–5854.

Lu, Y., Wang, L., Ge, Z., Zhou, Z., Deng, K., Zuo, S. (2020), "Fracture and pore structure dynamic evolution of coals during hydraulic fracturing", Fuel, Vol. 259, 116272.

Luo, S., Zhao, Y., Zhang, L., Chen, Z., Zhang, X. (2021), "Integrated simulation for hydraulic fracturing, productivity prediction, and optimization in tight conglomerate reservoirs", Energy & Fuels, Vol. 35 No. 18, pp. 14658–14670.

Manchanda, R., Sharma, M.M. (2014), "Impact of completion design on fracture complexity in horizontal shale wells", SPE Drilling & Completion, Vol. 29 No. 1, pp. 78–87.

Manríquez, A.L. (2018), "Stress behavior in the near fracture region between adjacent horizontal wells during multistage fracturing using a coupled stress-displacement to hydraulic diffusivity model", Journal of Petroleum Science and Engineering, Vol. 162, pp. 822–834.

Maulianda, B., Savitri, C.D., Prakasan, A., Atdayev, E., Yan, T.W., Yong, Y.K., Elrais, K.A., Barati, R. (2020), "Recent comprehensive review for extended finite element method (XFEM) based on hydraulic fracturing models for unconventional hydrocarbon reservoirs", Journal of Petroleum

Exploration and Production Technology, Vol. 10 No. 8, pp. 3319–3331.

Moës, N., Dolbow, J., Belytschko, T. (1999), "A finite element method for crack growth without remeshing. International Journal for Numerical Methods in Engineering, Vol. 46 No. 1, pp. 131–150.

Munjiza, A., Owen, D.R.J., Bicanic, N. (1995), "A combined finite‐discrete element method in transient dynamics of fracturing solids", Engineering computations, Vol. 12 No. 2, pp. 145–174.

Nagel, N., Zhang, M., Nagel, S., Lee, B., Agharazi, A. (2013), "Stress shadow evaluations for completion design in unconventional plays", SPE Unconventional Resources Conference Canada, SPE-167128-MS.

Nagel, N.B., Sanchez-Nagel, M. (2011), "Stress shadowing and microseismic events: a numerical evaluation", SPE Annual Technical Conference and Exhibition, SPE-147363-MS.

Olson, J.E. (2008), "Multi-fracture propagation modeling: applications to hydraulic fracturing in shales and tight gas sands", 42th US Rock Mechanics Symposium, American Rock Mechanics Association, ARMA-08-327.

Olson, J.E., Dahi, T.A. (2009), "Modeling simultaneous growth of multiple hydraulic fractures and their interaction with natural fractures", SPE Hydraulic Fracturing Technology Conference, SPE-119739-MS.

Ooi, E.T., Song, C., Tin‐Loi, F., Yang, Z. (2012), "Polygon scaled boundary finite elements for crack propagation modelling", International Journal for Numerical Methods in Engineering, Vol. 91 No. 3, pp. 319–342.

Pearson, C.M., Bond, A.J., Eck, M.E., Schmidt, J.H. (1992), "Results of stress-oriented and aligned perforating in fracturing deviated wells", Journal of Petroleum Technology, Vol. 44 No. 1, pp. 10–18.

Qian, Y., Guo, P., Wang, Y., Zhao, Y., Lin, H., Liu, Y. (2020), "Advances in laboratory-scale hydraulic fracturing experiments", Advances in Civil Engineering, Vol. 2020. 1386581.

Reynolds, M.A. (2020), "A technical playbook for chemicals and additives used in the hydraulic fracturing of shales", Energy & Fuels, Vol. 34 No. 12, pp. 15106−15125.

Roussel, N.P., Sharma, M.M. (2012), "Method for determining spacing of hydraulic fractures in a rock formation", United States Patent. 61501003.

Roussel, N.P., Sharma, M.M. (2011a), "Optimizing fracture spacing and sequencing in horizontal-well fracturing", SPE Production & Operations, Vol. 26 No. 2, pp. 173–184.

Roussel, N.P., Sharma, M.M. (2011b), "Strategies to minimize frac spacing and stimulate natural fractures in horizontal completions", SPE Annual Technical Conference and Exhibition, SPE-146104-MS.

Roussel, N.P., Manchanda, R., Sharma, M.M. (2012), "Implications of fracturing pressure data recorded during a horizontal completion on stage spacing design", SPE Hydraulic Fracturing Technology Conference, SPE-152631-MS.

Rui, Y., Yang, C. (2020), "Stress Shadow of fracture Law of Horizontal Well Seam Networks", IOP Conference Series: Earth and Environmental Science, Vol. 558 No. 2, 022083.

Rungamornrat, J., Wheeler, M.F., Mear, M.E. (2005), "A numerical technique for simulating nonplanar evolution of hydraulic fractures", Paper SPE, Vol. 96968, pp. 1–9.

Salimzadeh, S., Usui, T., Paluszny, A., Zimmerman, R.W. (2017), "Finite element simulations of interactions between multiple hydraulic fractures in a poroelastic rock", International Journal of Rock Mechanics and Mining Sciences, Vol. 99, pp. 9–20.

Sesetty, V., Ghassemi, A. (2013), "Numerical simulation of sequential and simultaneous hydraulic fracturing", ISRM International Conference for Effective and Sustainable Hydraulic Fracturing, ISRM-ICHF-2013-040.

Sesetty, V., Ghassemi, A. (2015a), "A numerical study of sequential and simultaneous hydraulic fracturing in single and multi-lateral horizontal wells", Journal of Petroleum Science and Engineering, Vol. 132, pp. 65–76.

Sesetty, V., Ghassemi, A. (2015b), "Simulation of simultaneous and zipper fractures in shale formations", 49th US Rock Mechanics/Geomechanics Symposium, American Rock Mechanics Association, ARMA-2015-558.

Sheng, Y., Sousani, M., Ingham, D., Pourkashanian, M. (2015), "Recent developments in multiscale and multiphase modelling of the hydraulic fracturing process", Mathematical Problems in Engineering, Vol. 2015.

Shi, X., Yang, L., Li, M., Cheng, Y. (2017), "Induced stress interaction during multi-stage hydraulic fracturing from horizontal wells using boundary element method", 4th ISRM Young Scholars Symposium on Rock Mechanics, ISRM-YSS-2017-084.

Sneddon, I. N. (1946), "The distribution of stress in the neighborhood of a crack in an elastic solid", Philosophical Transactions of the Royal Society of London, Series A, Vol. 187 No. 1009, pp. 1934–1990.

Sobhaniaragh, B., Mansur, W.J., Peters, F.C. (2018), "The role of stress interference in hydraulic fracturing of horizontal wells", International Journal of Rock Mechanics and Mining Sciences, Vol. 106, pp. 153–164.

Soliman, M.Y., East, L.E., Adams, D.L. (2008), "Geomechanics aspects of multiple fracturing of horizontal and vertical wells", SPE Drilling & Completion, Vol. 23 No. 3, pp. 217–228.

Tang, H., Winterfeld, P.H., Wu, Y.S., Huang, Z.Q., Di, Y., Pan, Z., Zhang, J. (2016), "Integrated simulation of multi-stage hydraulic fracturing in unconventional reservoirs", Journal of Natural Gas Science and Engineering, Vol. 36, pp. 875–892.

Tang, H., Wang, S., Zhang, R., Li, S., Zhang, L., Wu, Y. (2019b), "Analysis of stress interference among multiple hydraulic fractures using a fully three-dimensional displacement discontinuity method", Journal of Petroleum Science and Engineering, Vol. 179, pp. 378–393.

Tang, P., Wei, X., Liu, Y., Zhang, Y., Zhang, K., Yang, C., Li, C. (2019a), "Study on fracture interference for volume fracturing", International Field Exploration and Development Conference (pp. 2894–2909).

Tarasovs, S., Ghassemi, A. (2014), "Self-similarity and scaling of thermal shock fractures", Physical Review E, Vol. 90 No. 1, 012403.

Tian, W., Li, P., Dong, Y., Lu, Z., Lu, D. (2019), "Numerical simulation of sequential, alternate and modified zipper hydraulic fracturing in horizontal wells using XFEM", Journal of Petroleum

Science and Engineering, Vol. 183, 106251.

Vogler, D., Settgast, R.R., Annavarapu, C., Madonna, C., Bayer, P., Amann, F. (2018), "Experiments and simulations of fully hydro-mechanically coupled response of rough fractures exposed to high-pressure fluid injection", Journal of Geophysical Research: Solid Earth, Vol. 123 No. 2, pp. 1186–1200.

Wang, H. (2016), "Numerical investigation of fracture spacing and sequencing effects on multiple hydraulic fracture interference and coalescence in brittle and ductile reservoir rocks", Engineering Fracture Mechanics, Vol. 157, pp. 107–124.

Wang, X., Liu, C., Wang, H., Liu, H., Wu, H. (2016), "Comparison of consecutive and alternate hydraulic fracturing in horizontal wells using XFEM-based cohesive zone method", Journal of Petroleum Science and Engineering, Vol. 143, pp. 14–25.

Wang, X., Yu, P., Zhang, X., Yu, J., Hao, Q., Li, Q., Yu, Y. (2021d), "Simulation of three-dimensional tension-induced cracks based on cracking potential function-incorporated extended finite element method", Journal of Central South University, Vol. 28 No. 1, pp. 235–246.

Wang, T., Tian, S., Zhang, W., Ren, W., Li, G. (2021b), "Production model of a fractured horizontal well in shale gas reservoirs", Energy & Fuels, Vol. 35 No. 1, pp. 493–500.

Wang, Y. (2020a), "An h-version adaptive FEM for eigenproblems in system of second order ODEs: vector Sturm-Liouville problems and free vibration of curved beams", Engineering Computations, Vol. 37 No. 1, pp. 1210–1225.

Wang, Y. (2020b), Adaptive Analysis of Damage and Fracture in Rock with Multiphysical Fields Coupling. Springer Nature.

Wang, Y. (2021), "Adaptive finite element–discrete element analysis for stratal movement and microseismic behaviours induced by multistage propagation of three-dimensional multiple hydraulic fractures", Engineering Computations, Vol. 38 No. 5, pp. 1350–1371.

Wang, Y., Liu, X. (2021), "Stress-dependent unstable dynamic propagation of three-dimensional multiple hydraulic fractures with improved fracturing sequences in heterogeneous reservoirs: Numerical cases study via poroelastic effective medium model", Energy & Fuels, Vol. 35 No. 22, pp. 18543–18562.

Wang, Y., Li, X., Wang, J. B., Zheng, B., Zhang, B., Zhao, Z. (2015), "Numerical modeling of stress shadow effect on hydraulic fracturing", Natural Gas Geoscience, Vol. 26 No. 10, pp. 1941–1950.

Wang, Y., Ju, Y., Zhuang, Z., Li, C. (2018), "Adaptive finite element analysis for damage detection of non-uniform Euler–Bernoulli beams with multiple cracks based on natural frequencies", Engineering Computations, Vol. 35 No. 3, pp. 1203–1229.

Wang, Y., Ju, Y., Chen, J., Song, J. (2019), "Adaptive finite element–discrete element analysis for the multistage supercritical CO_2 fracturing and microseismic modelling of horizontal wells in tight reservoirs considering pre-existing fractures and thermal-hydro-mechanical coupling", Journal of Natural Gas Science and Engineering, Vol. 61, pp. 251–269.

Wang, Y., Duan, Y., Liu, X., Huang, J., Hao, N. (2021a), "Numerical analysis for dynamic propagation and intersection of hydraulic fractures and pre-existing natural fractures involving the sensitivity factors: orientation, spacing, length, and persistence", Energy & Fuels, Vol. 35 No. 19, pp. 15728–15741.

Wang, Y., Ju, Y., Zhang, H., Gong, S., Song, J., Li, Y., Chen, J. (2021c), "Adaptive finite element–discrete element analysis for the stress shadow effects and fracture interaction behaviours in three-dimensional multistage hydrofracturing considering varying perforation cluster spaces and fracturing scenarios of horizontal wells", Rock Mechanics and Rock Engineering, Vol. 54 No. 4, pp. 1815–1839.

Warpinski, N.R. (2000), "Analytic crack solutions for tilt fields around hydraulic fractures", Journal of Geophysical Research: Solid Earth, Vol. 105 No. B10, pp. 23463–23478.

Warpinski, N.R., Branagan, P.T. (1989), "Altered-stress fracturing", Journal of petroleum technology, Vol. 4 No.9, pp. 990–997.

Wu, K., Olson, J.E. (2015), "Simultaneous multifracture treatments: fully coupled fluid flow and fracture mechanics for horizontal wells", SPE Journal, Vol. 20 No. 2, pp. 337–346.

Wu, K., Olson, J.E. (2016), "Mechanisms of simultaneous hydraulic-fracture propagation from multiple perforation clusters in horizontal wells", SPE Journal, Vol. 21 No. 3, pp. 1000–1008.

Wu, K., Olson, J., Balhoff, M.T., Yu, W. (2017), "Numerical analysis for promoting uniform development of simultaneous multiple-fracture propagation in horizontal wells", SPE Production & Operations, Vol. 32 No. 1, pp. 41–50.

Wu, R., Kresse, O., Weng, X., Cohen, C.E., Gu, H. (2012), "Modeling of interaction of hydraulic fractures in complex fracture networks", SPE Hydraulic Fracturing Technology Conference, SPE-152052-MS.

Wong, S.W., Geilikman, M., Xu, G. (2013), "Interaction of multiple hydraulic fractures in horizontal wells", SPE Unconventional Gas Conference and Exhibition, SPE-163982-MS.

Xia, L., Zeng, Y. (2018), "Stress shadow effect of alternative fracturing based on numerical simulation of PFC2D", Rock and Soil Mechanics, Vol. 39, pp. 4269–4281.

Xie, J., Huang, H., Sang, Y., Fan, Y., Chen, J., Wu, K., Yu, W. (2019), "Numerical study of simultaneous multiple fracture propagation in Changning shale gas field", Energies, Vol. 12 No. 7, 1335.

Yamamoto, K., Shimamoto, T., Sukemura, S. (2004), "Multiple fracture propagation model for a three-dimensional hydraulic fracturing simulator", International Journal of geomechanics, Vol. 4 No. 1, pp. 46–57.

Yew, C.H., Schmidt, J.H., Li, Y. (1989), "On fracture design of deviated wells", SPE Annual Technical Conference and Exhibition, SPE-19722-MS.

Yu, Y., Zhu, W., Li, L., Wei, C., Dai, F., Liu, S., Wang, W. (2017), "Analysis on stress shadow of mutual interference of fractures in hydraulic fracturing engineering", Chinese Journal of Rock Mechanics and Engineering, Vol. 36 No. 12, pp. 2926–2939.

Yu, Y., Zhu, W., Li, L., Wei, C., Yan, B., Li, S. (2020), "Multi-fracture interactions during two-phase flow of oil and water in deformable tight sandstone oil reservoirs", Journal of Rock Mechanics and Geotechnical Engineering, Vol. 12 No. 4, pp.821–849.

Zhang, R., Zhang, L., Wang, R., Zhao, Y., Huang, R. (2016), "Simulation of a multistage fractured horizontal well with finite conductivity in composite shale gas reservoir through finite-element method", Energy & Fuels, Vol. 30 No. 11, pp. 9036–9049.

Zhang, S., Lei, X., Zhou, Y., Xu, G. (2015), "Numerical simulation of hydraulic fracture propagation in tight oil reservoirs by volumetric fracturing", Petroleum Science, Vol. 12 No. 4, pp. 674–682.

Zhang, H. (2010), "Shale gas: new bright point of the exploitation of the global oil-gas resources—the present status and key problems of the exploitation of shale gas", Bulletin of Chinese Academy of Science, Vol. 25 No. 4, pp. 406–410.

Zeng, Q., Tong, Y., Yao, J. (2019), "Production distribution in multi-cluster fractured horizontal wells accounting for stress interference", Journal of China University of Petroleum, Vol. 43 No. 1, pp. 99–107.

Zeng, S., Zhang, G., Han, J., Yuan, B., Wang, Y., Ji, Z. (2015), "Model of multi-fracture stress shadow effect and optimization design for staged fracturing of horizontal wells", Natural Gas Industry, Vol. 35 No. 3, pp. 55–59.

Zou, J., Chen, W., Jiao, Y. (2018), "Numerical simulation of hydraulic fracture initialization and deflection in anisotropic unconventional gas reservoirs using XFEM", Journal of Natural Gas Science and Engineering, Vol. 55, pp. 466–475.

Chapter 2   Dual bilinear cohesive zone model for fluid-driven propagation of multiscale tensile and shear fractures

2.1   Introduction

Many oil and gas reservoirs worldwide exist in low-permeability fractured reservoirs, and hydrofracturing technology has been extensively used to stimulate low-permeability unconventional oil and gas reservoirs to increase productivity (Bažant *et al.*, 2014). Many field monitoring data and laboratory tests have confirmed that in the actual reservoir fracturing process, fracture propagation is along the direction of a certain angle with the original fracture surface instead of the original direction, facilitating the formation of composite fractures (Chandler *et al.*, 2016; Wang *et al.*, 2021). Moreover, fracturing may cause microseismic and contact slip events (Wang *et al.*, 2018, 2019). Predicting, controlling, and optimising the efficiency of hydrofracturing is impossible if the mechanism of hydraulic fracture propagation in a fractured reservoir is not fully understood, eventually leading to a poor effect after fracturing and failure of a fracturing operation. Therefore, conducting theoretical research on the dynamic propagation of hydraulic fractures in fractured reservoirs is necessary.

Field monitoring and laboratory tests have been conducted to determine the morphology of hydraulic fractures in rock materials (Bohloli and De Pater, 2006; Bunger *et al.*, 2015). To study the effect of fracturing fluid viscosity on fracture propagation, researchers have conducted hydrofracturing experiments with fracturing fluids of different viscosities. In the experimental results, the fracture propagation path of hydrofracturing is jointly induced by shear stress and tensile stress (Chen *et al.*, 2015). To reveal the acoustic emission response characteristics of hydraulic fracture propagation, researchers have used a true triaxial fracturing simulation system to simulate the fracturing. The experimental results showed that a shear mechanism occurred around the perforation in the initial stage of fracturing (Ma *et al.*, 2017). To study the influence of geological and engineering factors on fracture propagation, on the basis of a hydrofracturing experiment in a laboratory, researchers have analysed the propagation law of hydraulic fractures in shale. The findings demonstrated that under the condition of high stress difference, hydraulic fractures tend to propagate along the predetermined direction and typically form a single tension fracture along the direction of the maximum principal stress (Hou *et al.*, 2014).

In summary, there are two types of hydraulic fractures: tension and shear. Tensile fractures primarily propagate along a direction parallel to the maximum principal stress. Shear fractures appear at the initial stage of fracture propagation, primarily around the perforation, and the number of shear fractures was less than that of tensile fractures. When there are natural fractures, the shear fractures propagate easily along the natural fracture direction (Wang *et al.*, 2019). Understanding and mastering the type and distribution of fractures in the study of fracture propagation are crucial. However, distinguishing types of tensile and shear fractures in physical experiments is difficult because observing the dynamic propagation of fracturing fractures in deep strata is difficult; therefore, numerical simulation has become an essential research method.

With the rapid development of computer technology, interest in complex hydrofracturing simulations is increasing. Several numerical methods have been used to study the propagation of various types of fractures. For example, using the finite element (FE) programme (Yang *et al.*, 2004), the flow, stress, and damage analyses are combined, and the maximum tensile stress criterion and Mohr–Coulomb fracture criterion are used to judge the damage. A numerical simulation of hydraulic fractures under the action of the internal pressure of a circular hole was conducted to study the influence of heterogeneity on the hydrofracturing of permeable rocks. The discrete element method (DEM) is used to establish a three-dimensional fracture propagation model of the target reservoir considering mechanical anisotropy, weak bedding plane, and vertical stress difference, and the maximum tensile stress criterion and Mohr–Coulomb criterion are used as the basis for fracture propagation (Deng *et al.*, 2014; Duan *et al.*, 2020). The distribution of hydraulic fractures under different bedding densities, mechanical properties, and fracturing engineering parameters (e.g. perforation clusters, injected flow rate, and fracturing fluid viscosity) were analysed (Zhou *et al.*, 2020). The finite difference method and numerical solution of a single element were used to determine the direction of the formation of hydraulic fractures. According to linear elastic theory, fractures occur when the tensile stress at the tip of the fracture reaches the tensile strength of the rock (Al-Rubaie and Ben

Mahmud, 2020). In addition, based on the extended FE method, a displacement enhancement scheme is proposed to simulate the action mechanism of the compression-shear fracture surface, and the bifurcation fracture treatment method is given. The Mohr–Coulomb criterion was used to determine the initiation and propagation of compression-shear fractures (Shi *et al.*, 2014). However, there are defects in the traditional numerical models and methods, namely, the mesh is not easy to refine in the modelling process, the network will actually be changed and moved because of the flexibility, and the shape of the material or the multiscale fracture propagation simulation cannot be conducted. Therefore, by combining the reliability of the stress solution of FE in continuum with the flexibility of fracture propagation in discrete element (DE) results (Munjiza *et al.*, 1995; Munjiza, 2004), this study overcomes the limitations of the traditional finite element method (FEM) in simulating three-dimensional fracture propagation, providing a potential technique for hydrofracturing (Wang *et al.*, 2018; Wang, 2020, 2021). Based on the three-dimensional engineering-scale numerical model, considering the crucial hydro-mechanical (HM) coupling and fracturing fluid leak-off, stimulation of three-dimensional multistage fracturing and fracture interaction behaviour can occur.

Achievements have been made in the study of the fracture criteria of hydrofracturing. The maximum tensile stress criterion considers that fractures propagate along the direction of the first principal normal stress, and the stress condition of fracture propagation is that the first principal normal stress at the fracture tip equals the tensile strength of the material (Palaniswamy and Knauss, 1978). The Mohr–Coulomb criterion considers that there is a certain functional relationship between the normal stress and the shear strength on the fracture surface; thus, whether the shear stress reaches the shear strength of the soil is regarded as the failure criterion (Ju *et al.*, 2018; Deng *et al.*, 2014; Duan *et al.*, 2020). The maximum tensile stress criterion and Mohr–Coulomb strength criterion are stress-based criteria that are simple in form and convenient for application. However, the application scope of these two criteria is relatively limited, and explaining the fracture propagation mechanism under complex stress states is difficult.

According to the fracture criterion based on energy, a certain amount of strain energy is released during fracture propagation. The fracture propagates only when the strain energy released per unit length of fracture propagation is greater than the work required to overcome the resistance. The critical energy release rate was used as the criterion for fracture (Griffith, 1920; Irwin, 1957). Based on the concept of energy, the theory can explain the result of brittle rock mass tensile failure. In any stress state, the tensile stress concentration at the fracture tip leads to fracture propagation. However, this theory fails to consider the role of plastic deformation and stress concentration in the energy balance, and it can only consider the aperture and propagation of the main fractures, ignoring the fracture closure and friction on the fracture surface under the action of compressive stress; it also fails to consider the propagation of multiple hydraulic fractures and their interactions.

According to the fracture criterion based on the stress intensity factor, brittle fracture occurs when the stress intensity factor is greater than the fracture toughness (Irwin et al., 1958). The stress intensity factor describes the stress state of the plastic zone around the fracture tip, which synthesises various factors that affect the behaviour of the fracture tip into the strength of the stress–strain field at the fracture tip, and numerically represents the severity of the fracture tendency at different fracture tips. However, when the size of the plastic zone at the fracture tip cannot be ignored, the stress–strain field in the plastic zone at the tip cannot be characterised by the K-field, and the fracture criterion based on the stress intensity factor is also not suitable for the propagation of composite fractures. When the value of KI/KII is small, the theoretical results are far from the experimental results. Damage refers to the progressive weakening of material cohesion under monotonic or repeated loading, leading to the failure of elements. On the basis of the fracture criterion of damage mechanics, damage variables were used to describe the damage degree of materials (Xie and Chen, 1988). The results show that the model based on damage mechanics is in good agreement with the experimental results of rock stress–strain under the condition of brittle material tension, and the stress–strain curve is divided into several sections, but the fitting degree with the actual curve is not ideal.

The cohesive zone model (CZM) has been widely used to study fracture growth (Liu et al., 2018b; Wu et al., 2020). The existence of cohesion keeps the fracture tip close and reduces or even eliminates the stress singularity to a certain extent. Compared with traditional fracture mechanics, the CZM has significant advantages in simulating the evolution of the plastic zone at the fracture front and during fracture initiation. The CZM is used to simulate the propagation of hydraulic fractures, and the nonlinear fracture behaviour near the fracture tip and their interaction with the existing natural fractures were considered (Dahi Taleghani et al., 2018; Khoei et al., 2018). In addition, according to the CZM, the element surface around the fracture tip is determined, which is closest to the normal direction of the maximum principal stress at the fracture tip. A three-dimensional hydrofracturing simulation method for fully saturated porous media was proposed (Secchi and Schrefler, 2012). In this study, the challenge of the traditional fracture criterion in fracture propagation can be overcome: the fracture criterion based on dual bilinear CZM is used to simulate fluid-driven fracture propagation. The coupling 'finite element-discrete element-finite volume' (FE-DE-FV) method is used to simulate hydrofracturing, and the propagation of multiple hydraulic fractures is studied in laboratory and engineering scales.

The remainder of this chapter is organized as follows. In Section 2, the governing partial differential equations of hydrofracturing are presented, including the governing equations of solid deformation and fluid flow in fractured porous media. Section 3 describes the bilinear CZM, and Section 4 summarises the steps of the numerical discretization of solids and fluids. Section 5 introduces the detection and separation of the DEM. Section 6 presents the overall algorithm and process of this study. In Section 7, the accuracy of the quasi-two-dimensional model is verified by comparing it with the analytical solutions in the Khristianovic–Geertsma–de Klerk (KGD) model (Barenblatt, 1962; Nordgren, 1972) and the Perkins–Kern–Nordgren (PKN) model (Perkins and Kern, 1961); subsequently, the fracture propagation behaviour and mechanism are analysed by numerical simulation at the laboratory scale and engineering scale, and the fracture types and distribution in these two scales are discussed. The concluding remarks are provided in Section 8.

2.2   Governing partial differential equations for hydrofracturing

In this study, the pore pressure and stress were dependent, and the permeability of the fluid flow in a hydraulic fracture was relative to the aperture of the hydraulic fracture. FEM and DEM were combined to analyse the solid deformation and fracture, and the FV method (FVM) was used for fluid flow. HM coupling is introduced by updating and iterating the solid and fluid pressure fields. A schematic of the FE-DE-FV algorithm for HM coupling is shown in Figure 2.1. A microstructural model of a rock material comprises two elements: block elements and jointed elements. The model was configured such that the FEM was used inside the block, and the DEM was adopted for the interface. Moreover, the strength of each element is related to its deformation modulus, and the deformation of each block element is obtained according to the state of stress and the constitutive relation of the materials. Various elements of the conventional FEM can be flexibly introduced into the models. This method has considerable advantages in simulating dynamic fracture processes, and it has various applications related to problems associated with continuous or discontinuous deformation under dynamic or static loads.

The governing equation for the movement of elements is the motion equation, which considers the deformation of elements. The following governing equations must be satisfied for each block within the model; few investigators have considered the seepage effect between hydraulic fractures and the pore matrix. In this study, Darcy's law was integrated into the model to incorporate the seepage effect and govern fluid leak-off. The fully coupled discretization of a single-phase flow in fractured porous media is implemented by the FVM. The details are presented in this section.

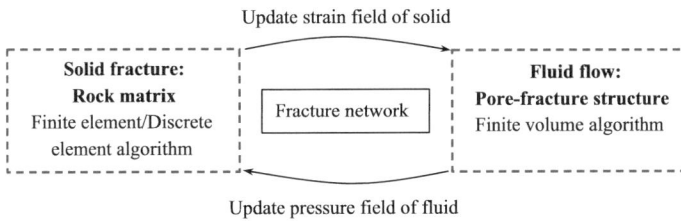

Figure 2.1. Iteration computation and combined finite element-discrete element-finite volume algorithm for hydro-mechanical coupling.

2.2.1   Governing equations of solid deformation

The mechanical equilibrium equation in Cartesian coordinates for solid deformation, considering the influence of dynamic inertia, is given as

$$\nabla \cdot \boldsymbol{\sigma} = \rho \ddot{\boldsymbol{u}} + c \dot{\boldsymbol{u}} - \boldsymbol{f} , \quad x, y, z \in \Omega , \tag{2.1}$$

where $\boldsymbol{u}(x, y, z) = (u(x, y, z), v(x, y, z), w(x, y, z),)^{\mathrm{T}}$ is the displacement vector; $\boldsymbol{\sigma}$ represents the stress field vector; $\boldsymbol{f}$ is the external load vector, including the body force, fluid pressure on the fracture surface, spring force, and force on the interface; ρ is the density; c is the damping coefficient; $\boldsymbol{u}$ is the displacement vector; $\dot{\boldsymbol{u}}$ and $\ddot{\boldsymbol{u}}$ denote the velocity and acceleration vectors, respectively, and are derivatives of displacement vectors to time t; and Ω is the solution domain. The meanings and values of the physical parameters in these equations are listed in Table 2.1. In Equation (2.1), the inertia term is used because the dynamic process must consider the inertia term. The damping terms are not considered in the hydrofracturing process; however, to fully provide all the functions that can be solved, the damping terms are considered in the governing equations. Therefore, both damping and inertia terms are given for the sake of completeness.

Table 2.1. Basic physical parameters of models for hydrofracturing in laboratory and engineering scales.

Parameter	Value
Young's modulus E /GPa	70
Poisson's ratio v	0.25
Tensile strength σ_{max} /MPa	10
Shear strength τ_{max} /MPa	14
Fracture energy G_f /(N·m)	165
Density ρ /(g/m^3)	2.5
Biot's coefficient α	0.75
Porosity n	0.4
Permeability k /nD	1×10^{-16}
Viscosity coefficient μ/(Pa·s)	1×10^{-11}
Damping coefficient c	0.8

The effective stress tensor $\boldsymbol{\sigma}^{\mathrm{e}}$ is given as

$$\boldsymbol{\sigma}^{\mathrm{e}} = \boldsymbol{\sigma} - \alpha p \boldsymbol{I} , \tag{2.2}$$

where α is Biot's coefficient, p is the pore water pressure, and $\boldsymbol{I}$ is the identity matrix.

The strain–displacement relationship is given as

$$\varepsilon = \frac{1}{2}(\nabla \boldsymbol{u} + (\nabla \boldsymbol{u})^{\mathrm{T}}) , \tag{2.3}$$

where ε is the strain tensor.

The constitutive relation law is given as

$$\boldsymbol{\sigma}^{\mathrm{e}} = \boldsymbol{D} : \varepsilon , \tag{2.4}$$

where $\boldsymbol{D}$ is the elastic matrix.

The boundary conditions are given as

$$\boldsymbol{u} = \bar{\boldsymbol{u}} , \text{ on } \Gamma_u \tag{2.5a}$$

$$\boldsymbol{\sigma} \cdot \boldsymbol{n} = \bar{\boldsymbol{\sigma}} , \text{ on } \Gamma_\sigma \tag{2.5b}$$

where the displacement $\bar{\boldsymbol{u}}$ is prescribed on the displacement boundary Γ_u, and the confining stress, $\bar{\boldsymbol{\sigma}}$, is prescribed on the external force boundary, Γ_σ.

The initial conditions are given as

$$u(t=0)=\bar{u}^0, \text{ on } \Gamma_u^0 \tag{2.6a}$$

$$\sigma(t=0)\cdot n=\bar{\sigma}^0, \text{ on } \Gamma_\sigma^0 \tag{2.6b}$$

where the displacement $\bar{u}^0$ is prescribed on the initial displacement boundary Γ_u^0, and the confining stress, $\bar{\sigma}^0$, is prescribed on the initial external force boundary, Γ_σ^0.

2.2.2　Governing equations of fluid flow in fractured porous media

The hydrofracturing fracture network and original porous media constituted the fractured porous media. In Equation (2.7), variables containing the subscript m represent the variables in porous media, and variables containing the subscript f represent the variables in porous media. Assuming single-phase flow in fractured porous media, Darcy's law for fluid flow in porous media and hydraulic fractures, in its simplified form and ignoring gravitational forces, can be used to obtain the velocity field:

$$v_m=-\frac{k_m}{\mu}\nabla p_m, \quad x,y,z\in\Omega, \tag{2.7a}$$

$$v_f=-\frac{k_f}{\mu}\nabla p_f, \quad x,y,z\in\Omega, \tag{2.7b}$$

where v_m and v_f denote the velocity field of fluid flow for porous media and hydraulic fractures, respectively; p_m and p_f represent the pressure values for porous media and hydraulic fractures, respectively; k_m and k_f are the permeability of the porous media and hydraulic fractures, respectively; and μ is the viscosity. The pressure equation for single-phase flow in porous media and hydraulic fractures in the absence of gravity and capillary forces can be written as follows:

$$S_m\dot{p}_m-\nabla\cdot v_m=-\alpha\frac{\partial\varepsilon_V}{\partial t}, \quad x,y,z\in\Omega, \tag{2.8a}$$

$$S_f\dot{p}_f-\nabla\cdot v_f=q_f, \quad x,y,z\in\Omega, \tag{2.8b}$$

where $S_m=n/K_f$ and $S_f=1/K_f$ are the water storage coefficients of porous media and hydraulic fractures, respectively; n is the porosity, and the porosity of the fractures is equal to unity; q_f denotes the external fluid sources; and ε_V is the volume strain of the rock matrix.

　　The permeability can be written as follows (Snow, 1965):

$$k_f=\frac{h^2}{12}, \tag{2.9}$$

where h denotes the aperture of the hydraulic fracture.

The boundary conditions are given as

$$\boldsymbol{v}_m \cdot \boldsymbol{n}_q = \overline{q} \text{ , on } \Gamma_q , \tag{2.10a}$$

$$p_m = \overline{p} \text{ , on } \Gamma_p , \tag{2.10b}$$

where the fluid source, $\overline{q}$, is prescribed on the external fluid source boundary, Γ_q , and $\boldsymbol{n}_q$ is the unit normal vector on this boundary; the fluid pressure, $\overline{p}$, is prescribed on the external fluid pressure boundary, Γ_p .

2.3   Dual bilinear cohesive zone model

Failure, slipping, and fracture of the solid occur at the interface between the elements. A dual bilinear CZM was used to judge the tensile and shear failures of the elements. This CZM is based on the elastic-plastic fracture mechanics method (Barenblatt, 1959; Dugdale, 1960), which considers the existence of a plastic deformation region at the fracture tip, and can effectively avoid the problem of infinite stress at the fracture tip.

Cohesion elements are typically regarded as isotropic materials with weak mechanical properties, and their mechanical properties are defined by a traction–separation relationship. In essence, cohesion elements belong to two separate surfaces. With the injection of fluid, tensile and shear fractures propagate, and stress concentration and damage zones appear around the fluid-driven fracture tip (Figure 2.2).

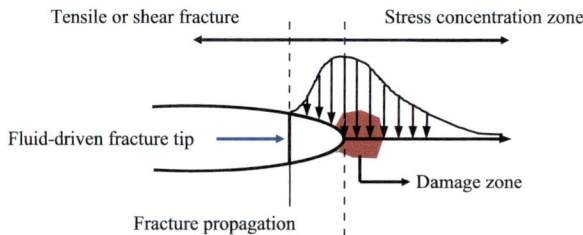

Figure 2.2. Stress concentration and damage zones around the fluid-driven fracture tip.

Based on the bilinear cohesive criteria (i.e. the tensile cohesive criterion and shear cohesive criterion in Figure 2.3) and the basic variables of the rock mass to be measured, the tensile or shear fracture conditions cause by the rock fracturing fluid are obtained, and the tensile and shear fracture propagations are judged by the tensile bilinear cohesive fracture criterion (i.e. the relationship between stress and strain is linear in the ascending and descending sections) and the shear bilinear cohesive fracture criterion, respectively; the corresponding types of fractures initiate and propagate when a certain point of the rock mass fulfils the fracture criterion.

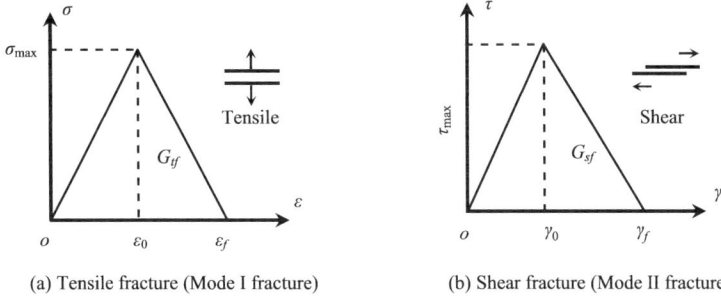

(a) Tensile fracture (Mode I fracture)   (b) Shear fracture (Mode II fracture)

Figure 2.3. Dual bilinear cohesive criteria for tensile and shear fracture of rock.

The dual bilinear cohesive criterion includes the tensile cohesive criterion and shear cohesive criterion. Each fracture criterion includes the evolution process of the stress–strain relationship in rock mass: in the initial stage, the stress and strain show a linear increasing trend driven by the fracturing fluid; as the stress continues to increase, the stress reaches tensile or shear strength (in Figure 2.3, σ_{max} is tensile strength and τ_{max} is shear strength), and the cohesion of rock material reaches its peak; subsequently, the rock mass begins to enter the stage of damage evolution (presented as plastic deformation), and the stress and strain show a linear decreasing trend; until the stress reduces to 0, the area surrounded by the stress–strain curve reaches the fracture energy of tensile or shear fracture of rock mass material (e.g. Equation (2.11)); at this time, the rock mass reaches the maximum damage, that is, tensile or shear fracture occurs.

This failure process is that of energy dissipation of material cohesion, and the damage evolution and plastic deformation at the fracture tip shown in Figure 2.3 can be effectively characterised. Mode I represents tensile fracture, and Mode II represents shear fracture, in Figures 2.3(a) and 2.3(b), respectively. In Figure 2.3(a), when $\varepsilon_0 < \varepsilon < \varepsilon_f$, the rock mass undergoes plastic deformation and enters the stage of damage evolution; when $\varepsilon > \varepsilon_f$, the rock mass undergoes tensile fracture. The evolution of Figure 2.3(b) was similar. The bilinear cohesive fracture criteria are expressed as follows:

$$\text{Tensile failure: } \int_0^{\varepsilon_f} \sigma \, d\varepsilon = \int_0^{\varepsilon_0} \sigma \, d\varepsilon + \int_{\varepsilon_0}^{\varepsilon_f} \sigma \, d\varepsilon = G_{tf} , \tag{2.11a}$$

$$\text{Shear failure: } \int_0^{\gamma_f} \tau \, d\gamma = \int_0^{\gamma_0} \tau \, d\gamma + \int_{\gamma_0}^{\gamma_f} \tau \, d\gamma = G_{sf} , \tag{2.11b}$$

where σ is the tensile stress, τ is the shear stress, ε_0 is the tensile strain value when the tensile stress reaches the maximum value, ε_f is the tensile strain value when tensile fracture occurs, γ_0 is the shear strain value when the shear stress reaches the maximum value, γ_f is the shear strain value when shear fracture occurs, and G_{tf} and G_{sf} are the tensile fracture energy and shear fracture energy, respectively.

2.4   Numerical discretization

2.4.1   Finite element discretization for solid

To solve the governing equations of solid deformation, based on the FEM and using the variation formulation, the equilibrium Equation (2.1) can be transformed into the following matrix form on element e:

$$\boldsymbol{M}^e \ddot{\boldsymbol{D}} + \boldsymbol{C}^e \dot{\boldsymbol{D}} + \boldsymbol{K}^e \boldsymbol{D} = \boldsymbol{F}^e , \quad x, y, z \in \Omega , \tag{2.12}$$

where $\boldsymbol{D}(t)$ is the displacement vector composed by assembling the nodal displacements of element e; $\boldsymbol{M}^e$, $\boldsymbol{C}^e$, and $\boldsymbol{K}^e$ are the mass, damping, and stiffness matrices, respectively; $\dot{\boldsymbol{D}}(t)$ and $\ddot{\boldsymbol{D}}(t)$ denote vectors containing the nodal velocities and accelerations at time t, respectively; and $\boldsymbol{F}^e$ is the external loading vector. $\boldsymbol{F}^e$ can be expressed as

$$\boldsymbol{F}^e = \boldsymbol{F}_b^e + \boldsymbol{F}_p^e + \boldsymbol{F}_s^e + \boldsymbol{F}_t^e , \tag{2.13}$$

where $\boldsymbol{F}_b^e$ is the body force, $\boldsymbol{F}_p^e$ is the fluid pressure on the fracture surface, $\boldsymbol{F}_s^e$ is the spring force, and $\boldsymbol{F}_t^e$ is the force on the traction boundary.

In the time domain, an explicit iteration technique is applied. In this technique, the acceleration is iterated by the central difference scheme, and the velocity is iterated by the unilateral difference scheme. The schemes can be written as

$$\ddot{u}_i^n = \frac{u_i^{n+1} - 2u_i^n + u_i^{n-1}}{(\Delta t)^2} , \tag{2.14a}$$

$$\dot{u}_i^{n+1} = \frac{u_i^{n+1} - u_i^n}{\Delta t} , \qquad (2.14b)$$

where, $\ddot{u}_i^n$, $\dot{u}_i^n$, and u_i^n, respectively, represent the acceleration, the velocity, and the displacement of the i-th node of one element at the nth time step.

2.4.2   Finite volume discretization for fluid

The cell-centred FVM is suitable for fluid flow in porous media and hydraulic fractures, assuming a single-phase flow in fractured porous media (Ju *et al.*, 2018). In the following descriptions, the unified symbol is used to represent the same type of variables in porous media and hydraulic fractures. When they need to be discussed independently, the variables containing the subscript m represent the variables in porous media, and the variables containing the subscript f represent the variables in hydraulic fractures. To derive the set of FV mass balance equations for the pressure equation, we consider that a cell Ω_i in the domain is denoted by Ω. Ω_j is one of the neighbouring cells of Ω_i, and they have the same interface $\partial\Omega_{ij} = \Omega_i \bigcap \Omega_j$. The fluid flow between the two FV cells (elements) through the centres of the cells and the interface is shown in Figure 2.4. The centres of the two cells were C_i and C_j respectively. C_0 is the midpoint of $\partial\Omega_{ij}$. The pressures in the two adjacent cells were p_i and p_j, respectively.

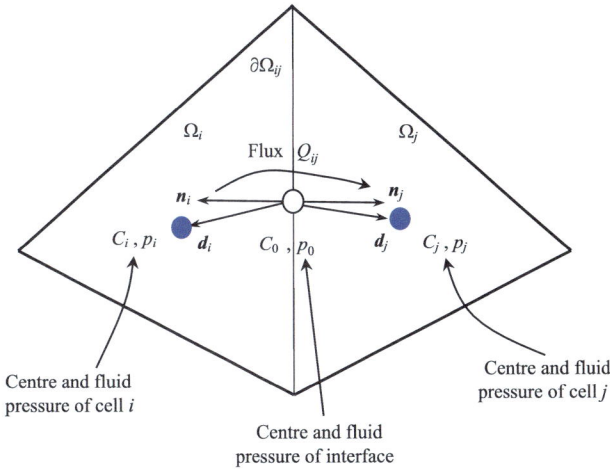

Figure 2.4. Fluid flow between two finite volume elements through the centres of elements and interface.

FVM is derived by obtaining the relationship between the flux Q_{ij} across the interface $\partial\Omega_{ij}$ and pressure p of the two adjacent cells. We let $K = k/\mu$ for simplicity. The flow velocities along the line segments $C_i C_0$ and $C_0 C_j$ can be obtained from Darcy's law by using Equation (2.7):

$$v_{i0} = -K_i \nabla p_{i0} = -K_i \frac{p_0 - p_i}{D_i}(-\boldsymbol{d}_i), \tag{2.15a}$$

$$v_{0j} = -K_j \nabla p_{0j} = -K_i \frac{p_j - p_0}{D_j}\boldsymbol{d}_j. \tag{2.15b}$$

Flux across the interface $\partial\Omega_{ij}$ can be computed by using the following integral:

$$Q_{ij} = Q_{i0} = \int_{\partial\Omega_{ij}} v_{i0} \cdot (-\boldsymbol{n}_i)\,\mathrm{d}S = AK_i \frac{p_i - p_0}{D_i}(\boldsymbol{d}_i \cdot \boldsymbol{n}_i), \tag{2.16a}$$

$$Q_{ij} = Q_{0j} = \int_{\partial\Omega_{ij}} v_{0j} \cdot (\boldsymbol{n}_j)\,\mathrm{d}S = AK_j \frac{p_0 - p_j}{D_j}(\boldsymbol{d}_j \cdot \boldsymbol{n}_j), \tag{2.16b}$$

where A is the area of the interface between the adjacent cells, K_i is the intrinsic permeability of cells i, D_i is the distance between the cell centre and the middle point of the interface, $\boldsymbol{n}_i$ is the unit normal vector to the interface, and $\boldsymbol{d}_i$ is the unit direction vector along $C_0 C_i$.

$p_i - p_0$ can be obtained from Equation (2.16a), and $p_0 - p_j$ can be obtained from Equation (2.16b). By adding the two terms, the following equation can be derived:

$$p_i - p_j = \left[\frac{D_i}{AK_i (\boldsymbol{d}_i \cdot \boldsymbol{n}_i)} + \frac{D_j}{AK_j (\boldsymbol{d}_j \cdot \boldsymbol{n}_j)} \right] Q_{ij}, \tag{2.17}$$

By denoting $\alpha_i = \dfrac{AK_i}{D_i}(\boldsymbol{d}_i \cdot \boldsymbol{n}_i)$, $\alpha_j = \dfrac{AK_j}{D_j}(\boldsymbol{d}_j \cdot \boldsymbol{n}_j)$, and $T_{ij} = \dfrac{\alpha_i \alpha_j}{\alpha_i + \alpha_j}$, Equation (2.17) can be simplified as follows:

$$Q_{ij} = T_{ij}(p_i - p_j), \tag{2.18}$$

where T_{ij} represents the geometric transmissibility between cell i and cell j.

2.5   Detection and separation of discrete elements

The fractures in this study were separated along the boundary of the elements to realise dynamic propagation. Therefore, the detection of elements to determine which element boundaries are separated is the quantitative basis for ensuring the propagation direction and length of the fractures. With the increasing pressure of the injected fluid, the contact force between the elements fulfils the bilinear fracture criterion shown in Equation (2.11), and the contact between the element failure and the elements is separated. Figure 2.5 shows a schematic of the fluid-driven tensile or shear failure of the nodes and fracture propagation. Figure 2.5(a) is the initial geometric domain, and Figure 2.5(b) shows the discrete FE domain discretized by FE elements (elements A, B, C, and D) in the initial geometric domain, and the elements are connected by FE nodes. Driven by fluid load, once some notes are detected by the current state of stress and energy that the dual bilinear cohesive fracture criteria are fulfilled, such as the node between elements A and C shown in Figure 2.5(c), the initial fracture initiation is induced by tensile or shear failure separation from the node. With the progressive increase in fluid load, more nodes around the fracture will be detected to fulfil the fracture criteria, such as the node between elements B and D shown in Figure 2.5(d), which will form fracture propagation induced by continuous failure and separation of nodes. Through this process, the detection and separation of element nodes will implement fluid-driven tensile or shear failure of nodes and fracture propagation.

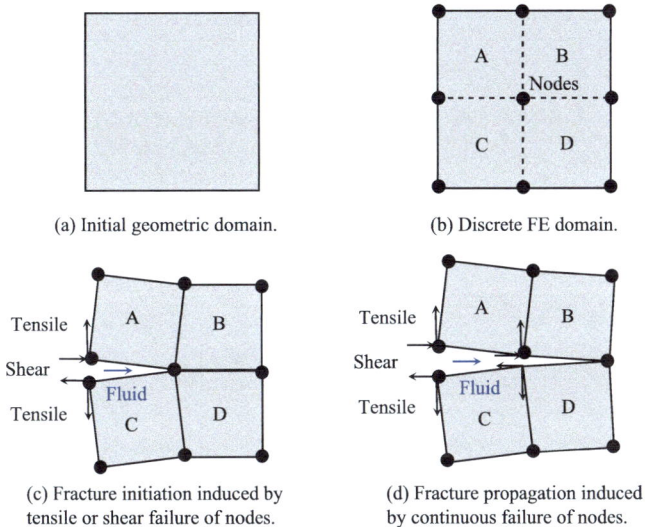

(a) Initial geometric domain.

(b) Discrete FE domain.

(c) Fracture initiation induced by tensile or shear failure of nodes.

(d) Fracture propagation induced by continuous failure of nodes.

Figure 2.5. Schematic of fluid-driven tensile or shear failure of nodes and fracture propagation.

2.6   Global algorithm and procedure

The microstructure model of rock materials includes two elements: blocks and joints. The construction method of this model allows FEM to be applied to the computation in the block, and the DE is used to compute the joint interface. Therefore, FEM and DEM were combined to analyse the deformation and fracture of the solid. The FVM is used to compute the fluid flow; thus, the HM coupling problem in the numerical simulation of hydrofracturing is introduced into the numerical computation.

The global methodology and procedure for studying hydraulic fracture propagation driven by fluid based on the dual bilinear cohesive fracture criterion is shown in Figure 2.6. First, we analyse the porous media with fluid as the HM coupling and convert it into FE and FV models. In the procedure, the computation of mechanical

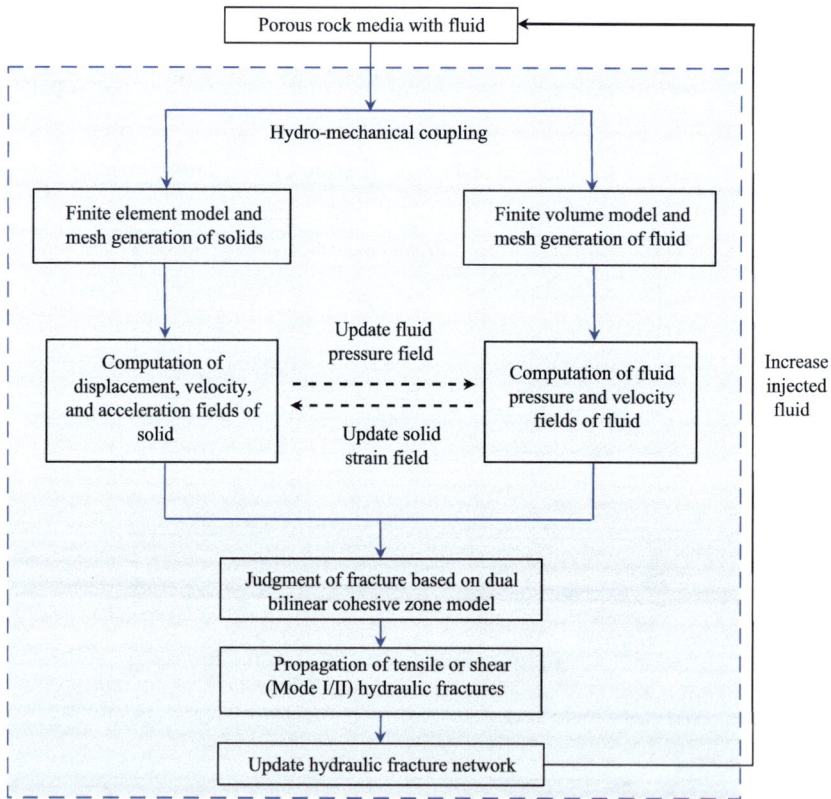

Figure 2.6. Global procedure of dual bilinear cohesive zone model-based fluid-driven propagation of tensile and shear hydraulic fractures.

deformation and fluid flow is shown on the left and right sides, respectively. The basic variables of the solid, such as displacement, velocity, and stress, are computed by FEM, and the basic variables of the fluid, such as fluid pressure and velocity, are computed by the FVM. The solid and fluid are in a state of balance by iteration, and the displacements, stresses, fluid flow, and fluid pressure in a stable state are obtained. Finally, based on the current stress state, the hydraulic tensile and shear fractures (Mode I/II) propagate once the fracture criterion is satisfied. Subsequently, the increase in the injected fluid flow is computed in the new fracture network for the next round to implement the propagation of hydraulic fractures in response to hydraulic pressure.

2.7   Results and discussion

In this section, the accuracy of the FE-DE-FV method is verified by comparing the analytical solution of the fracture length obtained by the quasi-two-dimensional model with the analytical solution in the KGD and PKN models. Next, the fracture propagation behaviour of multiscale rock mass is studied using this algorithm, and the reliability of the algorithm process and implementation scheme considering the HM coupling of hydrofracturing based on the FE-DE-FV method is verified. According to the simulation results, the dynamic propagation of multiple hydraulic fractures and the distribution of tensile and shear fractures in hydrofracturing were solved.

2.7.1   Verification of fracture propagation through analytical solutions in KGD and PKN models

The well-known theoretical models of hydrofracturing include the PKN model (Perkins and Kern, 1961) and the KGD model (Barenblatt, 1962; Nordgren, 1972), in which the leak-off effects are considered to be constant along the hydraulic fractures. The PKN and KGD models are relatively mature two-dimensional contour models for simulating fracture propagation in hydrofracturing, which have reasonable assumptions, complete mathematical models, and clear analytical solutions. The PKN and KGD models can be used to verify the accuracy of the numerical model, in addition to the computation of the hydrofracturing model. The numerical model based on the FE-DE-FV method can simulate the hydrofracturing process accurately and efficiently. By comparing the analytical solution of the fracture length obtained by the two-dimensional contour model with the numerical model, the numerical model can also be optimised to further increase the accuracy of the numerical model.

The KGD model has the following assumptions: the fracture height remains constant during the fracture propagation; the horizontal plane where the fracture extends is in a plane strain condition; the fracturing fluid is a Newtonian fluid, and the fluid flow in the fracture satisfies the cubic law; the fracture tip fulfils the Barenblatt fracture tip condition; the rock matrix is in an elastic state; and leakage of fracturing fluid is not considered. The analytical solution of the fracture length $L(t)$ in the KGD model can be obtained by solving

$$L(t) = 0.68 \left[\frac{Gq_0^3}{\mu(1-v)h^3} \right]^{\frac{1}{6}} t^{\frac{2}{3}}, \tag{2.19}$$

where h is the fracture height, q_0 is the flow rate of the fracturing fluid injection, μ is the viscosity coefficient of the fracturing fluid, G is the shear modulus of the rock matrix, v is the Poisson's ratio of the rock matrix. The PKN model has the following assumptions: the fracture height remains constant in the process of fracture propagation; the vertical plane where the fracture extends is under plane strain condition; the liquid pressure p on the cross section of the Newtonian fluid perpendicular to the fracture propagation direction is a constant, the cross section of vertical fractures is oval, the rock matrix is in an elastic state, and the fracture toughness does not affect the fracture geometry deformation. The analytical solution of the fracture length $L(t)$ in the PKN model can be obtained by solving

$$L(t) = 0.605 \left[\frac{Gq_0^3}{\mu(1-v)h^3} \right]^{\frac{1}{5}} t^{\frac{4}{5}}, \tag{2.20}$$

where h is the fracture height, q_0 is the flow rate of the fracturing fluid injection, μ is the viscosity coefficient of the fracturing fluid, G is the shear modulus of the rock matrix, v is the Poisson's ratio of the rock matrix. All parameters for the KGD and PKN models are the same as those for the model proposed in this study (Table 2.2).

Table 2.2. Parameters for verification of fracture propagation through analytical solutions in KGD and PKN models.

Parameter	Value
Fracture height h /m	1
Injected flow rate q /(m³/s)	1×10^{-5}
Viscosity coefficient μ /(Pa·s)	1×10^{-11}
Shear modulus G /GPa	28
Poisson's ratio ν	0.25

In this section, a quasi-two-dimensional model is used to simulate the propagation behaviour of rock fractures. The model conforms to the assumption of a two-dimensional contour model (KGD and PKN model), and the numerical solution can be compared with the analytical solution of the two-dimensional contour model to verify the accuracy of the method used in this study. The model for the verification of fracture propagation using analytical solutions is shown in Figure 2.7. The geometric model is shown in Figure 2.7(a): the size of the quasi-two-dimensional model is $100 \text{ m} \times 100 \text{ m} \times 1 \text{ m}$, and the perforation is located at the midpoint of the edge of the model. The FE model is shown in Figure 2.7(b), and the eight-node hexahedron element is used to divide the mesh into 6000 elements and 12,322 nodes. The mesh is encrypted in the region of $y = 40$ m to $y = 60$ m , and the element size is $1 \text{ m} \times 1 \text{ m} \times 1 \text{ m}$, which makes the results more accurate and can be compared with the two-dimensional fracture propagation model. The boundary condition of the load is shown in Figure 2.7(c), where the two sides perpendicular to the x axis are constrained in the y direction, and two sides perpendicular to the y axis are constrained in the x direction. The *in-situ* stress is loaded on the model boundary to better simulate the actual engineering conditions and speed up the fracture propagation. The main *in-situ* stress parallel to the x axis was set to S_h , and the main *in-situ* stress parallel to the y axis was set to S_H , as is shown in Figure 2.7(d). In this computational model, we set $S_h = 40$ MPa and $S_H = 60$ MPa . In this study, the FE-DE-FV method was used, and 100,000 time steps were set to reach a fracture length of 100 m. The numerical solution obtained was compared with two-dimensional contour fracture propagation models.

The analytical solution of the fracture propagation length of the rock mass computed by the model is compared with the analytical solution of the fracture propagation length of the KGD model (Figure 2.8). The analytical solution of the fracture propagation length of the rock mass computed by the model is compared with the analytical solution of the fracture propagation length of the PKN model (Figure 2.9). In addition, a comparison is conducted between the numerical solution and analytical solutions of the KGD and PKN models of fracture propagation length when $t =$ 0–600 s is shown in Figure 2.10. Through the numerical solution of the model, it can be found that the speed of rock fracture propagation decreases with the increase of propagation time, which is basically the same as the analytical solution of KGD and PKN models. This verifies the accuracy of the FE-DE-FV method, which can be used to study multiscale rock fracture propagation behaviour. Compared with the analytical solution of the KGD and PKN models, the numerical solution of the model is closer to the analytical solution of the PKN model. This difference is because the PKN model assumes that the vertical plane of fracture propagation is a plane strain state, whereas the KGD model assumes that the horizontal plane of fracture propagation is a plane strain state. The accuracy of the PKN model increases when the difference between the fracture length and fracture height is large, whereas that of the KGD model increases when the difference between the fracture length and fracture height is small.

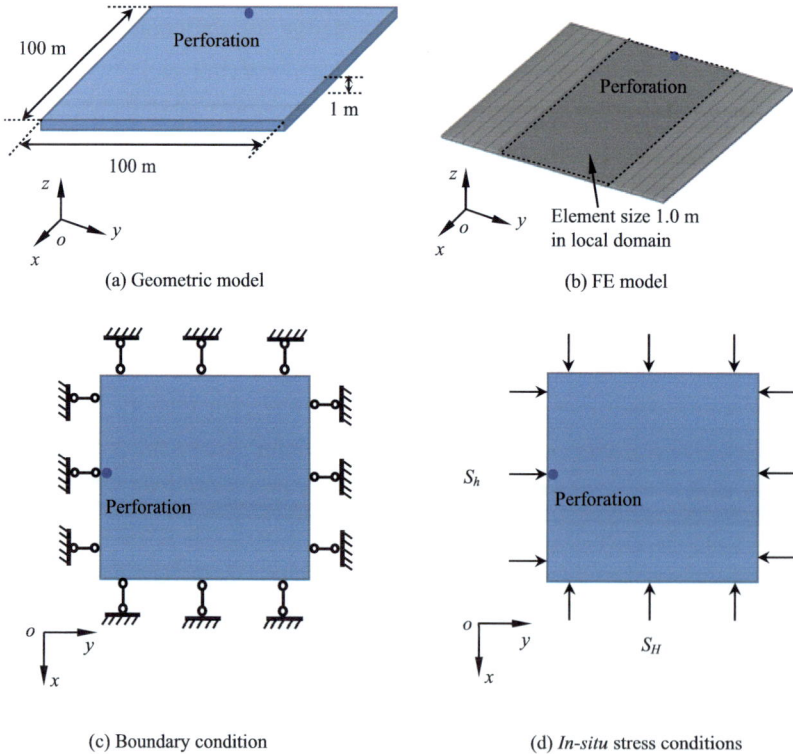

(a) Geometric model   (b) FE model

(c) Boundary condition   (d) *In-situ* stress conditions

Figure 2.7. Model for verification of fracture propagation through analytical solutions.

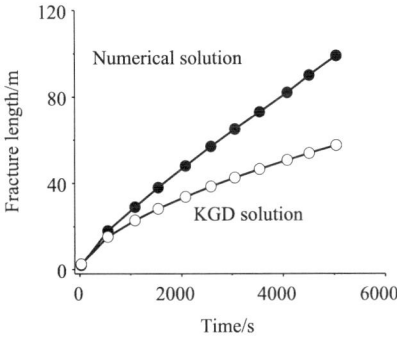

Figure 2.8. Comparison between the numerical solution and analytical solution of the KGD model of fracture propagation length.

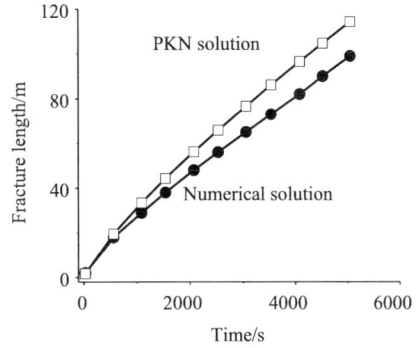

Figure 2.9. Comparison between the numerical solution and analytical solution of the PKN model of fracture propagation length.

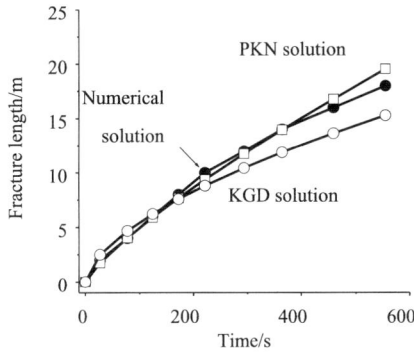

Figure 2.10. Comparison among the numerical solution and analytical solutions of the KGD and PKN models of fracture propagation length ($t = 0$–600 s).

2.7.2   Laboratory scales: dynamic propagation of small-size hydraulic fractures

According to the size of the laboratory-scale model, a quasi-two-dimensional fracturing model was designed in Figure 2.11. According to the geometric model shown in Figure 2.11(a), the perforation of the model is in the geometric centre of the model, and the size of the model is $30\,\text{mm} \times 30\,\text{mm} \times 0.3\,\text{mm}$. Figure 2.11(b) shows the FE model: the number of elements is 1600 (eight-node hexahedral element), and the size is $0.75\,\text{mm} \times 0.75\,\text{mm} \times 0.75\,\text{mm}$, and the number of nodes is 3362. The boundary condition of the load was the same as that of the quasi-two-dimensional model. Under the loading condition of *in-situ* stress, the main *in-situ* stress parallel to the x direction is $S_h = 40\,\text{MPa}$, and the main *in-situ* stress parallel to the y direction is $S_H = 44\,\text{MPa}$. The injected conditions were as follows: the geometric centre of the model was considered the perforation, and the injected flow rate was q.

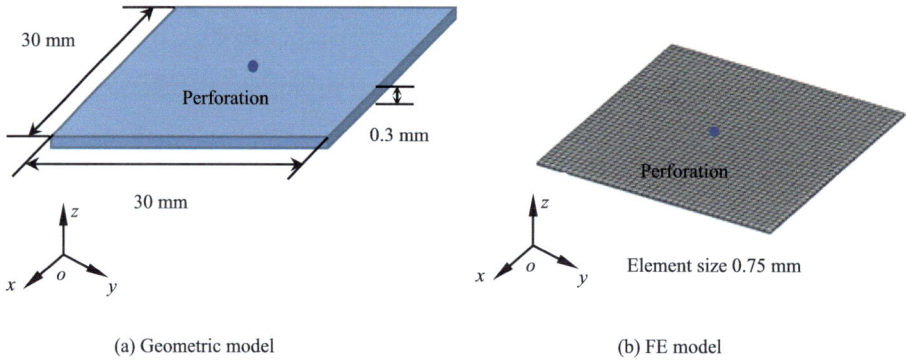

<table>
<tr><td>(a) Geometric model</td><td>(b) FE model</td></tr>
</table>

Figure 2.11. Laboratory-scale model for dynamic propagation of small-size hydraulic fractures.

Considering the actual situation of the laboratory-scale injected flow rate, the injected flow rate is set as $q = 0.0000001\,\text{m}^3/\text{s}$. The fracture propagation morphology and fracture aperture under different loading steps are shown in Figure 2.12. In Figure 2.12(a), the directivity of fracture propagation is not obvious at the beginning of fracturing, and shear fractures appear around the perforation. With the injection of fracturing fluid, the tensile fracture gradually propagates along the direction of the maximum *in-situ* stress. At the end of fracturing, as shown in Figures 2.12(c) and 2.12(d), a large aperture occurs in the middle of the fracture.

2.7.3   Engineering scales: dynamic propagation of large-size hydraulic fractures

According to the size of the engineering-scale model, a quasi-two-dimensional fracturing model is designed (Figure 2.13). According to the geometric model shown in Figure 2.13(a), the perforation of the model is in the geometric centre of the model, and the size of the model is $100\,\text{m} \times 100\,\text{m} \times 1\,\text{m}$. Figure 2.13(b) shows the FE model: the number of elements is 1600 (eight-node hexahedral elements), the size is $2.5\text{m} \times 2.5\text{m} \times 2.5\text{m}$, and the number of nodes is 3,362. The boundary condition of the load was set to be the same as that of the quasi-two-dimensional mode. The injection conditions were as follows: the geometric centre of the model was considered the perforation, and the injected flow rate was q.

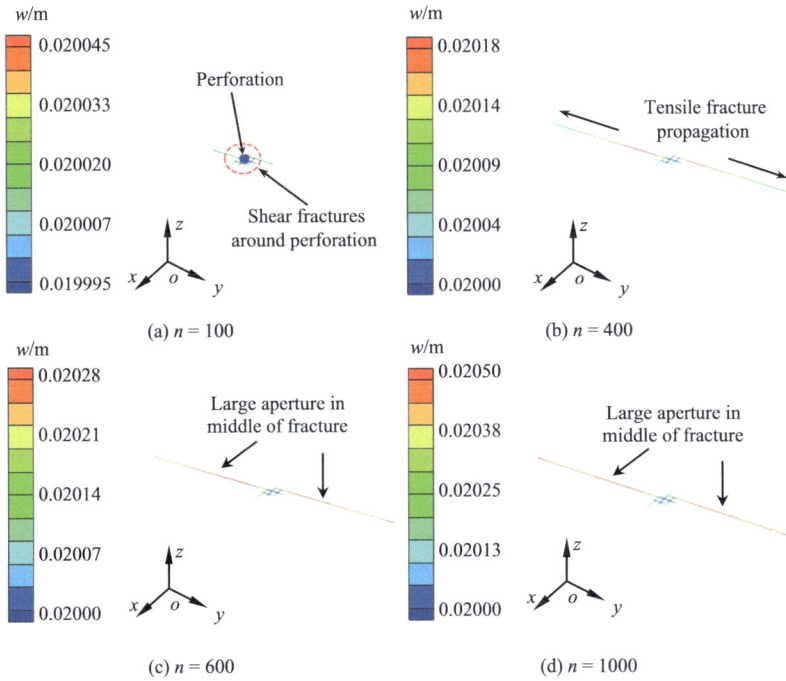

Figure 2.12. Fracture propagation morphology and fracture aperture under different loading steps n.

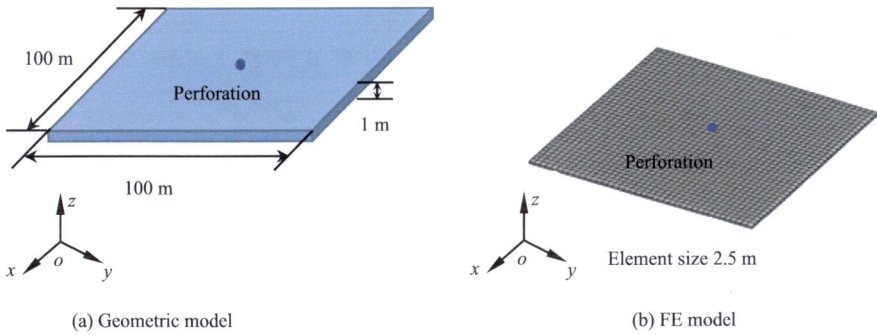

Figure 2.13. Engineering-scale model for dynamic propagation of large-size hydraulic fractures.

Due to the directionality of the horizontal *in-situ* stress distribution, based on the *in-situ* stress measurement data of actual projects, the influence of different injected flow rates on fracture propagation morphology was qualitatively studied under the conditions of three types of horizontal main *in-situ* stress ratios— $S_h : S_H =$ 40MPa:44MPa, 40MPa:60MPa, and 40 MPa : 80 MPa —and under 100000 loading steps (with a certain injected time). The representative numerical cases with different injected flow rates and *in-situ* stresses in the engineering scale are listed in Table 2.3.

Table 2.3. Numerical cases with different injected flow rate and *in-situ* stress in the engineering scale.

Cases	Injected flow rate q /(m³/s)	S_h	S_H
I	0.000001		
		40	44
II	0.00001		
		40	60
III	0.0001		
		40	80
IV	0.001		

The results are shown in Figures 2.14–2.16, from which the following conclusions are obtained: (1) When $q = 0.000001$ m³/s and $S_h : S_H = 40$ MPa: 44 MPa, the maximum horizontal *in-situ* stress has little influence on the fracture propagation direction, and the fracture propagation morphology presents cross-type fractures, as shown in Figure 2.14(a). (2) When $q = 0.00001$ m³/s or $q = 0.0001$ m³/s $S_h : S_H = 40$ MPa:60 MPa, the maximum horizontal *in-situ* stress has little influence

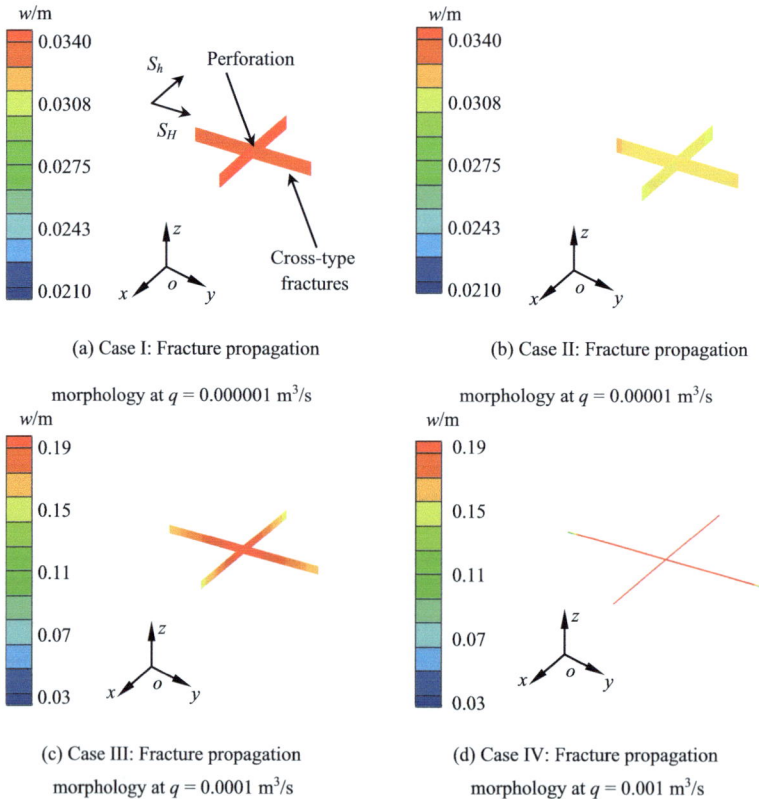

(a) Case I: Fracture propagation

morphology at $q = 0.000001$ m³/s

(b) Case II: Fracture propagation

morphology at $q = 0.00001$ m³/s

(c) Case III: Fracture propagation

morphology at $q = 0.0001$ m³/s

(d) Case IV: Fracture propagation

morphology at $q = 0.001$ m³/s

Figure 2.14. Effects of different injected flow rates on fracture propagation morphology under $S_h : S_H = 40$ MPa : 44 MPa .

on the fracture propagation direction, and the trend of fracture propagation in the direction of maximum *in-situ* stress is slightly obvious, as shown in Figures 2.15(b) and (c). (3) When $q = 0.001 \text{ m}^3/\text{s}$ and $S_h : S_H = 40 \text{ MPa} : 80 \text{ MPa}$, the maximum horizontal *in-situ* stress has a large influence on the fracture propagation direction, and the fracture propagation extends along the direction of maximum *in-situ* stress to form the double wing-type fractures, as shown in Figure 2.16(d).

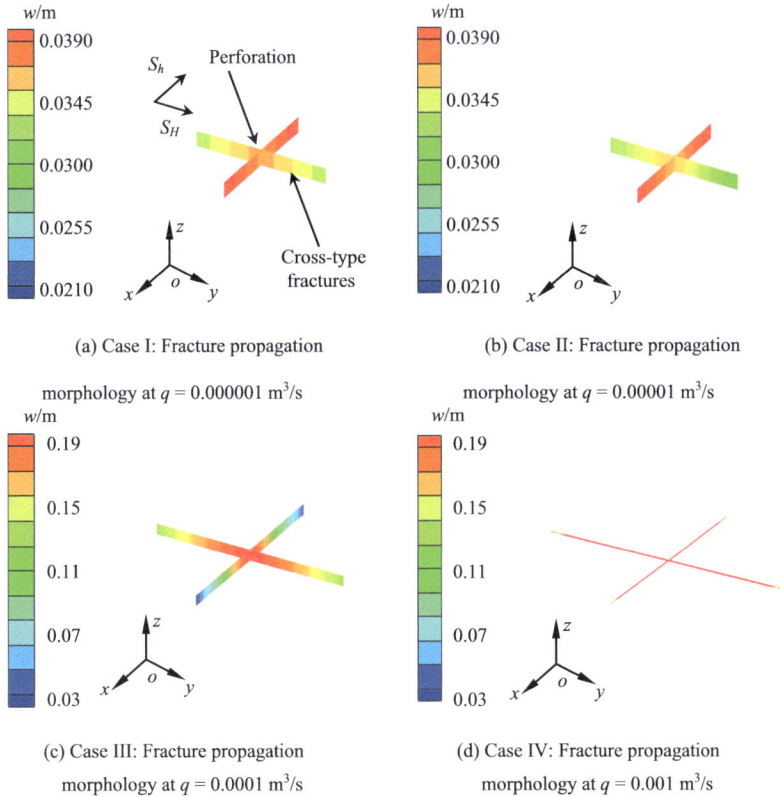

(a) Case I: Fracture propagation

morphology at $q = 0.000001 \text{ m}^3/\text{s}$

(b) Case II: Fracture propagation

morphology at $q = 0.00001 \text{ m}^3/\text{s}$

(c) Case III: Fracture propagation

morphology at $q = 0.0001 \text{ m}^3/\text{s}$

(d) Case IV: Fracture propagation

morphology at $q = 0.001 \text{ m}^3/\text{s}$

Figure 2.15. Effects of different injected flow rates on fracture propagation morphology under $S_h : S_H = 40 \text{ MPa} : 60 \text{ MPa}$.

To test the effectiveness of the proposed method on the computed fracture propagation pattern, we compared and analysed the experimental results (Liu *et al.*, 2018a). In Table 2.4, in the physical experiment of hydrofracturing, there are primarily two types of fracture networks under different *in-situ* stresses: the first type is double wing-type fracture (Table 2.4(a)), in which fracture occurs when the *in-situ* stress ratio is small ($S_h : S_H = 40 \text{ MPa} : 44 \text{ MPa}$), and the second type is cross-type fracture

(a) Case I: Fracture propagation
morphology at $q = 0.000001$ m^3/s

(b) Case II: Fracture propagation
morphology at $q = 0.00001$ m^3/s

(c) Case III: Fracture propagation
morphology at $q = 0.0001$ m^3/s

(d) Case IV: Fracture propagation
morphology at $q = 0.001$ m^3/s

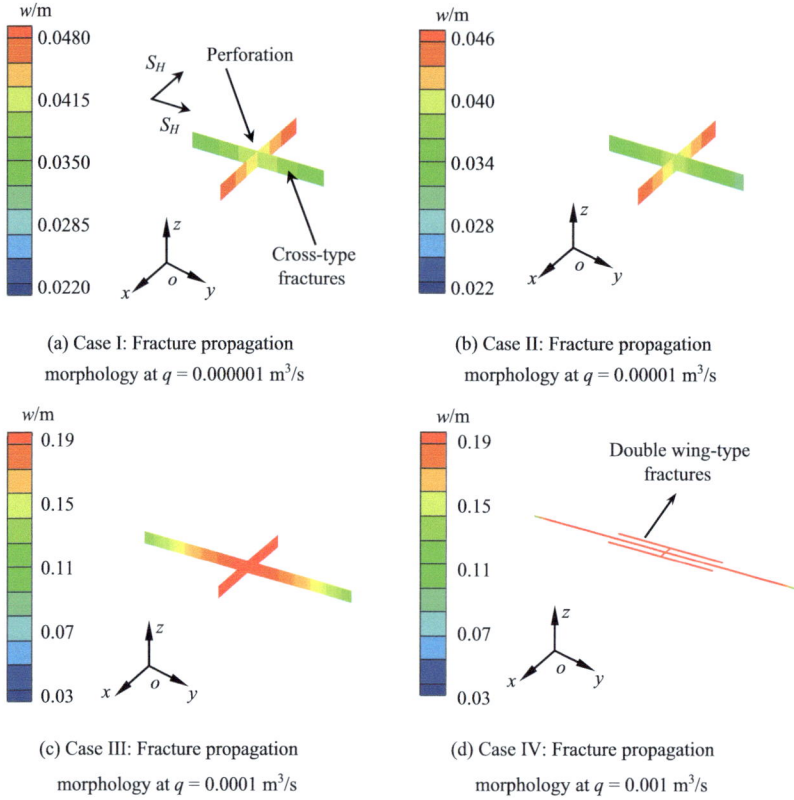

Figure 2.16. Effects of different injected flow rates on fracture propagation morphology under $S_h : S_H = 40$ MPa : 80 MPa.

Table 2.4. Comparison of experimental and numerical simulation results of fracture morphology.

Fracture morphology	Experimental results (Liu *et al.*, 2018a)	Numerical results (Proposed method)
(a) Double wing-type fractures		

Continued

Fracture morphology	Experimental results (Liu *et al.*, 2018a)	Numerical results (Proposed method)
(b) Cross-type fractures		

(Table 2.4(b)), which occur when the *in-situ* stress ratio is larger ($S_h : S_H = 40\,\text{MPa} : 80\,\text{MPa}$). From the numerical simulation of hydrofracturing in this study, the two types of fracture propagation were effectively simulated under different *in-situ* stress cases. The morphology of fracture distribution computed by the numerical models is consistent with the experimental results, which verifies the effectiveness of the proposed method for the simulation of various types of fractures.

2.7.4   Distribution of tensile and shear fractures in hydrofracturing process

According to the conventional theory, hydrofracturing typically occurs via tensile fracture generation. However, shear-type mechanisms have been observed in most AE events recorded during laboratory and *in-situ* hydrofracturing experiments (Falls *et al.*, 1992). A comparison between the experimental and numerical results of the shear fractures is shown in Figure 2.17. The geometric shape of the hydraulic fracture propagation obtained using CT scanning technology is shown in Figure 2.17(a). The shear failure mechanism around the perforation is obviously caused by hydrofracturing (Ma *et al.*, 2017). Based on the DEM, the fracture propagation evolution law and fracture mode of dual-porosity rock synchronous hydrofracturing were numerically simulated. The fracturing results show that when the stress ratio decreases, small shear fractures appear around the perforation (Yang *et al.*, 2021), as shown in Figure 2.17(b). In this study, the fracture propagation diagram obtained by simulating hydrofracturing at the laboratory scale is shown in Figure 2.17(c). The *in-situ* stress condition was $S_h : S_H = 40\,\text{MPa} : 44\,\text{MPa}$. Shear fractures occur around perforations during hydrofracturing.

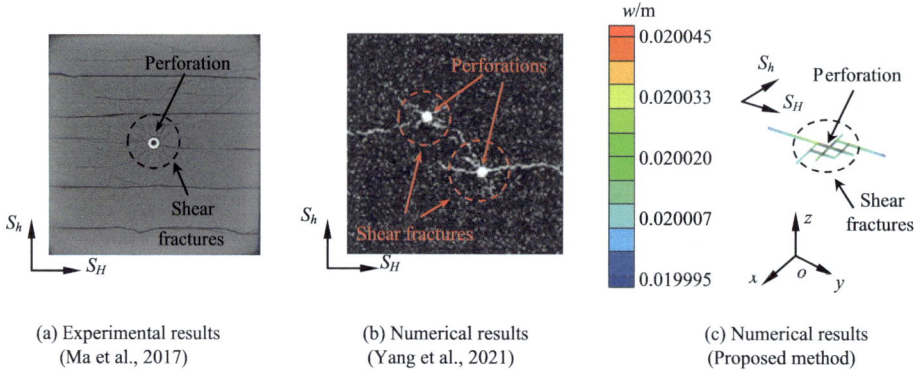

| (a) Experimental results (Ma et al., 2017) | (b) Numerical results (Yang et al., 2021) | (c) Numerical results (Proposed method) |

Figure 2.17. Comparison of experimental and numerical results of shear fractures around perforation.

The initiation and propagation of hydrofracturing fractures under hydraulic pressure are important factors in determining the distribution pattern of a hydrofracturing network. In the hydrofracturing model, fractures initiate around the perforation if the fluid pressure exceeds the strength of bonding and continues to appear, propagate, and break down under the constantly injected fluid flow. In this study, the distribution of tensile and shear fractures was studied using hydrofracturing simulation results at the laboratory and engineering scales.

The fracture propagation patterns at the laboratory scale are shown in Figure 2.12. The fracture propagation pattern of the engineering scale is shown in Figures 2.14–2.16. In the process of fracturing, fractures initially appear around the perforation, and with the increase in loading steps, continues to propagate under the influence of fluid pressure. Moreover, the initial fractures around the perforation diminished with an increase in the *in-situ* stress ratio. There are more tensile fractures than shear fractures, and the shear fractures primarily appear near the hole, where the effect of *in-situ* stress differences is most obvious. The tensile fracture direction is parallel to the maximum *in-situ* stress direction, which conforms to the typical experimental results (Liu *et al.*, 2016). The fracture propagation direction is also related to the injected velocity. When the injected flow rate is small or large, the minimum *in-situ* stress has little influence on the fracture propagation direction; only when the injected flow rate reaches a certain value, the minimum *in-situ* stress has a significant influence on the fracture propagation direction. At this time, the fractures primarily extend along the direction perpendicular to the minimum *in-situ* stress. When the ratio of the injected flow rate to the *in-situ* stress is appropriate, the fracture propagation bifurcates.

2.8   Conclusions

Based on this study, our primary concluding remarks are summarised as follows:

(1) A dual bilinear CZM based on energy evolution was introduced to detect the initiation and propagation of fluid-driven tensile and shear fractures. The model overcomes the limitations of classical linear fracture mechanics, such as the stress singularity at the fracture tip, and considers the important role of fracture surface behaviour in the shear activation. The bilinear cohesive criterion based on the energy evolution criterion can reflect the formation mechanism of complex fracture networks objectively and accurately. Considering the HM coupling and leak-off effects, the combined FE-DE-FV approach was introduced and implemented successfully, and the results showed that the models considering HM coupling and leak-off effects could form a more complex fracture network.

(2) The accuracy and applicability of the proposed algorithm were verified by comparing the analytical solution of the fracture length computed by the quasi-two-dimensional fracture propagation model with the KGD and PKN models. The comparison of the numerical solution of the model with the analytical solution of the KGD and PKN models showed that the speed of rock fracture propagation decreased with the increase in propagation time. Therefore, the algorithm can be used to study multiscale rock fracture propagation behaviour.

(3) The effects of different *in-situ* stresses and flow rates on the dynamic propagation of hydraulic fractures at multiple scales were investigated. At the laboratory scale, because of the injected flow rate, the fracture propagation direction was not obvious at the beginning of fracturing, and as time went on, the fracture propagated along the direction parallel to the maximum *in-situ* stress. Moreover, a large fracture aperture appeared in the middle of the fracture. At the engineering scale, when the ratio of *in-situ* stress is small, the fracture propagation direction is not affected, and the fracture morphology is a cross-type fracture. When the ratio of *in-situ* stress is relatively large, the propagation direction of the fracture is affected by the maximum *in-situ* stress, and it is more inclined to propagate along the direction of the maximum *in-situ* stress, forming double wing-type fractures.

(4) Hydrofracturing tensile and shear fractures were identified, and the

distribution and number of each type were obtained. In the fracturing process, the total number and distribution of fractures are closely related to the *in-situ* stress ratio. With an increase in the *in-situ* stress ratio, the total number of fractures decreases, and it is more likely to have a double wing-type fracture. There are fewer hydraulic shear fractures than tensile fractures, and shear fractures appear in the initial stage of fracture propagation and then propagate and distribute around the perforation. With the injection of fluid, tensile fractures form a straight main fracture, and the direction of the fracture is parallel to the direction of the maximum *in-situ* stress.

This work introduces a dual bilinear CZM for detecting the initiation and propagation of fluid-driven tensile and shear fractures. The FE-DE-FV method involved HM coupling and leak-off effects and verified its applicability. Practical fracturing process involves the multi-type and multiscale fluid-driven fracture propagation. This study introduces general fluid-driven fracture propagation, which can be extended to the fracture propagation analysis of fluid fracturing, such as other liquids or supercritical gases, which will be the subject of further research.

References

Al-Rubaie, A., Ben Mahmud, H. K. (2020), "A numerical investigation on the performance of hydraulic fracturing in naturally fractured gas reservoirs based on stimulated rock volume", Journal of Petroleum Exploration and Production Technology, Vol. 10 No. 8, pp. 3333–3345.

Barenblatt, G. I. (1959), "The formation of equilibrium cracks during brittle fracture. General ideas and hypotheses. Axially-symmetric cracks", J.Appl.Math.Mech, Vol. 23 No. 3, pp. 622–636.

Barenblatt, G. I. (1962), "The mechanical theory of equilibrium cracks in brittle fracture", Advances in Applied Mechanics, Vol. 7, pp. 55–129.

Bažant, Z. P., Salviato, M., Chau, V. T., Viswanathan, H., Zubelewicz, A. (2014), "Why fracking works", Journal of Applied Mechanics, Vol. 81 No. 10, pp. 1–10.

Bohloli, B., De Pater, C. J. (2006), "Experimental study on hydraulic fracturing of soft rocks: influence of fluid rheology and confining stress", Journal of Petroleum Science and Engineering, Vol. 53 No. 1–2, pp. 1–12.

Bunger, A. P., Kear, J., Dyskin, A. V., Pasternak, E. (2015), "Sustained acoustic emissions following tensile crack propagation in a crystalline rock", International Journal of Fracture, Vol. 193 No. 1, pp. 87–98.

Chandler, M. R., Meredith, P. G., Brantut, N., Crawford, B. R. (2016), "Fracture toughness anisotropy in shale", Journal of Geophysical Research: Solid Earth, Vol. 121 No. 3, pp. 1706–1729.

Chen, Y., Nagaya, Y., Ishida, T. (2015), "Observations of fractures induced by hydraulic fracturing in anisotropic granite", Rock Mechanics and Rock Engineering, Vol. 48 No. 4, pp. 1455–1461.

Dahi Taleghani, A., Gonzalez-Chavez, M., Yu, H., Asala, H. (2018), "Numerical simulation of hydraulic fracture propagation in naturally fractured formations using the cohesive zone model", Journal of Petroleum Science and Engineering, Vol. 165, pp. 42–57.

Deng, S., Li, H., Ma, G., Huang, H., Li, X. (2014), "Simulation of shale–proppant interaction in hydraulic fracturing by the discrete element method", International Journal of Rock Mechanics and Mining Sciences, Vol. 70, pp. 219–228.

Duan, K., Li, Y., Yang, W. (2020), "Discrete element method simulation of the growth and efficiency of multiple hydraulic fractures simultaneously-induced from two horizontal wells", Geomechanics and Geophysics for Geo-Energy and Geo-Resources, Vol. 7 No. 1.

Dugdale, D. S. (1960), "Yielding of steel sheets containing slits", Journal of the Mechanics and Physics of Solids, Vol. 8 No. 2, pp. 100–104.

Falls, S. D., Young, R. P., Carlson, S. R., Chow, T. (1992), "Ultrasonic tomography and acoustic emission in hydraulically fractured Lac du Bonnet Grey granite", Journal of Geophysical Research, Vol. 97 No. B5.

Griffith, A. A. (1920), "The phenomena of rupture and flow in solids", Philosophical Transactions of the Royal Society A: Mathematical, Physical and Engineering Sciences, Vol. A221 No. 4, pp. 163–198.

Hou, B., Chen, M., Li, Z., Wang, Y., Diao, C. (2014), "Propagation area evaluation of hydraulic fracture networks in shale gas reservoirs", Petroleum Exploration and Development, Vol. 41 No. 6, pp. 833–838.

Irwin, G. R. (1957), "Analysis of stresses and strains near the end of a crack traversing a plate", Journal of Applied Mechanics, Vol. 24, pp. 361–364.

Irwin, G. R., Kies, J. A., Smith, H. L. (1958), "Fracture strengths relative to onset and arrest of crack propagation", Proceedings of the American Society for Testing Materials, Vol. 58, pp. 640–657.

Ju, Y., Chen, J., Wang, Y., Gao, F., Xie, H. (2018), "Numerical analysis of hydrofracturing behaviours and mechanisms of heterogeneous reservoir glutenite, using the continuum-based discrete element method while considering hydro-mechanical coupling and leak-off effects", Journal of Geophysical Research: Solid Earth, Vol. 123 No. 5, pp. 3621–3644.

Khoei, A. R., Vahab, M., Hirmand, M. (2018), "An enriched–FEM technique for numerical simulation of interacting discontinuities in naturally fractured porous media", Computer Methods in Applied Mechanics and Engineering, Vol. 331, pp. 197–231.

Liu, J., Yao, Y., Liu, D., Xu, L., Elsworth, D., Huang, S., Luo, W. (2018a), "Experimental simulation of the hydraulic fracture propagation in an anthracite coal reservoir in the southern Qinshui basin, China", Journal of Petroleum Science and Engineering, Vol. 168, pp. 400–408.

Liu, P., Ju, Y., Ranjith, P. G., Zheng, Z., Chen, J. (2016), "Experimental investigation of the effects of heterogeneity and geostress difference on the 3D growth and distribution of hydrofracturing cracks in unconventional reservoir rocks", Journal of Natural Gas Science and Engineering, Vol. 35, pp. 541–554.

Liu, W., Zeng, Q., Yao, J. (2018b), "Numerical simulation of elasto-plastic hydraulic fracture propagation in deep reservoir coupled with temperature field", Journal of Petroleum Science and Engineering, Vol. 171, pp. 115–126.

Ma, X., Li, N., Yin, C., Li, Y., Zou, Y., Wu, S., He, F., Wang, X., Zhou, T. (2017), "Hydraulic fracture propagation geometry and acoustic emission interpretation: A case study of Silurian Longmaxi Formation shale in Sichuan Basin, SW China", Petroleum Exploration and Development, Vol. 44 No. 6, pp. 1030–1037.

Munjiza, A. (2004), The combined finite-discrete element method. John Wiley & Sons Press.

Munjiza, A., Owen, D.R.J., Bicanic, N. (1995), "A combined finite-discrete element method in transient dynamics of fracturing solids", Engineering Computations, Vol. 12 No. 2, pp. 145–174.

Nordgren, R. P. (1972), "Propagation of a vertical hydraulic fracture", Society of Petroleum Engineers Journal, Vol. 12 No. 04, pp. 306–314.

Palaniswamy, K., Knauss, W. G. (1978), "On the problem of crack extension in brittle solids under general loading", Mechanics Today, Vol. 4, pp. 87 148.

Perkins, T. K., Kern, L. R. (1961), "Widths of hydraulic fractures", Journal of Petroleum Technology, Vol. 13 No. 09, pp. 937–949.

Secchi, S., Schrefler, B. A. (2012), "A method for 3-D hydraulic fracturing simulation", International Journal of Fracture, Vol. 178 No. 1–2, pp. 245–258.

Shi, F., Gao, F., Li, X. R., Shen, X. M. (2014), "Propagation finite element method for simulating primary and secondary crack initiation and cracking in rock under compression and shear", Geotechnical Mechanics, Vol. 35 No. 6, pp. 1809–1817.

Snow, D. T. (1965), "A parallel plate model of fractured permeable media", Ph.D. Dissertation University of California, USA.

Wang, Y. (2020), Adaptive analysis of damage and fracture in rock with multiphysical fields coupling, Springer Press.

Wang, Y. (2021), "Adaptive finite element–discrete element analysis for stratal movement and microseismic behaviours induced by multistage propagation of three-dimensional multiple hydraulic fractures", Engineering Computations, Vol. 38 No. 5, pp. 1350–1371.

Wang, Y., Ju, Y., Yang, Y. (2018), "Adaptive finite element-discrete element analysis for microseismic modelling of hydraulic fracture propagation of perforation in horizontal well considering pre-existing fractures", Shock and Vibration, Vol. 2018, pp.1–14.

Wang, Y., Ju, Y., Chen, J., Song, J. (2019), "Adaptive finite element-discrete element analysis for the multistage supercritical CO_2 fracturing and microseismic modelling of horizontal wells in tight reservoirs considering pre-existing fractures and thermal-hydro-mechanical coupling", Journal of Natural Gas Science and Engineering, Vol. 61, pp. 251–269.

Wang, Y., Ju, Y., Zhang, H., Gong, S., Song, J., Li, Y., Chen, J. (2021), "Adaptive finite element-discrete element analysis for the stress shadow effects and fracture interaction behaviours in three-dimensional multistage hydrofracturing considering varying perforation cluster spaces and fracturing scenarios of horizontal wells", Rock Mechanics and Rock Engineering, Vol. 54 No. 4, pp. 1815–1839.

Wu, M. Y., Zhang, D. M., Wang, W. S., Li, M. H., Liu, S. M., Lu, J., Gao, H. (2020), "Numerical simulation of hydraulic fracturing based on two-dimensional surface fracture morphology reconstruction and combined finite-discrete element method", Journal of Natural Gas Science and Engineering, Vol. 82.

Xie, H. P., Chen, Z. D. (1988), "Discussion on continuous damage mechanical model of rock", Journal of China Coal Society, Vol. 01, pp. 33–42.

Yang, T. H., Tham, L. G., Tang, C. A., Liang, Z. Z., Tsui, Y. (2004), "Influence of heterogeneity of mechanical properties on hydraulic fracturing in permeable rocks", Rock Mechanics and Rock Engineering, Vol. 37 No. 4, pp. 251–275.

Yang, W., Li, S., Geng, Y., Zhou, Z., Li, L., Gao, C., Wang, M. (2021), "Discrete element numerical simulation of two-hole synchronous hydraulic fracturing", Geomechanics and Geophysics for Geo-Energy and Geo-Resources, Vol. 7 No. 3.

Zhou, T., Wang, H., Li, F., Li, Y., Zou, Y., Zhang, C. (2020), "Numerical simulation of hydraulic fracture propagation in laminated shale reservoirs", Petroleum Exploration and Development, Vol. 47 No. 5, pp. 1117–1130.

Chapter 3　Multi-thread parallel computation method for dynamic propagation of hydraulic fracture networks

3.1　Introduction

The exploitation of conventional oil and gas energy is being gradually exhausted worldwide, such that unconventional oil and gas resources are beginning to occupy an important position in the global energy resource structure due to the increasing global demand for energy (Ahmed and Meehan, 2019; Song *et al.*, 2017; Wang *et al.*, 2016). China has huge reserves of unconventional oil and gas resources; hence, reasonable exploitation of tight oil, gas, and shale gas can reduce the consumption of conventional energy and optimize the energy structure (Gao *et al.*, 2021; Song *et al.*, 2020; Zou *et al.*, 2010). Hydrofracturing refers to the method where fracturing fluid is injected into a tight reservoir using a high-pressure pump group on the ground. Fluid-driven dynamic fracture propagation in a porous elastic rock mass is the main process of hydrofracturing. When the fracturing fluid pressure is greater than the strength of the reservoir rock, the rock will produce fractures and then form a hydraulic fracture network, such that oil and gas can flow out of the tight reservoir along the hydraulic fracture network to improve the production of tight oil and gas (Ingraffea, 1983; Weng *et al.*, 2010). Since the unconventional energy exploitation mode by hydrofracturing was proposed, the demand for the exploitation of unconventional natural gas resources has progressed, and hydrofracturing technology has been increasingly used (Montgomery and Smith, 2010; Beckwith, 2010; Chen *et al.*, 2021). Hydrofracturing has greatly improved the efficiency of conventional oil and gas energy exploitation. Hydrofracturing is still an effective method to improve its productivity, however the fracture growth, proppant transport and fluid flow, and induced reservoir characteristics damaged in unconventional oil and gas energy exploitation are extremely complex (Cipolla *et al.*, 2009; Ajayi *et al.*, 2013; Nicot *et al.*, 2014; Davies *et al.*, 2013; Barboza *et al.*, 2021; Sun *et al.*, 2021). Unconventional oil and gas production methods are completely different from conventional oil and gas production techniques, which has led to restrictions on the use of conventional research methods and technical means (Scanlon *et al.*, 2014; Rahm, 2011).

Some key factors (such as the coupling between multiphysical fields including fluid and solid, large-scale mining modes on engineering-scale structures, and multi-scale fracture evolution behaviours) involved in the hydrofracturing process, which make the process exceedingly complex (Vaziri, 1996; Bower and Zyvoloski, 1997; Figueiredo *et al.*, 2015; Li *et al.*, 2017; Gao *et al.*, 2020; Li *et al.*, 2017; Su *et al.*, 2015). These problems make the fracturing production process a complex physical procedure. It is difficult to detect and control detailed information through physical methods and to obtain the entire process of fracture propagation and evolution through field monitoring and indoor physical experiments; therefore, the control and optimization of the fracture network is a challenge. A comprehensive analysis of the behaviour and mechanism of fracture network propagation is unlikely without understanding the physical process (Fu *et al.*, 2013; McClure *et al.*, 2016). Numerical simulations can simulate the entire process of the generation and expansion of hydraulic fractures. However, the current numerical simulation method for the hydrofracturing process has certain limitations for efficiently capturing the evolution process of fracture propagation. The slowness of the solution process is precisely because of the difficulty to solve the problems of multiphysical fields and large-scale models; therefore, the computation mode needs to be updated to improve the computation speed (Li *et al.*, 2018; Cai *et al.*, 2007). One of the key methods for improvement in this regard is parallel computation (Byeon and Lee, 2020; Becker and Noels, 2013; Gao and Ghassemi, 2020; Wheeler *et al.*, 2020). However, because the numerical model on the engineering scale often contains a large number of elements and degrees of freedom, the numerical simulation is limited to small-scale problems. With the development of parallel computation, it is possible to introduce parallel computation algorithms and models into the study of rock fracture propagation (Gebali, 2011; Dimakopoulos, 2014).

In the numerical simulation of hydrofracturing, some typical models and methods have been developed, such as finite element method (FEM) (Giovanardi *et al.*, 2020; Settgast *et al.*, 2016), extended finite element method (XFEM) (Sukumar *et al.*, 2000; Wang *et al.*, 2018), discrete element method (DEM) (Lisjak and Grasselli, 2014; Deng *et al.*, 2014), boundary element method (BEM) (Mi and Aliabadi, 1992; Ke *et al.*, 2008), and phase field method (PFM) (Borden *et al.*, 2012; Ziaei-Rad and Shen, 2016). To solve the efficient computation of fracture propagation in an engineering-scale rock reservoir model, some parallel computation schemes have been developed when these methods are applied. What needs to be emphasized here is the combined finite element and discrete element coupling methods for continuous and discontinuous analysis (Lei et al., 2014; Lukas et al., 2014; Rougier et al., 2014; Fukuda et al., 2021). Based on this method, the parallel computation of the fracture and failure process of the rock mass is carried out. Through this method, the continuous physical field (such as displacement and stress field) and discontinuous fracture behaviour in rock mass can be well simulated, and the fracture behaviour results of some three-dimensional large-scale models can be analysed quickly (Lisjak et al., 2018; Fukuda et al., 2020). In particular, some parallel acceleration methods of CPU, GPU, and GPGPU, as well as some methods based on MPI and OpenMP, have been applied in rock mass fracture analysis, but these methods are rarely used to solve the problem of multiphysical field coupling (Owen and Feng, 2001; Mohammadnejad et al., 2020a; Mohammadnejad et al., 2020b). Moreover, considering the fluid and solid coupling behaviours in the fracturing process, computational efficiency can be improved in the solution of each physical field by a parallel computation method, which is an effective way to improve the simulation of fracture propagation in hydrofracturing networks.

The author of this book has carried out preliminary research on the hydrofracturing process of deep tight rock reservoirs using finite element and discrete element methods. The coupling effect of the fluid and solid was considered by using an engineering-scale model (Wang, 2021a, 2021b; Ju et al., 2018). Furthermore, cohesive zone model was applied (Wang et al., 2021), and dual bilinear cohesive zone model-based fluid-driven tensile and shear propagation of multiscale fractures in tight reservoirs were proposed and implemented. However, the method used is a serial computation technique, which is time consuming. The aim of this study is to model and analyse the fluid structure coupling of tight rock reservoirs at an engineering scale. Based on previous research methods, parallel computation methods have been developed to improve the computation efficiency. This chapter introduces the progress obtained regarding parallel computation in this study.

The remainder of this chapter is organized as follows. Section 3.2 introduces the governing partial differential equations and fracture criteria for hydrofracturing. Section 3.3 introduces the multi-thread parallel computation scheme for solid and fluid analyses. Section 3.4 introduces the global algorithm and the procedure. Section 3.5 introduces the results and discussion for the verification of multi-thread parallel computation solutions of hydraulic fracture propagation, multi-thread parallel computation efficiency for fluid-driven fracture propagation, parallel computation using multi-type elements and meshes, and dynamic propagation behaviours of fractures under *in-situ* stresses and external fluid drive. Section 3.6 presents the conclusions of this study.

3.2   Governing partial differential equations and fracture criteria for hydrofracturing

In this study, the governing partial differential equations for hydrofracturing (Governing equations of solid deformation and fluid flow in fractured porous media), numerical discretization (Finite element discretization for solid and finite volume discretization for fluid), and fracture criteria via dual bilinear cohesive zone model are the same as the introductions in Chapter 2. The basic physical parameters of models for hydrofracturing are also extended to this multi-threaded parallel computation, and the number of threads in parallel computation n_t used in this study is set as 8 in the following examples.

3.3   Multi-thread parallel computation scheme for solid and fluid analysis

The OpenMP programming technique can provide an application interface that provides a portable and extensible model for developers of shared memory parallel computation programs. While programming, only the specific pre-compiling instructions need to be added in front of the specific code; the compiler will automatically start parallel computation processing of the code and add thread synchronization and data communication mechanisms. If the compiler chooses to ignore the pre-compiled instruction of parallel computation, or the compiler does not support parallel computation, the program naturally degenerates into a common serial program. This feature makes parallel computing easy to implement. At this point, the code can still operate normally, but it cannot use multi-threading to accelerate the execution of the program. This parallel computation is designed for shared memory parallel computation programming; therefore, it is largely limited to single-node parallelism. Generally, the number of processing elements on a node can determine the extent to which parallelism is achieved. This parallel programming uses and executes the fork-join model. The multi-thread parallel computation scheme and procedure based on the OpenMP technique are illustrated in Figure 3.1. The implementation of the fork-join model first requires task segmentation from the serial domain: that is, dividing large tasks into smaller subtasks for each thread in the parallel domain, and further dividing subtasks if they are still large. Then, the subtasks are merged, that is, the subtasks are placed in two-terminal queues. Subsequently, several parallel threads obtain the task execution from the queue, and the results after the execution of the subtasks are stored in another queue. Then, another thread starts to read the computation data from the queue and finally merges all the computation data, enabling the fork-join model to complete the parallel computation process. In the multi-thread parallel computation procedure used in this study, parallel computing processing can be implemented according to the above-mentioned processes for solid and fluid analysis involved in the hydrofracturing process. A parallel computation scheme is used to compute the solid deformation by FEM, compute the fluid pressure by FVM, and analyse element detection and separation by DEM. In this way, parallel computation and analysis of hydro-mechanical coupling in hydrofracturing is achieved.

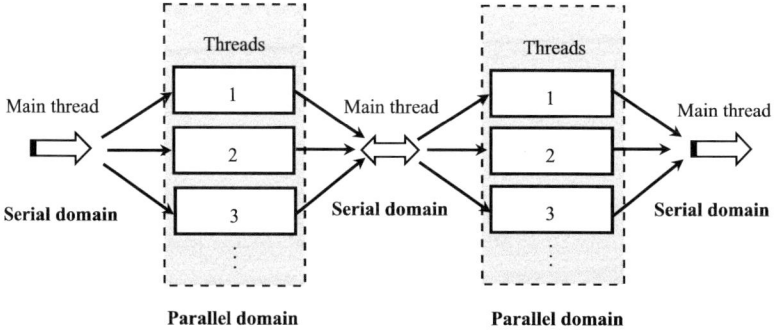

Figure 3.1. Illustration of multi-thread parallel computation scheme and procedure based on OpenMP technique.

3.4   Global algorithm and procedure

The microstructure model of porous rock media includes two elements: blocks and joints. The construction method of this model allows for the FEM to be applied to the computation in the block, and the DE is used to compute the joint interface. Therefore, FEM and DEM were combined to analyse the deformation and fracture of the solid.

The FVM is used to compute the fluid flow; thus, the HM coupling problem in the numerical simulation of hydrofracturing is introduced into the numerical computation.

The global procedure of multi-thread parallel combined finite element-discrete element-finite volume (FE-DE-FV) method for the dynamic propagation of hydraulic fractures considering hydro-mechanical coupling in tight reservoirs is shown in Figure 3.2. Parallel computation is used in the entire solid and fluid coupling process. First, we analyse the porous media with fluid as the HM coupling and convert it into FE and FV models. In the procedure, the computation of mechanical deformation and fluid flow is shown on the left and right sides, respectively. The basic variables of the solid, such as displacement, velocity, and stress, are computed by FEM in parallel, and the basic variables of the fluid, such as fluid pressure and velocity, are computed by the FVM in parallel. The solid field will pass strain to fluid, and the fluid field passes fluid pressure to solid. By iterative computations, the solid and fluid flow gradually become

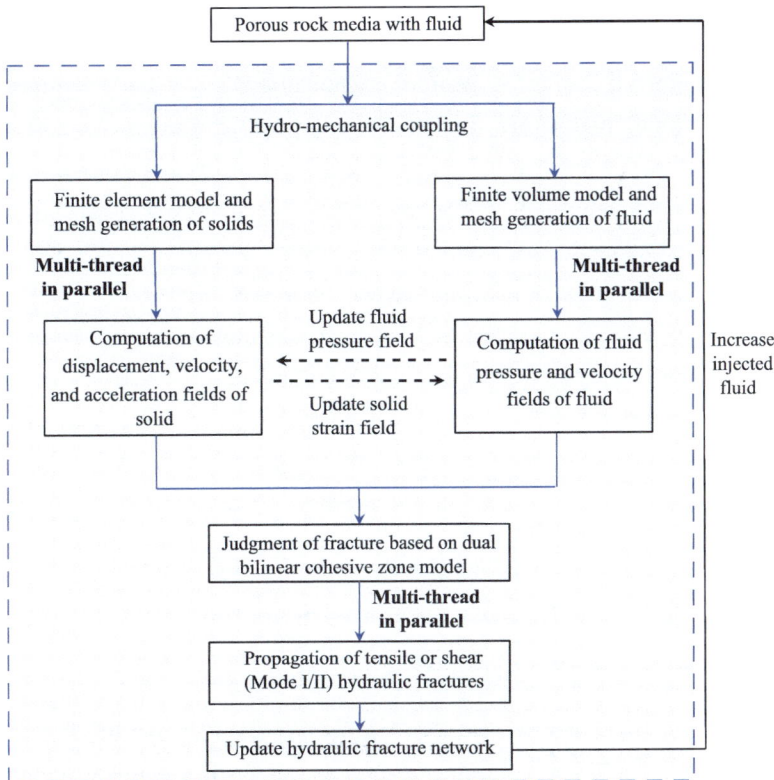

Figure 3.2. Global procedure of multi-thread parallel combined FE-DE-FV method for dynamic propagation of hydraulic fractures considering hydro-mechanical coupling in tight reservoirs.

a state of balance and stability, resulting in stable displacement, stress, fluid flow, and fluid pressure. Once the DE meets the fracture criterion for fractures, hydraulic tensile fractures (Mode I) and shear fractures (Mode II) will propagate. Furthermore, an increase in fluid flow rate was injected into the new fracture network for the next round of computations. Through this continuous increase in fluid load, parallel computation is performed to achieve high-efficiency propagation simulation of hydraulic fractures.

3.5　Results and discussion

In this section, through the results of typical numerical examples, the verification for multi-thread parallel computing solutions of hydraulic fracture propagation was executed using the solutions of the propagation length of hydraulic fractures in the analytical KGD model (Barenblatt, 1962; Nordgren, 1972) and the PKN model (Perkins and Kern, 1961). For testing the computational efficiency, different threads are used to validate the improvement of computational efficiency by multi-thread parallel computation, and some threads are adopted to simulate the fluid-driven fracture propagation. To test parallel computing using multi-type elements and meshes, the effectiveness of solutions under pentahedral and hexahedral elements was successfully implemented, and the mesh sensitivity and convergence for the propagation direction of the fracture were analysed. To analyse the dynamic propagation behaviours of engineering-scale hydraulic fractures under *in-situ* stresses and external fluid drive using an efficient parallel computation method, the differences in each component among the *in-situ* stresses on hydraulic fractures are analysed, and the dynamic viscosity of the fracturing fluid for simulating the liquid and gas are used as fracturing fluids to drive the fracture propagation. Each example and its results are introduced below.

3.5.1　Example 1: Verification for multi-thread parallel computation solutions of hydraulic fracture propagation

The numerical solution obtained by the parallel FE-DE-FV computation method is compared with the analytical solution, which can verify the accuracy of the parallel computation method and optimize the numerical computation model to further increase the accuracy of the numerical model. The well-known theoretical models of hydrofracturing include the PKN and KGD models, in which the leak-off effects are considered to be constant along the hydraulic fractures. The PKN and KGD models are relatively mature two-dimensional contour models for simulating fracture propagation in hydrofracturing, which are characterized by reasonable assumptions, complete mathematical models, and clear analytical solutions. The PKN and KGD models can be used to verify the accuracy of the numerical model, in addition to the computation of the hydrofracturing model. The numerical model based on the parallel FE-DE-FV method can simulate the hydrofracturing process accurately and efficiently. By comparing the analytical solution of the fracture length obtained by the two-dimensional model with the numerical model, the numerical model can also be optimized to further increase the accuracy of the numerical model.

The KGD model has the following assumptions: the fracture height remains constant during the fracture propagation; the horizontal plane where the fracture extends is in a plane strain condition; the fracturing fluid is a Newtonian fluid; and the fluid flow in the fracture satisfies the cubic law; the fracture tip fulfills the Barenblatt fracture tip condition; the rock matrix is in an elastic state; and leakage of fracturing fluid is not considered. The analytical solution of the fracture length $L(t)$ in the KGD model can be obtained by solving (Barenblatt, 1962; Nordgren, 1972).

$$L(t) = 0.68 \left[\frac{G q_0^3}{\mu(1-v)h^3} \right]^{\frac{1}{6}} t^{\frac{2}{3}} \qquad (3.1)$$

where h is the fracture height, q_0 is the flow rate of the fracturing fluid injection, μ is the viscosity coefficient of the fracturing fluid, G is the shear modulus of the rock matrix, and v is the Poisson's ratio of the rock matrix. The PKN model contains the following assumptions: the fracture height remains constant in the process of fracture propagation; the vertical plane where the fracture extends is under plane strain condition; the liquid pressure p on the cross section of the Newtonian fluid perpendicular to the fracture propagation direction is a constant; the cross section of vertical fractures is oval; the rock matrix is in an elastic state; and the fracture toughness does not affect the fracture geometry deformation. The analytical solution of the fracture length $L(t)$ in the PKN model can be obtained by solving (Perkins and Kern, 1961)

$$L(t) = 0.605 \left[\frac{Gq_0^3}{\mu(1-v)h^3} \right]^{\frac{1}{5}} t^{\frac{4}{5}} \tag{3.2}$$

where h is the fracture height, q_0 is the flow rate of the fracturing fluid injection, μ is the viscosity coefficient of the fracturing fluid, G is the shear modulus of the rock matrix, and v is the Poisson's ratio of the rock matrix. All parameters for the KGD and PKN models were the same as those for the model proposed in this study (Table 3.1). The parallel computation in this example adopts eight threads.

Table 3.1. Example 1: Basic physical parameters of models for verification for multi-thread parallel computation solutions of hydraulic fracture propagation.

Parameter	Value
Young's modulus E/GPa	70
Shear modulus G/GPa	28
Fracture height h/m	1
Viscosity coefficient of fracturing fluid μ/(Pa·s)	1×10^{-11}
Poisson's ratio v	0.25
Number of threads in parallel computation n_t	8

In this section, a quasi-two-dimensional model is used to simulate the propagation behaviour of rock fractures. The model conforms to the assumption of a two-dimensional contour model (KGD and PKN model), and the numerical solution can be compared with the analytical solution of the two-dimensional model to verify the accuracy of the method used in this study. The model for the verification of fracture propagation using analytical solutions is shown in Figure 3.3. The geometric model is shown in Figure 3.3(a), where the size of the quasi-two-dimensional model is $100\,\mathrm{m} \times 100\,\mathrm{m} \times 1\,\mathrm{m}$, and the perforation is located at the midpoint of the edge of the model. The FE model is shown in Figure 3.3(b), and an eight-node hexahedron element was used to divide the mesh into 6000 elements and 12,322 nodes. The mesh is encrypted in the region of $y = 40\,\mathrm{m}$ to $y = 60\,\mathrm{m}$, and the element size is $1\,\mathrm{m} \times 1\,\mathrm{m} \times 1\,\mathrm{m}$, which provides more accurate results and can be compared with the two-dimensional fracture propagation model. The main *in-situ* stress parallel to the x-axis was set to S_h, and the main *in-situ* stress parallel to the y-axis was set to S_H. In this computational model, we set $S_h = 40\,\mathrm{MPa}$ and $S_H = 60\,\mathrm{MPa}$. In this study, the parallel FE-DE-FV method was used, and 100,000 time steps were set to reach a fracture length of 100 m. The obtained numerical solutions were compared with two-dimensional fracture propagation models.

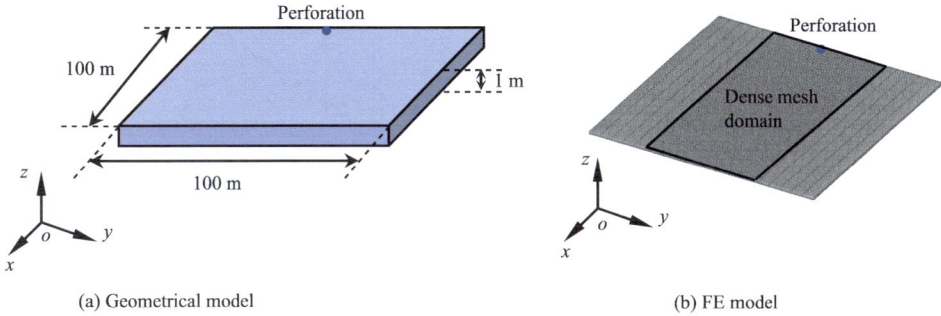

(a) Geometrical model   (b) FE model

Figure 3.3. Example 1: Geometrical and FE models for verification for multi-thread parallel computation solutions of hydraulic fracture propagation.

   The analytical solution of the fracture propagation length of the rock mass computed by the model was compared with the analytical solution of the fracture propagation length of the KGD model (Figure 3.4(a)). The analytical solution of the fracture propagation length of the rock mass computed by the model was compared with the analytical solution of the fracture propagation length of the PKN model (Figure 3.4(b)). In addition, a comparison is conducted between the numerical solution and analytical solutions of the KGD and PKN models of the fracture propagation length when $t = 0$–600 s, as shown in Figure 3.4(c). Through the numerical solution of the model, we found that the speed of rock fracture propagation decreases with the increase in propagation time, which is basically the same as the analytical solution of the KGD and PKN models. This verifies the accuracy of the parallel FE-DE-FV method, which can be used to study multiscale rock fracture propagation behaviour. Compared with the analytical solution of the KGD and PKN models, the numerical solution of the model is closer to the analytical solution of the PKN model. This is because the PKN model assumes that the vertical plane of fracture propagation is a plane strain state, whereas the KGD model assumes that the horizontal plane of fracture propagation is a plane strain state. The accuracy of the PKN model increases when the difference between the fracture length and fracture height is large, whereas that of the KGD model increases when the difference between the fracture length and fracture height is small.

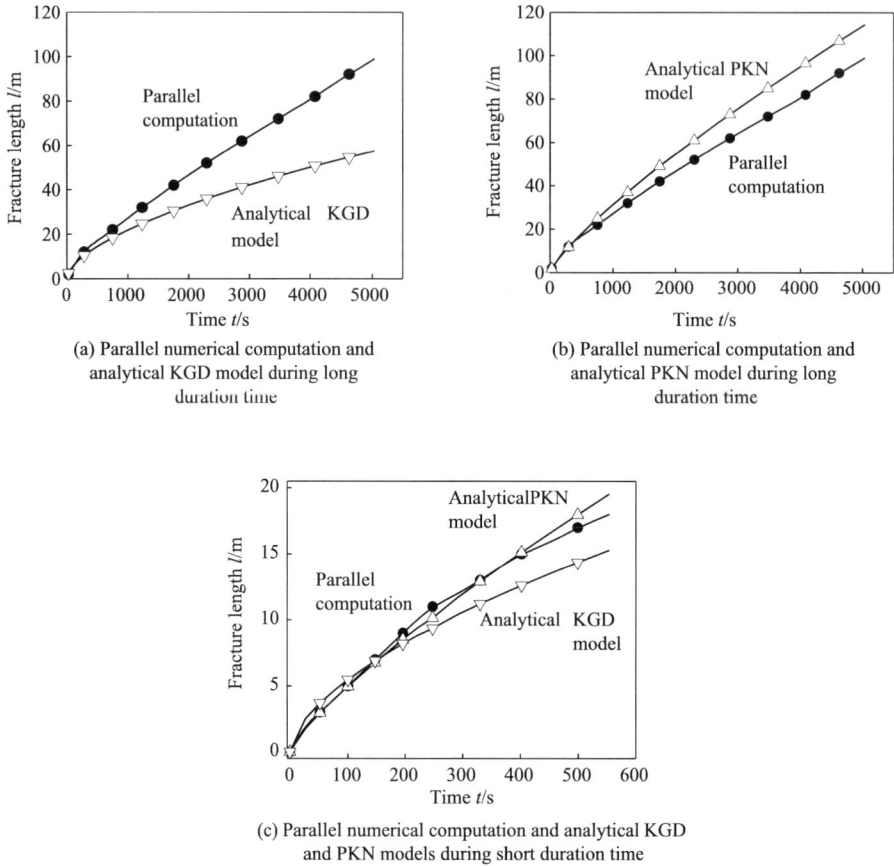

(a) Parallel numerical computation and
analytical KGD model during long
duration time

(b) Parallel numerical computation and
analytical PKN model during long
duration time

(c) Parallel numerical computation and analytical KGD
and PKN models during short duration time

Figure 3.4. Example 1: Comparison between numerical solutions of analytical solutions for propagation length of hydraulic fractures.

3.5.2 Example 2: Multi-thread parallel computation efficiency for fluid-driven fracture propagation

To test the performance of the proposed method and procedure, in this section, we used the same computation model by setting different parallel threads and analysed the efficiency of the parallel algorithm using a quasi-two-dimensional rock fracture propagation model. The model designed in this section is illustrated in Figure 3.5. The size of the quasi-two-dimensional model was 100 m × 100 m × 1 m. The fracturing fluid injection perforation was located at the geometric center of the model. An eight-node hexahedron element was used, which was divided into 1600 elements and 3362 nodes. The main minimum horizontal *in-situ* stress parallel to the *x*-axis was set to S_h, and the main maximum *in-situ* stress parallel to the *y*-axis was set to S_H. In this computation model, the size of S_h was set to 40 MPa, and the size of S_H was set to 60 MPa.

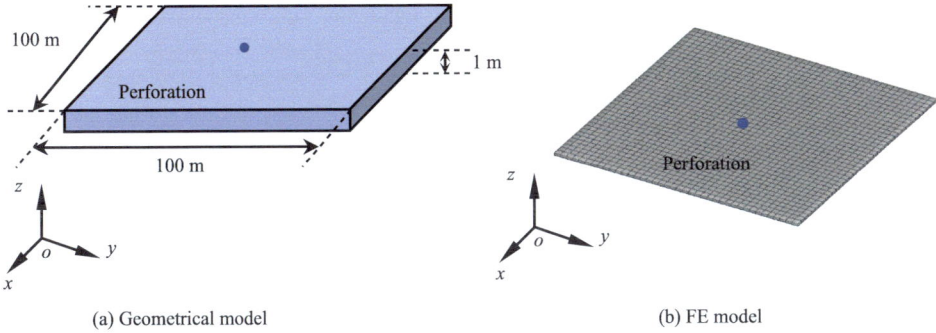

<table>
<tr><td>(a) Geometrical model</td><td>(b) FE model</td></tr>
</table>

Figure 3.5. Example 2: Geometrical and FE models for multi-thread parallel computation efficiency for fluid-driven fracture propagation.

The results of serial computing (single thread) and parallel computing (1-8 threads) are compared to assess the accuracy of parallel computing solutions. Figure 3.6 shows the computation results under 30000 and 100000 load steps, that is, identical fracture morphology and pore pressure (MPa) under different fluid load steps in serial and multi-threaded parallel computation. The same crack propagation morphology and the results of each physical field were obtained by serial and parallel computations. This result shows that parallel computing maintains serial computation results in solution accuracy.

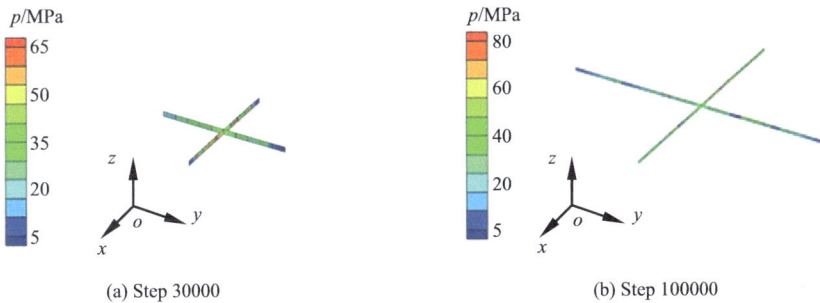

<table>
<tr><td>(a) Step 30000</td><td>(b) Step 100000</td></tr>
</table>

Figure 3.6. Example 2: Identical fracture morphology and pore pressure (MPa) under different fluid load steps in serial and multi-threaded parallel computation.

Table 3.2 shows the computation time statistics of multiple threads in parallel computation and the computation duration time for fluid-driven fracture propagation. As shown in the results, when multiple threads are used, the computing time is continuously shortened, and the time ratio (the ratio of parallel computing time to serial computing time) gradually increases as shown in the table.

Table 3.2. Example 2: Number of threads in parallel computation and computation duration time for fluid-driven fracture propagation.

Number of threads in parallel computation n_t	Computation duration time t /s	Time ratio r_t
1	2414.78	1.00
2	1490.61	1.62
3	1123.15	2.15
4	856.30	2.82
5	759.36	3.18
6	686.02	3.52
7	677.63	3.56
8	676.96	3.56

To illustrate a more detailed improvement of efficiency by parallel computing, Figure 3.7 shows the computation duration time and time ratio for different numbers of threads in parallel computation. As shown in this figure, with an increase in the number of threads, the calculation time gradually decreases and the time ratio gradually increases; however, when the number of threads increases to a certain number, such as 6, the computing time is no longer reduced, that is, the efficiency of parallel computing has reached the maximum. At this time, parallel computing has achieved the maximum improvement in computing efficiency, and the factor affecting the improvement of computing time has become its own computing algorithm. For the hydrofracturing problem calculated here, the time ratio of the parallel computation is approximately 3.56.

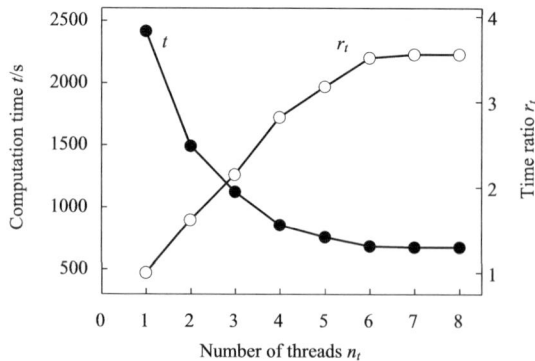

Figure 3.7. Example 2: Computation duration time and time ratio for different number of threads in parallel computation.

3.5.3 Example 3: Parallel computation using multi-type elements and meshes

3.5.3.1 Effectiveness of solutions under pentahedral and hexahedral elements

These examples verify the validity of the solution by comparing pentahedral elements and hexahedral elements to simulate the fracture propagation behaviour of the rock mass. In this section, the pentahedral model with the same size, *in-situ* stress conditions, and model parameters as the hexahedral model in Example 2 are set as shown in Figure 3.8 (a). Relatively dense meshes were used in the finite element models to avoid the mesh sensitivity of the results. The FE model was divided into 20000 pentahedral elements and 20402 nodes, and the FE model was divided into 1600 hexahedral pentahedral elements and 3362 nodes. The established FE models are shown in Figure 3.8 (b) and 3.8 (c). To better illustrate the difference between pentahedral elements and hexahedral elements, local enlarged views of the meshes in the FE models are shown in Figure 3.9.

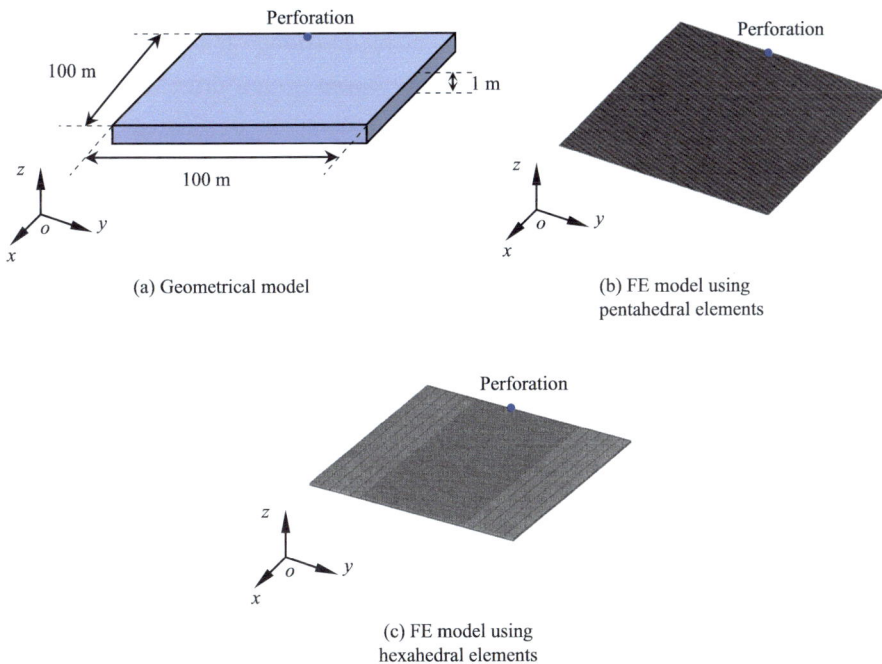

(a) Geometrical model

(b) FE model using pentahedral elements

(c) FE model using hexahedral elements

Figure 3.8. Example 3: Geometrical and FE models for parallel computation using multi-type elements and meshes.

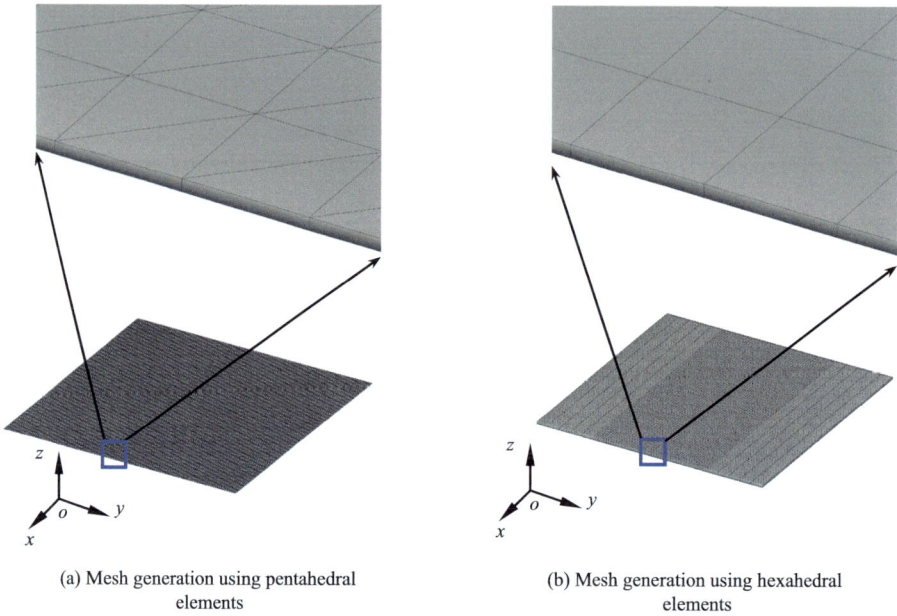

(a) Mesh generation using pentahedral
elements

(b) Mesh generation using hexahedral
elements

Figure 3.9. Example 3: Local enlarged view of meshes in finite element models.

To detect the influence of the two element divisions on fracture propagation, Figure 3.10 shows the morphologies of fracture propagation driven by fluids using different elements. In parallel computation, eight threads were used. Because the fracture of this method propagates along the edge of the element, we see from Figure 3.10(a) that near the initial perforation, the fracture propagates along the two edges of

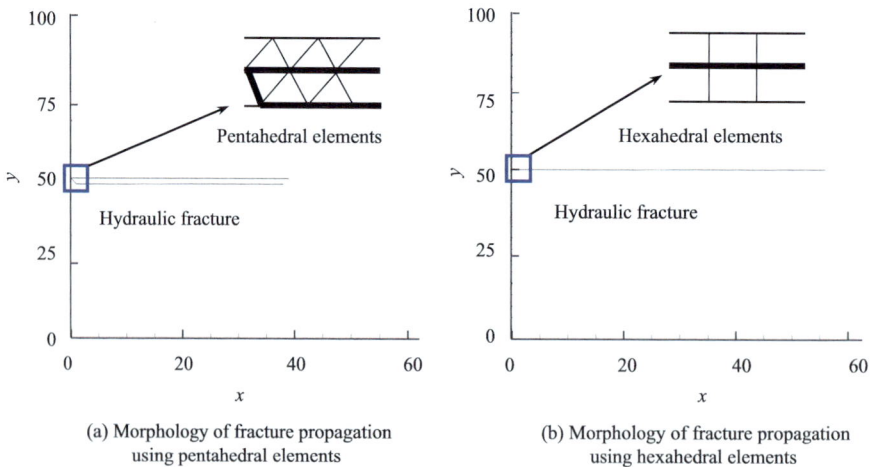

(a) Morphology of fracture propagation
using pentahedral elements

(b) Morphology of fracture propagation
using hexahedral elements

Figure 3.10. Example 3: Morphologies of fracture propagation using different elements.

the pentahedral elements to form two adjacent parallel fractures. As shown in Figure 3.10(b), near the initial perforation, the fracture propagates along one edge of the hexahedral element to form a single fracture. Because the flow rate of the injected fluid is the same in the above two cases, a single fracture formed by the hexahedral element model is longer than that formed by the pentahedral element model.

3.5.3.2   Mesh sensitivity for propagation length of fracture

To determine the mesh sensitivity for the propagation length of the fracture, two groups of hexahedral elements with different mesh densities were set up to simulate the fracture propagation behaviour of the rock mass. The mesh sizes (lengths) of the models were 1 m, 5 m, 10 m, 15 m, and 20 m. In this example, the hexahedral model with different sizes, *in-situ* stress conditions, and model parameters as those in Example 2 were set. In parallel computation, eight threads were used. In this study, the fracture length under different element sizes was computed, and statistical analysis was performed. From the statistical results, we can see that the fracture propagation length is gradually stable with mesh densification. When a larger element length is used, the variation in the fracture propagation length with time is evidently different. When the smaller unit lengths are 1 m and 5 m, the fracture propagation length tends to be the same over time. Figure 3.11 shows the mesh sensitivity for stable fracture lengths with time for small element sizes (element lengths of 1 m and 5 m). We observe that the fracture length tends to be the same at this time, and this fracture propagation behaviour can be considered as a more reliable result. This also reflects a problem: fracture

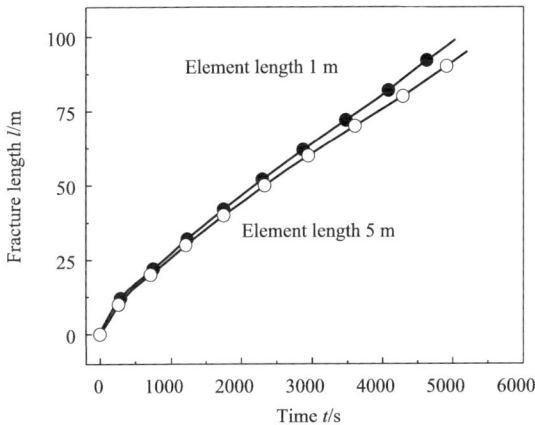

Figure 3.11. Example 3: Mesh sensitivity for stable fracture lengths with time under small element sizes (element lengths 1 m and 5 m).

propagation is related to the density of the mesh generation. Only when the mesh is fine to a certain extent can reliable fracture propagation be obtained. Reasonable optimal meshing is a challenge to be solved in the future.

3.5.4   Example 4: Dynamic propagation behaviours of fractures under *in-situ* stresses and external fluid drive

In engineering practice, fracture propagation mainly depends on the external *in-situ* stress conditions and the external fluid drive, and an important factor of the driving force is the dynamic viscosity coefficient of the fluid. In this example, the parallel computation method proposed in this chapter is used to study the fracture propagation behaviour under the difference of each component among the *in-situ* stresses and dynamic viscosity of the fracturing fluid in the engineering-scale reservoir model. Based on the size of the engineering-scale model, a quasi-two-dimensional fracturing model was designed (Figure 3.12). According to the geometric model shown in Figure 3.12(a), the perforation of the model is in the geometric center of the model, and the size of the model is 400 m × 600 m × 1 m. Figure 3.12(b) shows the FE model, where the number of elements is 60000 (eight-node hexahedral elements), and the number of nodes is 121002. The states of the *in-situ* stresses are $S_h : S_H = 40$ MPa : 40 MPa, $S_h : S_H = 40$ MPa : 60 MPa, $S_h : S_H = 40$ MPa : 80 MPa, respectively. High viscosity coefficients $\mu = 1 \times 10^{-11}$ Pa·s and low viscosity coefficient $\mu = 1 \times 10^{-13}$ Pa·s were used for different types of fracturing fluids.

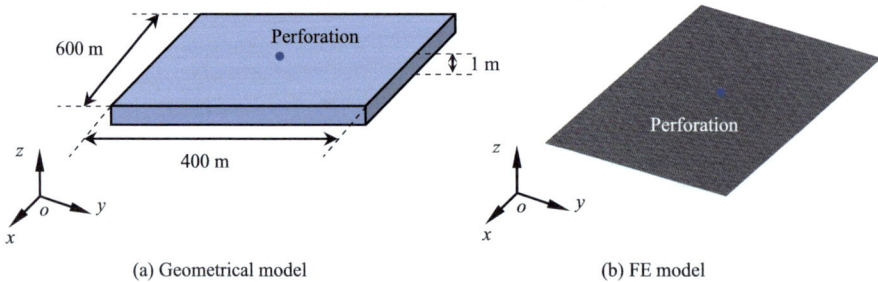

Figure 3.12. Example 4: Geometrical and FE models for dynamic propagation behaviours of fractures under different *in-situ* stresses and external fluid drive.

The 8-thread parallel computation time of the engineering-scale model example here is approximately 10 h, which is significantly improved in terms of efficiency compared with 52 h of serial single-thread computation, and it can effectively and accurately study the fracture propagation behaviour of rock mass at the engineering scale. Table 3.3 shows the computed results of the dynamic propagation behaviours and pore pressures of hydraulic fractures under different *in-situ* stresses and external fluid drives. When $S_h : S_H = 40\,\text{MPa} : 40\,\text{MPa}$ and $S_h : S_H = 40\,\text{MPa}:60\,\text{MPa}$, the maximum horizontal *in-situ* stress has little influence on the fracture propagation direction, and the fracture propagation morphology presents cross-type fractures. When $S_h : S_H = 40\,\text{MPa}:80\,\text{MPa}$, the maximum horizontal *in-situ* stress has a large influence on the fracture propagation direction, and the fracture propagation extends along the direction of the maximum *in-situ* stress to form the double wing-type fractures. By comparing the results of fracturing fluid with high and low dynamic viscosity coefficients, it can be seen that the shape of the fracture network propagation has little effect on the dynamic difficulty coefficient; however, when the dynamic viscosity coefficient is smaller, the flow rate and pressure distribution of the fracturing fluid are more uniform, and a lower fluid pressure is formed in the fracture. This type of fracturing fluid is more conducive to entering small fractures and forming a complex multiscale fracture network.

Table 3.3. Example 4: Computed results of dynamic propagation behaviours and pore pressure (MPa) of hydraulic fractures under different *in-situ* stresses and external fluid drives.

State of *in-situ* stresses	High viscosity coefficient $\mu = 1 \times 10^{-11}\,\text{Pa·s}$	Low viscosity coefficient $\mu = 1 \times 10^{-13}\,\text{Pa·s}$
$S_h : S_H = 40\,\text{MPa} : 40\,\text{MPa}$		
$S_h : S_H = 40\,\text{MPa} : 60\,\text{MPa}$		

Continued

State of *in-situ* stresses	High viscosity coefficient $\mu =1\times10^{-11}$ Pa·s	Low viscosity coefficient $\mu =1\times10^{-13}$ Pa·s
$S_h : S_H$ =40 MPa : 80 MPa		

p/MPa (High viscosity): 120, 90, 70, 30, 10

p/MPa (Low viscosity): 100, 70, 50, 30, 10

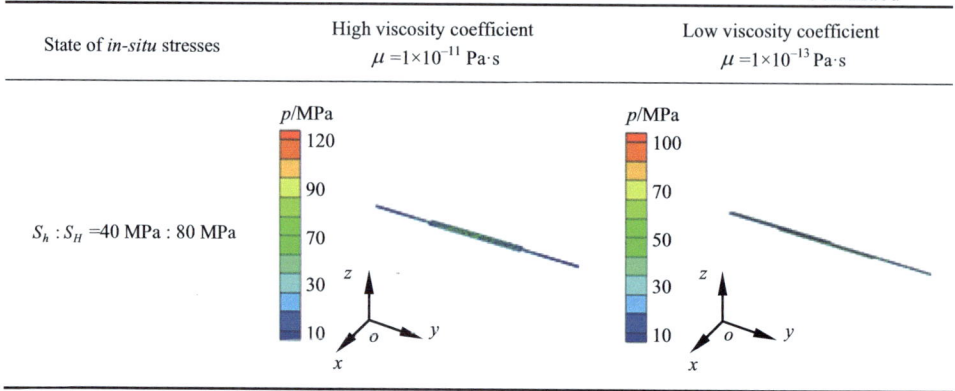

We observe from the above computation results that when the local *in-situ* stress amplitude is small, the fracture morphology under a single perforation in a homogeneous reservoir is adequately simple, and cross-type fractures and double wing-type fractures appear. When the reservoir depth continues to increase and the *in-situ* stress amplitude increases, the fracture network becomes extremely complex, driven by *in-situ* stress and fracturing fluid. Figure 3.13 shows the morphology of the complex fracture network under high *in-situ* stress fracture network = 70 MPa and 70 MPa. This is because the *in-situ* stress is very large. Once the fracturing disturbance occurs, it is easy to reach the limit of the fracture and form a complex stimulated reservoir volume phenomenon.

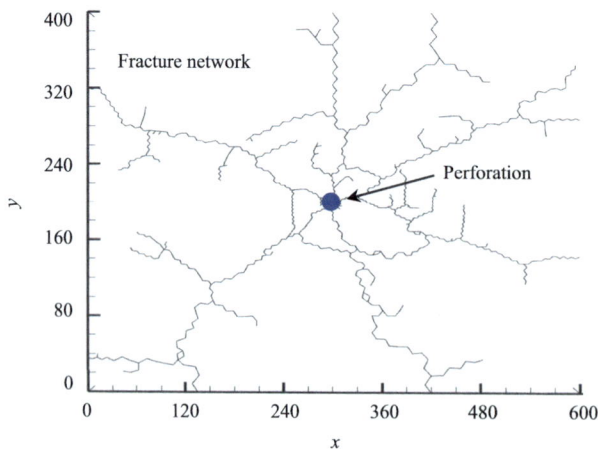

Figure 3.13. Example 4: Morphology of complex fracture network under high *in-situ* stress $S_h : S_H$ =70 MPa : 70 MPa.

3.6   Conclusions

Based on this study, our primary concluding remarks can be summarized as follows:

(1) By comparing the analytical solutions of the KGD and PKN models with the numerical solution of the proposed method, it can be seen that the parallel FE-DE-FV method based on OpenMP has effective consistency and high accuracy in simulating the dynamic propagation of hydraulic fractures considering hydro-mechanical coupling in tight reservoirs. Using different numbers of threads for parallel computation, the results of rock fracture propagation are identical, which verifies the reliability of the parallel algorithm, and multi-threaded parallel computation has a significant improvement in computational efficiency compared with single-thread serial computation.

(2) To test parallel computing using multi-type elements and meshes, the effectiveness of solutions under pentahedral and hexahedral elements was successfully executed, and the mesh sensitivity and convergence for the propagation direction of the fracture were analysed. Hexahedral and pentahedral elements were used to simulate the fracture propagation of the rock mass. The fracture propagation patterns solved by the hexahedral element model and pentahedral element model were similar, which confirms the validity of the solutions under different element forms. Two groups of hexahedral elements with different mesh densities were used to simulate the fracture propagation of the rock mass. The fracture propagation velocities under the two mesh densities were similar, which verifies the convergence of the solution under different mesh densities.

(3) To analyse the dynamic propagation behaviours of engineering-scale hydraulic fractures under *in-situ* stresses and external fluid drive using an efficient parallel computation method, the differences in each component among the *in-situ* stresses on the hydraulic fractures were analysed. In the engineering-scale model, when the ratio of the horizontal principal stresses was small, the fracture was a bifurcation fracture. However, when the ratio of horizontal principal stresses increased, the fractures mainly propagated along the direction of the maximum principal stress, which is consistent with the experimental results.

(4) The dynamic viscosity of the fracturing fluid for simulating liquid and gas is used to drive the fracture propagation in the proposed parallel computation method. In the engineering-scale model, under the same *in-situ* stress conditions, changing the

viscosity coefficient of the fracturing fluid had an impact on the speed and the final length of hydraulic fracture propagation. Within a certain range, the fracture propagation speed decreased with an increase in the viscosity coefficient of the fracturing fluid, but changing the viscosity coefficient of the fracturing fluid did not have a significant impact on the fracture propagation morphology.

(5) A multi-thread parallel computation scheme for solid and fluid analysis was developed to improve the computational efficiency in large-scale models and multi-physical field coupling (hydro-mechanical) problems, and fracture criteria via dual bilinear cohesive zone model was introduced. Considering the hydro-mechanical coupling and leak-off effects, the combined FE-DE-FV approach was introduced and implemented successfully, and a multi-thread parallel combined method and global procedure were proposed.

This chapter proposed a multi-thread parallel combined FE-DE-FV method for the dynamic propagation of hydraulic fractures considering hydro-mechanical coupling in a tight reservoir. Through numerical examples, this study analysed the influence of computation accuracy, efficiency, mesh applicability and sensitivity, influence of *in-situ* stress, and dynamic viscosity coefficient of the engineering-scale model. To better implement and test the method proposed in this study, we adopted a simplified two-dimensional model (i.e., the value in the thickness direction is one). The follow-up research is to extend the proposed method to an actual three-dimensional engineering reservoir, study the spatial propagation of hydraulic fractures in a large-scale three-dimensional model, and provide better parallel computation techniques.

References

Ahmed, U., Meehan, D.N. (2019), "Unconventional oil and gas resources: exploitation and development", CRC Press, USA.

Ajayi, B., Aso, I.I., Terry, I.J., Walker, K., Wutherich, K., Caplan, J., Gerdorn, D.W., Clark, B.D., Gunguly, U., Li, X., Xu, Y., Yang, H., Liu, H., Luo, Y., Waters, G. (2013), "Stimulation design for unconventional resources", Oilfield Review, Vol. 25 No. 2, pp. 34–46.

Barboza, B.R., Chen, B., Li, C.F. (2021), "A review on proppant transport modelling", Journal of Petroleum Science and Engineering, Vol. 204, pp. 108753.

Barenblatt, G.I. (1962), "The mechanical theory of equilibrium cracks in brittle fracture", Advances in Applied Mechanics, Vol. 7, pp. 55–129.

Becker, G., Noels, L. (2013), "A full-discontinuous Galerkin formulation of nonlinear Kirchhoff–Love shells: elasto-plastic finite deformations, parallel computation, and fracture applications", International Journal for Numerical Methods in Engineering, Vol. 93 No. 1, pp. 80–117.

Beckwith, R. (2010), "Hydraulic fracturing: the fuss, the facts, the future", Journal of Petroleum Technology, Vol. 62, pp. 34–41.

Borden, M.J., Verhoosel, C.V., Scott, M.A., Hughes, T.J.R., Landis, C.M. (2012), "A phase-field description of dynamic brittle fracture", Computer Methods in Applied Mechanics and Engineering, Vol. 217, pp. 77–95.

Bower, K.M., Zyvoloski, G. (1997), "A numerical model for thermo-hydro-mechanical coupling in fractured rock", International Journal of Rock Mechanics and Mining Sciences, Vol. 34 No. 8, pp. 1201–1211.

Byeon, S.P., Lee, D.Y. (2020), "Method for real-time simulation of haptic interaction with deformable objects using GPU-based parallel computing and homogeneous hexahedral elements", Computational Mechanics, Vol. 65, pp. 1205–1218.

Cai, M., Kaiser, P.K., Morioka, H., Minami, M., Maejima, T., Tasaka, Y., Kurose, H. (2007), "FLAC/PFC coupled numerical simulation of AE in large-scale underground excavations", International Journal of Rock Mechanics and Mining Sciences, Vol. 44 No. 4, pp. 550–564.

Chen, B., Barboza, B., Sun, Y., Bai, J., Thomas, H.R., Dutko, M., Cottrell, M., Li, C. (2021), "A review of hydraulic fracturing simulation", Archives of Computational Methods in Engineering, Vol. 29, pp. 1–58.

Cipolla, C.L., Mayerhofer, M.J., Warpinski, N.R. (2009), "Fracture design considerations in horizontal wells drilled in unconventional gas reservoirs", SPE Hydraulic Fracturing Technology Conference. Society of Petroleum Engineers, SPE-119366-MS.

Davies, R., Foulger, G., Bindley, A., Styles, P. (2013), "Induced seismicity and hydraulic fracturing for the recovery of hydrocarbons", Marine and Petroleum Geology, Vol. 45 No. 4, pp. 171–185.

Deng, S., Li, H., Ma, G., Huang, H., Li, X. (2014), "Simulation of shale–proppant interaction in hydraulic fracturing by the discrete element method", International Journal of Rock Mechanics and Mining Sciences, Vol. 70, pp. 219–228.

Dimakopoulos, V.V. (2014), "Parallel Programming Models", Springer, New York, USA.

Figueiredo, B., Tsang, C.F., Rutqvist, J., Niemi, A. (2015), "A study of changes in deep fractured rock permeability due to coupled hydro-mechanical effects", International Journal of Rock Mechanics and Mining Sciences, Vol. 79, pp. 70–85.

Fu, P., Johnson, S., Carrigan, C. (2013), "An explicitly coupled hydro-geomechanical model for simulating hydraulic fracturing in arbitrary discrete fracture networks", International Journal for Numerical and Analytical Methods in Geomechanics, Vol. 37 No. 14, pp. 2278–2300.

Fukuda, D., Mohammadnejad, M., Liu, H.Y., Zhang, Q.B., Zhao, J., Dehkhoda, S., Chan, A., Kodama, J.I., Fujii, Y. (2020), "Development of a 3D hybrid finite-discrete element simulator based on GPGPU-parallelized computation for modelling rock fracturing under quasi-static and dynamic loading conditions", Rock Mechanics and Rock Engineering, Vol. 53, pp. 1079–1112.

Fukuda, D., Liu, H.Y., Zhang, Q.B., Zhao, J., Kodama, J.I., Fujii, Y., Chan, A.H.C. (2021), "Modelling of dynamic rock fracture process using the finite-discrete element method with a novel and efficient contact activation scheme", International Journal of Rock Mechanics and Mining Sciences, Vol. 138, pp. 104645.

Gao, Q., Ghassemi, A. (2020), "Finite element simulations of 3D planar hydraulic fracture propagation using a coupled hydro-mechanical interface element", International Journal for Numerical and Analytical Methods in Geomechanics, Vol. 44 No. 15, pp. 1999–2024.

Gao, S.K., Dong, D.Z., Tao, K., Guo, W., Li, X. J., Zhang, S. (2021), "Experiences and lessons learned from China's shale gas development: 2005-2019", Journal of Natural Gas Science and Engineering, Vol. 85, pp. 103648.

Gao, X., Gu, D., Huang, D., Zhang, W., Zheng, Y. (2020), "Development of a DEM-based method for modeling the water-induced failure process of rock from laboratory-to engineering-scale", International Journal of Geomechanics, Vol. 20 No. 7, pp. 04020080.

Gebali, F. (2011), "Algorithms and Parallel Computing", John Wiley & Sons, Inc, Canada.

Giovanardi, B., Serebrinsky, S., Radovitzky, R. (2020), "A fully-coupled computational framework for large-scale simulation of fluid-driven fracture propagation on parallel computers", Computer Methods in Applied Mechanics and Engineering, Vol. 372, pp. 113365.

Ingraffea, A. R. (1983), "Numerical modeling of fracture propagation", Rock Fracture. Mechanics, Vol. 275, pp. 151–208.

Ju, Y., Chen, J., Wang, Y., Gao, F., Xie, H. (2018), "Numerical analysis of hydrofracturing behaviors and mechanisms of heterogeneous reservoir glutenite, using the continuum-based discrete element method while considering hydromechanical coupling and leak-off effects", Journal of Geophysical Research: Solid Earth, Vol. 123 No. 5, pp. 3621–3644.

Ke, C., Chen, C., Tu, C. (2008), "Determination of fracture toughness of anisotropic rocks by boundary element method", Rock Mechanics and Rock Engineering, Vol. 41, pp. 509–538.

Lei, Z., Rougier, E., Knight, E. E., Munjiza, A. (2014), "A framework for grand scale parallelization of the combined finite discrete element method in 2D", Computational Particle Mechanics, Vol. 1, pp. 307–319.

Li, G., Tang, C., Liang, Z. (2017), "Development of a parallel FE simulator for modeling the whole trans-scale failure process of rock from meso-to engineering-scale", Computers & Geosciences, Vol. 98, pp. 73–86.

Li, X., Zhang, Q., Li, H., Zhao, J. (2018), "Grain-based discrete element method (GB-DEM) modelling of multi-scale fracturing in rocks under dynamic loading", Rock Mechanics and Rock Engineering, Vol. 51 No. 12, pp. 3785–3817.

Lisjak, A., Grasselli, G. (2014), "A review of discrete modeling techniques for fracturing processes in discontinuous rock masses", Journal of Rock Mechanics and Geotechnical Engineering, Vol. 6, pp. 301–314.

Lisjak, A., Mahabadi, O. K., He, L., Tatone, B.S.A., Kaifosh, P., Haque, S.A., Grasselli, G. (2018), "Acceleration of a 2D/3D finite-discrete element code for geomechanical simulations using General Purpose GPU computing", Computers & Geotechnics, Vol. 100, pp. 84–96.

Lukas, T., D'Albano, G. G.S., Munjiza, A. (2014), "Space decomposition based parallelization solutions for the combined finite–discrete element method in 2D", Journal of Rock Mechanics and Geotechnical Engineering, Vol. 6, pp. 607–615.

McClure, M., Babazadeh, M., Shiozawa, S., Huang, J. (2016), "Fully coupled hydromechanical simulation of hydraulic fracturing in 3D discrete-fracture networks", SPE Journal, Vol. 21 No.

04, pp. 1302–1320.

Mi, Y., Aliabadi, M.H. (1992), "Dual boundary element method for three-dimensional fracture mechanics analysis", Engineering Analysis with Boundary Elements, Vol. 10, pp. 161–171.

Mohammadnejad, M., Dehkhoda, S., Fukuda, D., Liu H.Y., Chan A. (2020a), "GPGPU-parallelised hybrid finite-discrete element modelling of rock chipping and fragmentation process in mechanical cutting", Journal of Rock Mechanics and Geotechnical Engineering, Vol. 12, pp. 310–325.

Mohammadnejad, M., Fukuda, D., Liu, H.Y., Dehkhoda, S., Chan, A. (2020b), "GPGPU-parallelized 3D combined finite-discrete element modelling of rock fracture with adaptive contact activation approach", Computational Particle Mechanics, Vol. 7, pp. 849–867.

Montgomery, C.T., Smith, M. B. (2010), "Hydraulic fracturing: History of an enduring technology", Journal of Petroleum Technology, Vol.62, pp.26–40.

Nicot, J. P., Scanlon, B. R., Reedy, R. C., Costley, R. A. (2014), "Source and fate of hydraulic fracturing water in the barnett shale: a historical perspective", Environmental Science & Technology, Vol. 48 No. 4, pp. 2464–2471.

Nordgren, R.P. (1972), " Propagation of a vertical hydraulic fracture", Society of Petroleum Engineers Journal, Vol. 12 No. 4, pp. 306–314.

Owen, D. R. J., Feng, Y. T. (2001), "Parallelised finite/discrete element simulation of multi-fracture solids and discrete systems", Engineering Computations, Vol. 18, pp. 557–576.

Perkins, T.K., Kern, L.R. (1961), " Widths of hydraulic fractures", Journal of Petroleum Technology, Vol.13 No.9, pp.937–949.

Rahm, D. (2011), "Regulating hydraulic fracturing in shale gas plays: the case of texas", Energy Policy, Vol. 39 No. 5, pp. 2974–2981.

Rougier, E., Knight, E.E., Broome, S.T., Sussman, A.J., Munjiza, A. (2014), "Validation of a three-dimensional Finite-Discrete Element Method using experimental results of the Split Hopkinson Pressure Bar test", International Journal of Rock Mechanics and Mining Sciences, Vol. 70, pp. 101–108.

Scanlon, B., Reedy, R., Nicot, J. (2014), "Comparison of water use for hydraulic fracturing for unconventional oil and gas versus conventional oil", Environmental Science & Technology, Vol. 48 No. 20, pp. 12386–12393.

Settgast, R.R., Fu, P.C., Walsh, S. D.C., White, J. A., Annavarapu, C., Ryerson, F. J. (2016), "A fully coupled method for massively parallel simulation of hydraulically driven fractures in 3-dimensions", International Journal for Numerical and Analytical Methods in Geomechanics, Vol. 41, pp. 627–653.

Song, J. Y., Huo, Z. P., Fu, G., Hu, M., Sun, T. W., Liu, Z., Wang, W., Liu, L. F. (2020), "Petroleum migration and accumulation in the Liuchu area of Raoyang Sag, Bohai Bay Basin, China", Journal of Petroleum Science and Engineering, Vol. 192, pp. 107276–107291.

Song, Y., Li, Z., Jiang, Z., Luo, Q., Gao, Z. (2017), "Progress and development trend of unconventional oil and gas geological research", Petroleum Exploration and Development, Vol. 44 No. 4, pp. 638–648.

Su, Y., Zhang, Q., Wang, W., Sheng, G. (2015), "Performance analysis of a composite dual-porosity model in multi-scale fractured shale reservoir", Journal of Natural Gas Science and Engineering, Vol. 26, pp. 1107–1118. 1.

Sukumar, N., Moës, N., Moran, B., Belytschko, T. (2000), "Extended finite element method for three-dimensional crack modelling", International Journal for Numerical Methods in Engineering, Vol. 48, pp. 1549–1570.

Sun, Y. N., Edwards, M.G., Chen, B., Li, C.F. (2021), "A state-of-the-art review of crack branching", Engineering Fracture Mechanics, Vol. 257, pp. 108036.

Vaziri, H.H. (1996), "Theory and application of a fully coupled thermo-hydro-mechanical finite element model", Computers & Structures, Vol. 61 No. 1, pp. 131–146.

Wang, H., Ma, F., Tong, X., Liu, Z., Zhang, X., Wu, Z., Li, D., Wang, B., Xie, Ye., Yang, L. (2016), "Assessment of global unconventional oil and gas resources", Petroleum Exploration and Development, Vol. 43 No. 6, pp. 925–940.

Wang, X. L., Shi, F., Liu, C., Lu, D. T., Liu, H., Wu, H. A. (2018), "Extended finite element simulation of fracture network propagation in formation containing frictional and cemented natural fractures", Journal of Natural Gas Science and Engineering, Vol. 50, pp. 309–324.

Wang, Y. (2021a), "Adaptive finite element-discrete element analysis for stratal movement and microseismic behaviours induced by multistage propagation of three-dimensional multiple hydraulic fractures", Engineering Computations, Vol. 38 No. 5, pp. 1350–1371.

Wang, Y. (2021b), "Adaptive analysis of damage and fracture in rock with multiphysical fields coupling. Springer Press", Springer Nature.

Wang, Y., Ju, Y., Zhang, H., Gong, S., Song, J., Li, Y., Chen, J. (2021), " Adaptive finite element-discrete element analysis for the stress shadow effects and fracture interaction behaviours in three-dimensional multistage hydrofracturing considering varying perforation cluster spaces and fracturing scenarios of horizontal wells", Rock Mechanics and Rock Engineering, Vol. 54, pp. 1815–1839.

Weng, X., Kresse, O., Cohen, C., Wu, R., Gu, H. (2010), "Modeling of hydraulic-fracture-network propagation in a naturally fractured formation", SPE Production & Operations, Vol. 26 No. 04, pp. 368–380.

Wheeler, M. F., Wick, T., Lee, S. (2020), "IPACS: Integrated phase-field advanced crack propagation simulator. An adaptive, parallel, physics-based-discretization phase-field framework for fracture propagation in porous media", Computer Methods in Applied Mechanics and Engineering, Vol. 367, 113124.

Ziaei-Rad, V., Shen, Y.X. (2016), "Massive parallelization of the phase field formulation for crack propagation with time adaptivity", Computer Methods in Applied Mechanics and Engineering, Vol. 312, pp. 224–253.

Zou, C., Dong, D., Wang, S., Li, J., Cheng, K. (2010), "Geological characteristics and resource potential of shale gas in China", Petroleum Exploration and Development Vol. 37 No. 6, pp. 641–653.

Chapter 4 Heterogeneous continuum-discontinuum computation method for dynamic diversion and penetration of hydraulic fractures contacting multi-layers and granules

4.1 Introduction

As the development of conventional oil and gas resources is becoming increasingly complex, the development of tight oil and gas from unconventional resources has attracted growing attention (Wang *et al.*, 2014; Zou *et al.*, 2012). Hydrofracturing is one of the leading technologies to realize the efficient development of tight oil and gas reservoirs. Shale contains numerous layer bedding structures and several granules are present in the formation process of rock stratum. Figure 4.1 shows a schematic simulation of hydraulic fracture propagation in a tight reservoir embedded multi-layers and granules, in which the hydraulic fractures may display diversion or penetration behaviours when they contact with the bedded interface and granules. The high compressibility and weak cementation of bedding lead to the typical anisotropic characteristics of deformation characteristics, permeability evolution, tensile strength, fracture toughness, and fracture evolution of the reservoir (Niandou *et al.*, 1997; Wang *et al.*, 2018; Chen *et al.*, 2015). The initiation and propagation of hydraulic fractures depend on the relative mechanical properties between bedding and rock matrix. The cementation property of the bedding plane is weak and often disintegrates the rock matrix, which may cause hydraulic fracture propagation along the bedding plane during tight gas development, interfere with the trend and path of fractures, and deflect and affect their propagation mechanisms under the action of the stress field. On encountering rock granules, hydraulic fractures may divert or break them. The distribution and property of granules also affects the trend and final development of hydraulic fractures. The morphology of the fracture network formed by hydrofracturing directly affects oil and gas output. To obtain a highly complex fracture network, investigating the propagation mechanisms of hydraulic fracture is crucial; ignoring the influence of heterogeneous factors in reservoir rock, such as bedding and granules, it will be difficult to accurately evaluate and optimize the fracture growth and evolution behaviours, and oil and gas production may be not improved.

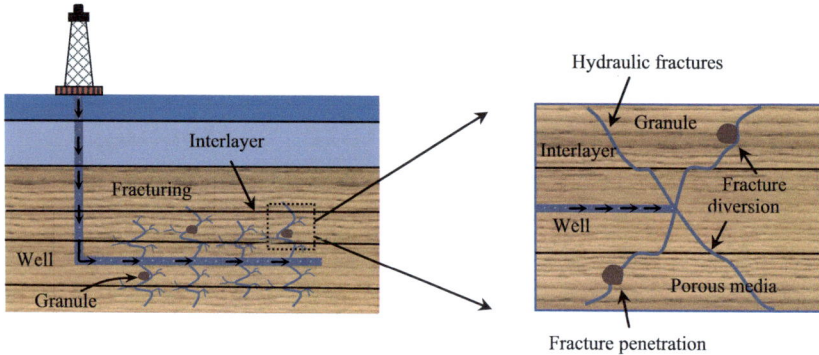

Figure 4.1. Schematic simulation of hydraulic fractures propagation in tight reservoir embedded multi-layers and granules.

Currently, some studies are implemented on the influence of bedding surfaces on hydraulic fracture propagation. Some three-dimensional fracture network models have been established to explore the impact of engineering-scale hydraulic fracture propagation in layered shale reservoirs (Zou et al., 2016a, 2017). Additionally, with the increase in bedding deviation angle, the elastic modulus of shale increases, compressive strength first decreases and later increases, and tensile strength first increases and later decreases. Shear failure can also easily occur along the bedding direction during Brazilian splitting (Cho et al., 2012). When hydraulic fractures intersect with bedding, the weak surfaces and bedding orientation become important influencing factors of the intersection behaviour of hydraulic fractures and natural fractures (Daneshy, 1974; Norman and Jessen, 1963). With the increase of deviation angle of the bedding plane, the maximum principal stress and bedding plane primarily control the initiation and propagation of fractures (Men et al., 2013). Several scholars have studied the change mechanisms of the angle between the maximum principal stress of the wellbore and normal direction of bedding and quantitatively evaluated its influence on the formation collapse pressure (Okland and Cook, 1998; Ong and Roegiers, 1993; Al-Bazali et al., 2009; Lee et al., 2012). However, research on the influence of bedding deviation angle, various mechanical parameters of bedding and shale matrix on the mechanisms of hydraulic fracture propagation, and the effect of deviation angle on various mechanical parameters is still scarce. Several researchers have made some achievements in researching this issue through experimental methods. For example, the influence of shale bedding on hydraulic fracture propagation in laboratory hydrofracturing tests were directly observed (Suarez-Rivera *et al.*, 2013; Guo *et al.*, 2014; Zou *et al.*, 2016b; Lin *et al.*, 2017). Further, experimental methods were employed to study the effects of bedding on shale mechanical properties, pore pressure, and fracture propagation during supercritical CO_2 fracturing (Zhang *et al.*, 2019). In the physical simulation experiment of large-scale true triaxial hydraulic fractures, the arrest, branching, penetration, and steering of hydraulic fractures when they encountered weak planes were simulated (Hou *et al.*, 2016). Some researchers also carried out triaxial fracturing experiments on natural rock samples with different bedding densities, studied the influence of bedding properties on fracture propagation, and established three types of fracture propagation modes; the formation of complex fracture networks is considered as the result of weak bedding property and high bedding density (Guo *et al.*, 2018). Besides, shale samples with different angles of bedding were found that the sample anisotropy primarily affected the fracture propagation behaviours (He *et al.*, 2015).

Deep reservoir rock is a natural geomaterial, which generally has granules of different sizes, that have an impact on rock's mechanical behaviour (Shao *et al.*, 2014; Hofmann *et al.*, 2015). Current researches show that when the fracture encounters granules, it will display different behaviours, such as termination, deflection, bifurcation, and penetration (Hou *et al.*, 2017; Li *et al.*, 2013; Ma *et al.*, 2017; Rui *et al.*, 2018). An experimental and numerical study on three granite samples with different granule sizes shows that under the same ligament angle, the angle between the fractures and principal stress direction is inversely proportional to grain size (Tian *et al.*, 2018). A series of experiments on quartz, mudstone and sub-quartz sand mudstone show that two fracture propagation mechanisms exist in sandstone: intragranular fracture and intergranular fracture (Li *et al.*, 2020); the experiments also showed that when fracture encounters granules, it could either cut across them or divert them. Some numerical results of hydrofracturing in heterogeneous sandy conglomerate show that the fracture can easily propagate along the weak surfaces around the granule, and the gravel hinders the hydraulic fracture propagation (Ju *et al.*, 2018). In the process of hydrofracturing in a glutenite reservoir, granules may lead to the deflection of the original fracture propagation, form complex fractures, and increase the difficulty of fracture shape prediction, making it difficult to implement fracturing design effectively (Rui *et al.*, 2012; Hu and Zhang, 2017).

Due to the limitations of laboratory experiments and the limited data collected, determining fracture propagation pattern and evolution process in the heterogeneous rock mass is challenging. Numerical simulation is alternatively used to study the influence of bedding on the mechanisms of hydraulic fracture propagation. Some researchers have studied how the bedding plane affects hydraulic fracture network propagation through numerical models (Zou et al., 2016a). To simulate the bedding plane in shale, the distributed parallel smooth joints are embedded into the discrete element (DE) model, which can adequately simulate transverse isotropy in shale (Zou et al., 2016a; Chong et al., 2017a, 2017b; Tan et al., 2015; Zangeneh et al., 2015). On geomaterial heterogeneity of reservoir, the bedding rock was simulated to investigate the influence of strength anisotropy on fracture initiation and propagation (Shen et al., 2015). On geometrical heterogeneity of reservoir, the effects of bedding deviation angle on the initiation and propagation mechanisms of fracture were studied (Chong et al., 2017a); the Brazilian splitting failure process of bedding shale was simulated, and the effects of bedding deviation angle, bedding strength, and other parameters on the failure mechanisms of shale were studied (Chong et al., 2017b). To investigate the influence of various kinds of heterogeneity, some novel heterogeneous computation methods and models have been developed. The cohesion model is used to study the fracturing behaviour of joint surfaces (Alfano et al., 2009). Based on the energy evolution, assuming that the fracture propagated in the predetermined weak plane, a variational model of quasi-static fracture propagation in hydrofracturing was proposed (Almia et al., 2014). To reflect the bedding effect by different mechanical parameters, the interface elements representing the matrix and bedding bonding are developed in bonding element method (Li and Zhang, 2019). A continuous medium-based discrete element method (DEM) was used to simulate the hydrofracturing of heterogeneous sandy conglomerate embedded granule and bedding (Hu and Zhang, 2017). In addition, the DEM can directly represent the granule size microstructure characteristics of granular materials and form a bonded granule model by considering each granule as a discrete granule (Potyondy and Cundall, 2004). Moreover, different mechanical parameters can be assigned to the granules according to the granule size distribution to study the heterogeneity (Shimizu et al., 2011). From the above summary, it can be seen that there is still a lack of the specialized and reliable numerical methods in the investigation of the dynamic propagation behaviours of hydraulic fractures contacting the multi-layers and granules, especially the systematic investigation of geometrical and geomaterial heterogeneity is urgently needed. In this study, the heterogeneous continuum-discontinuum computation method and models are proposed to investigate the effects of deviation angle, distribution, and geomaterial property of bedding and granule on diversion and penetration behaviours of hydraulic fractures.

The remainder of this chapter is organized as follows. Section 4.2 introduces the governing partial differential equations and fracture criteria in fractured porous media. Section 4.3 describes the combined finite element (FE)-DE-finite volume (FV) method and the algorithm for multi-materials. Section 4.4 presents the numerical models and schemes, including the models and fracturing schemes for tight reservoirs with different deviation angles and properties in bedding and granules. Sections 4.5 and 4.6 introduce the results and discussions of dynamic propagation behaviours of hydraulic fractures in multilayered and multi-granules reservoirs, respectively. Finally, Section 4.7 summarizes the main conclusions.

4.2   Governing partial differential equations and fracture criteria in fractured porous media

In this study, the governing partial differential equations for hydrofracturing (Governing equations of solid deformation and fluid flow in fractured porous media) and numerical discretization (FE discretization for solid and FV discretization for fluid) are the same as the introductions in Chapter 2 and Chapter 3. To compute the governing equations of solid deformation and fluid flow in fractured porous media, finite element method (FEM) and DEM are used for solid deformation, and finite volume method (FVM) is used for fluid flow computation. Using the above numerical methods, a combined FE-DE-FV method is formed to realize hydrofracturing considering hydro-mechanical coupling (Wang and Zhang, 2022; Wang et al., 2022).

The maximum tensile stress criterion and Mohr Coulomb strength criterion are used to judge the tensile and shear failure of the elements:

$$\text{Tensile failure criterion:} \quad \sigma_n = \bar{\sigma}_n \tag{4.1a}$$

$$\text{Shear failure criterion:} \quad \tau \geqslant \sigma_n \tan \varphi + C \tag{4.1b}$$

where, σ_n and τ are normal stress and tangential stress respectively; $\bar{\sigma}_n$ is tensile strength; C is cohesion; φ is the angle of internal friction.

Once the normal or tangential stress contacts the fracture criteria in Equation (4.1), the interface of the element will separate. Due to the continuous deformation of hydraulic pressure, the newly separated unit interface finally evolved into a typical hydraulic fracture. The aperture of hydraulic fracture is determined by the normal displacement of the separation interface caused by the separation node, as follows:

$$w = \left\| \left(\boldsymbol{u}_1^a - \boldsymbol{u}_2^b \right) \cdot \boldsymbol{n}_{12} \right\|, \tag{4.2}$$

where $\boldsymbol{u}_1^a$ and $\boldsymbol{u}_2^b$ are the displacement increments of the separated nodes 1 and 2 of the two elements a and b, and $\boldsymbol{n}_{12}$ is the normal unit vector of the contact between nodes 1 and 2.

4.3   Combined finite element-discrete element-finite volume method and algorithm for multi-materials

The global algorithm and program for dynamic diversion and penetration behaviours of hydraulic fractures contacting multiple layers and granules are shown in Figure 4.2. Firstly, a tight heterogeneous reservoir model with bedding or granules is established. In this study, the basic physical parameters of bedding and granule members (Young's modulus E, Cohesion C, tensile strength $\bar{\sigma}_n$, internal friction angle φ, and Poisson's ratio v) are set to be different from the geomaterial properties of the rock matrix. When the fluid is injected, the solid displacement, velocity, and acceleration fields were computed, and solid strain field was updated; fluid pressure and velocity field were determined, and fluid pressure field was updated. Next, according to the fracture criteria, the hydraulic fracture would propagate and contact the bedding or granules. Finally, according to the computation results, the hydraulic fracture network was updated, and fluid injection continued. To briefly illustrate the implementation of the above formula in the program, the pseudo-code algorithm for dynamic diversion and penetration behaviours of hydraulic fractures contacting multi-layers and granules in tight reservoir rock is shown in Algorithm 4.1. According to the above comprehensive processes and algorithms, a heterogeneous continuum-discontinuum computation method was formed.

Algorithm 4.1. Pseudo code algorithm for dynamic diversion and penetration behaviours of hydraulic fractures contacting multi-layers and granules in tight reservoir rock.

Input: Basic physical parameters of bedding and granule members (Young's modulus E, Cohesion C, tensile strength $\bar{\sigma}_n$, internal friction angle φ, and Poisson's ratio v), physical parameters of rock matrix, and fracturing fluid.

Geometric and numerical models of tight reservoirs with bedding and granules:

Bedding deviation angle, geometric endpoint coordinates of bedding lines;

Granule deviation angle, geometric central coordinates, and length of granules;

Initial mesh π_0.   !   Region division and parameter assignment for multi-materials

Heterogeneous continuum-discontinuum computation method

do $k=1, n$! k is the current step of injection fluid

  # **Update models of tight reservoirs with bedding and granules:**

  Update geometric model with new hydraulic fracture network

  Increase injection of fracturing fluid

  # **Combined finite element-discrete element-finite volume method considering hydromechanical coupling:**

  Finite element method   →   Solid deformation   ! Update solid strain field
  $$M^e \ddot{D}(t) + C^e \dot{D}(t) + K^e D(t) = F^e, \qquad x,y,z \in \Omega$$

  Discrete element method → Fracture propagation

  Finite volume method   →   Fluid flow   ! Update fluid pressure field
  $$v_{i0} = -K_i \nabla p_{i0} = -K_i \frac{p_0 - p_i}{D_i}(-d_i), \quad v_{0j} = -K_j \nabla p = -K_{0j} \frac{p_j - p_0}{D_j} d_j$$

  # **The failure, sliding and fracture of solids occur at the interface between elements:**

  Investigate fracture based on fracture criteria   ! Tensile and shear fractures

  Hydraulic fractures propagate and contact bedding or granule

  # **Update hydraulic fracture network**

end do

Output: Dynamic diversion and penetration of hydraulic fractures contacting multi-layers and granules in tight reservoir rock, fracture network morphology, and pore pressure.

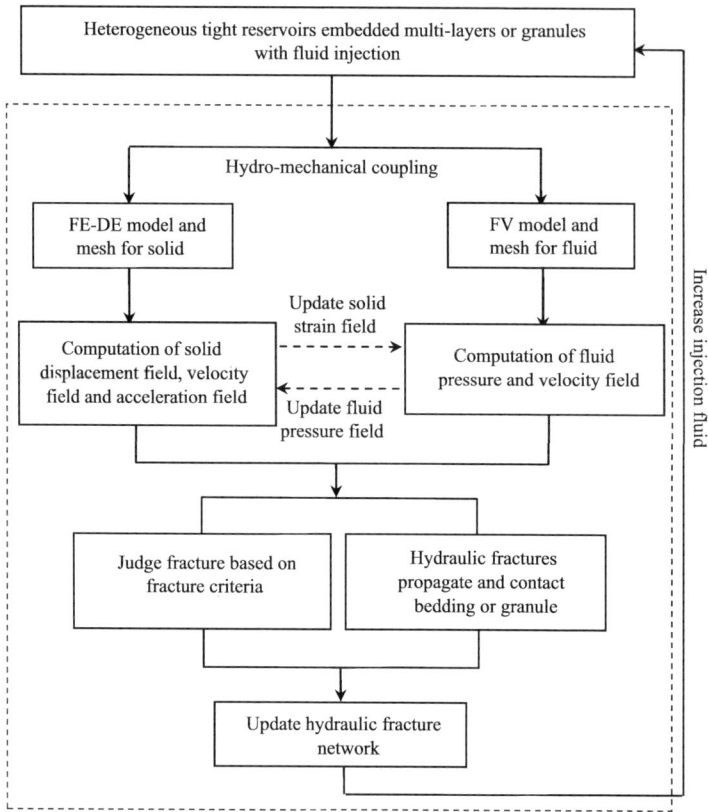

Figure 4.2. Global algorithm and procedure of dynamic diversion and penetration behaviours of hydraulic fractures contacting multi-layers and granules in tight reservoir rock.

4.4   Numerical models of tight heterogeneous reservoirs with bedding and granules

Based on the proposed heterogeneous continuum-discontinuum computation method and algorithm, this study primarily investigated the influence of two heterogeneous factors in reservoir rocks on hydraulic fracture propagation behaviour: the geometrical and geomaterial of heterogeneity. To study the dynamic diversion and penetration behaviour of heterogeneous factors (such as bedding and granule) and hydraulic fractures, two-dimensional numerical models were established. Figure 4.3(a) is a schematic representation of the geometric model of a tight reservoir with bedding. A total of six bedding members were set and numbered from 1 to 6. The bedding width was 1 m. The model size was 100 m × 100 m, the water injection point was set at the model's center, and the horizontal and vertical *in-situ* stresses are shown this figure. The basic physical parameters of the rock matrix are shown in Table 4.1, and the basic physical parameters of bedding members are shown in Table 4.2. Additionally, four types of geomaterial properties (strong, medium, weak, and extra-weak) of beddings were selected to analyse the diversion and penetration behaviours of hydraulic fractures. In this chapter, the stronger the geomaterial property, the higher the modulus and strength values, and vice versa; these expression and statement will be used in the following contents. The bedding deviation angle β is a vital parameter to describe bedding distribution, which is necessary to study the influence of bedding with different deviation angles on hydraulic fracture propagation. As shown in Table 4.3, the bedding deviation angles β of bedding models Layer-I, Layer-II, Layer-III, Layer-IV, and Layer-V are set to 0°, 30°, 45°, 60°, and 90° respectively; and Figure 4.4 shows the geometric models of different bedding deflection angles and the distribution of the multilayered interface. The distance from the water injection point to the bedding in each model is along a 10 m gradient, and the distance between adjacent beddings is equal. The FE model and initial mesh of different bedding deviation angles are shown in Figure 4.5; to balance the computation accuracy and efficiency, the meshes around the beddings were moderately densified, and the meshes were coarsened away from the beddings.

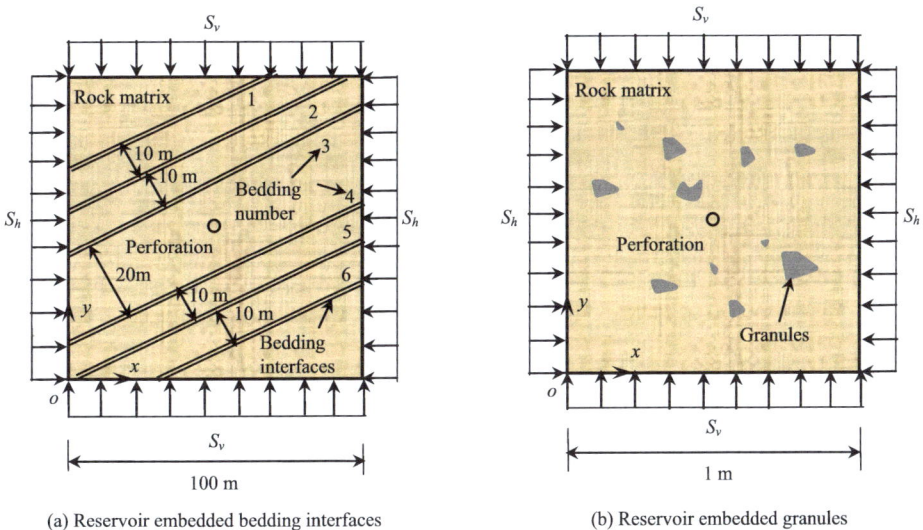

(a) Reservoir embedded bedding interfaces   (b) Reservoir embedded granules

Figure 4.3. Geometric models of heterogeneous tight reservoirs embedded bedding interfaces and granules.

Table 4.1. Basic physical parameters of the rock matrix.

Parameter	Value
Horizontal *in-situ* stress in x direction S_h/MPa	60
Vertical *in-situ* stress in y direction S_v/MPa	60
Fluid injection rate Q/(m^3/s)	0.001
Poisson's ratio of rock matrix v	0.08
Young's modulus of rock matrix E/GPa	43.88
Tensile strength of rock matrix $\bar{\sigma}_n$ /MPa	25
Porosity ϕ /%	0.4
Internal friction angle of rock matrix φ	26.13
Cohesion of rock matrix C/MPa	52.44
Density ρ_b /(kg/m^3)	2.5×10^3
Permeability k/m^2	1×10^{-16}
Gravity g/(m/s^2)	9.81
Damping coefficient c	0.8
Biot's coefficient α	0.7

Table 4.2. Basic physical parameters of the bedding and granule members.

Geomaterial Properties	Young's modulus E /GPa	Cohesion C /MPa	Tensile strength $\bar{\sigma}_n$ /MPa	Internal friction angle φ/(°)	Poisson's ratio v
Strong	438.8	524.4	250	22.9	0.1
Medium	43.88	52.44	25	22.9	0.1
Weak	4.388	5.244	2.5	22.9	0.1
Extra-weak	0.4388	0.5244	0.25	22.9	0.1

Table 4.3. Bedding deviation angle in each bedding model.

Bedding model	Layer-I	Layer-II	Layer-III	Layer-IV	Layer-V
Deviation angle/(°)	0	30	45	60	90

Figure 4.3(b) shows a schematic diagram of the geometric model for the tight reservoir with granules, in which the coarse granules are obtained by CT experimental scanning of tight reservoir rock samples. The model size was set to 1 m × 1 m; and the water injection point was set at the model's center. The basic physical parameters of the rock matrix and granule members are shown in Tables 1 and 2, respectively. Three

types of geomaterial properties (strong, medium, and weak) of granules were selected to analyse the diversion and penetration behaviours of hydraulic fractures contacting granules. The coordinates of bedding endpoints in each model are shown in Table 4.4. Furthermore, to study the influence of different granule distribution forms on hydraulic

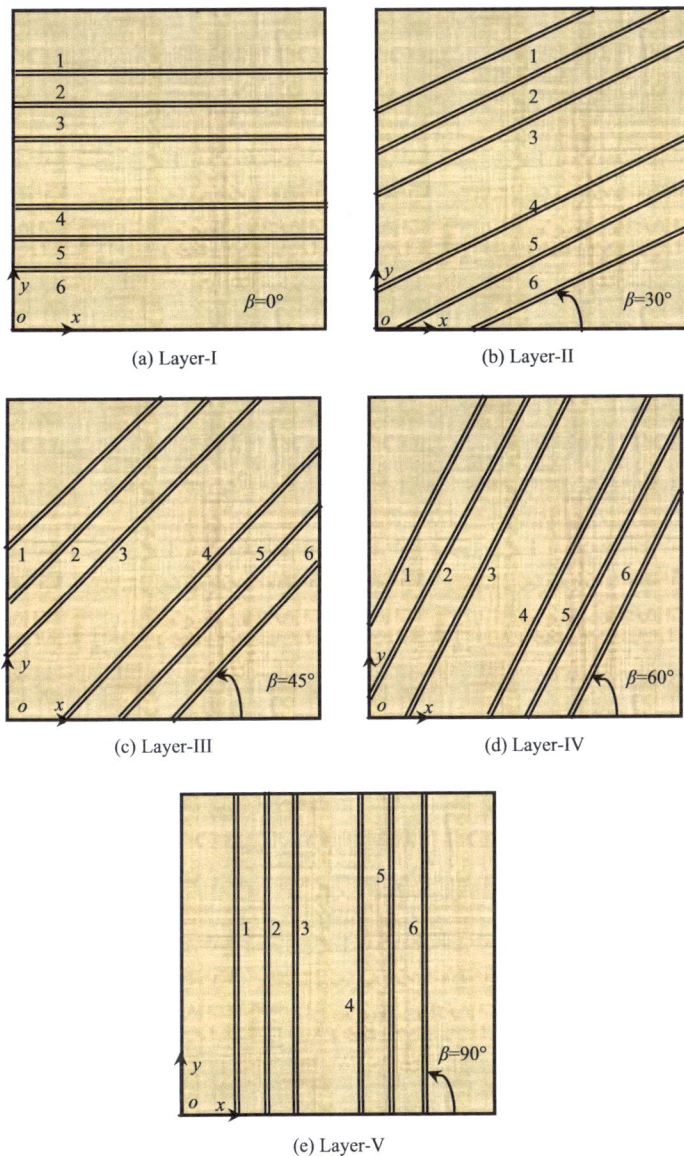

(a) Layer-I　　　　　　　　　　　　(b) Layer-II

(c) Layer-III　　　　　　　　　　　(d) Layer-IV

(e) Layer-V

Figure 4.4. Geometric models of different deviation angles β and distribution of bedded interfaces.

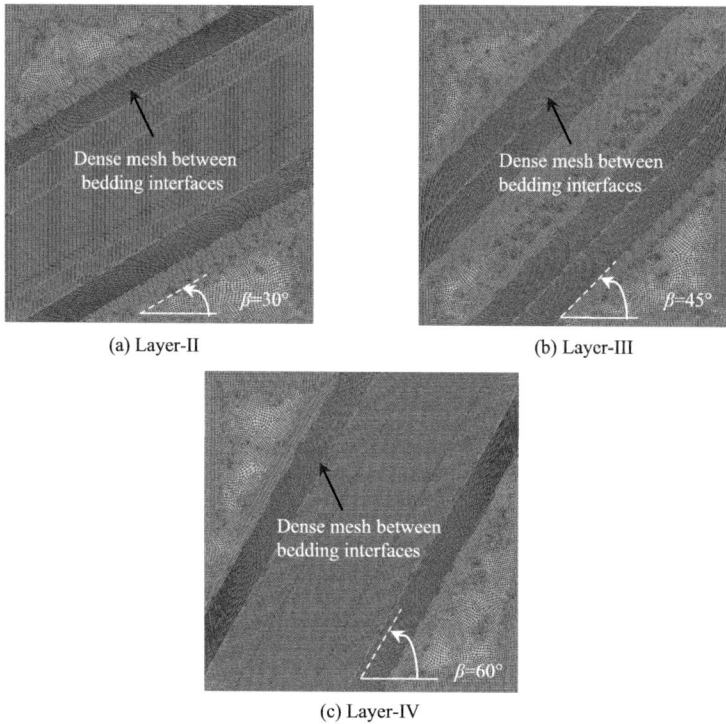

(a) Layer-II

(b) Layer-III

(c) Layer-IV

Figure 4.5. FE model and initial mesh of different bedding deviation angles.

fracture propagation, three granule models with different granule distribution forms, Granule-VI, Granule-VII, and Granule-VIII, were established as 0°, 45°, and 90° respectively; the granule deviation angle in each granule model is shown in Table 4.5. To facilitate the description, the main granules are numbered, and each granule's position and size information is shown in Table 4.6; and the models of different granule distributions and meshes are shown in Figure 4.6.

Table 4.4. Geometric endpoint coordinates (Unit: m) of bedding lines in bedding models.

Bedding number	Layer-I	Layer-II	Layer-III	Layer-IV	Layer-V
1	(0, 80)	(0, 61.2)	(0, 42.4)	(0, 23.4)	(20, 0)
	(100, 80)	(76.6, 100)	(57.6, 100)	(44.2, 100)	(20, 100)
2	(0, 70)	(0, 52.8)	(0, 28.3)	(0, 3.4)	(30, 0)
	(100, 70)	(81.6, 100)	(71.7, 100)	(55.8, 100)	(30, 100)
3	(0, 60)	(0, 44.2)	(0, 14.2)	(9.6, 0)	(40, 0)
	(100, 60)	(86.6, 100)	(85.9, 100)	(67.3, 100)	(40, 100)

Continued

Bedding number	Layer-I	Layer-II	Layer-III	Layer-IV	Layer-V
4	(0, 40)	(13.4, 0)	(14.2, 0)	(32.7, 0)	(60, 0)
	(100, 40)	(100, 55.8)	(100, 85.9)	(90.4, 100)	(60, 100)
5	(0, 30)	(18.4, 0)	(28.3, 0)	(44.2, 0)	(70, 0)
	(100, 30)	(100, 47.2)	(100, 71.7)	(100, 96.6)	(70, 100)
6	(0, 20)	(23.4, 0)	(42.4, 0)	(55.8, 0)	(80, 0)
	(100, 20)	(100, 38.5)	(100, 57.6)	(100, 76.7)	(80, 100)

Table 4.5. Granule deviation angle in each granule model.

Granule model	Granule-VI	Granule-VII	Granule-VIII
Deviation angle/(°)	0	45	90

Table 4.6. Geometric central coordinates and length (Unit: m) of granules in granule models.

Granule number	Granule-VI	Granule-VII	Granule-VIII
1	(0.31, 0.62)	(0.59, 0.70)	(0.51, 0.84)
	0.030	0.045	0.030
2	(0.45, 0.65)	(0.49, 0.59)	(0.50, 0.65)
	0.025	0.035	0.040
3	(0.62, 0.62)	(0.59, 0.60)	(0.38, 0.55)
	0.040	0.030	0.025
4	(0.65, 0.44)	(0.64, 0.43)	(0.62, 0.56)
	0.030	0.030	0.040
5	(0.78, 0.51)	(0.39, 0.44)	(0.45, 0.44)
	0.045	0.035	0.040
6	(0.64, 0.43)	(0.50, 0.35)	(0.59, 0.41)
	0.040	0.040	0.020
7	(0.28, 0.47)	(0.24, 0.34)	(0.36, 0.35)
	0.025	0.040	0.025
8	(0.41,0.52)	(0.40, 0.30)	(0.54, 0.31)
	0.035	0.030	0.025
9	(0.45, 0.45)	(0.08, 0.09)	(0.43, 006)
	0.040	0.025	0.045

(a) Granule-VI

(b) Granule-VII

(c) Granule-VIII

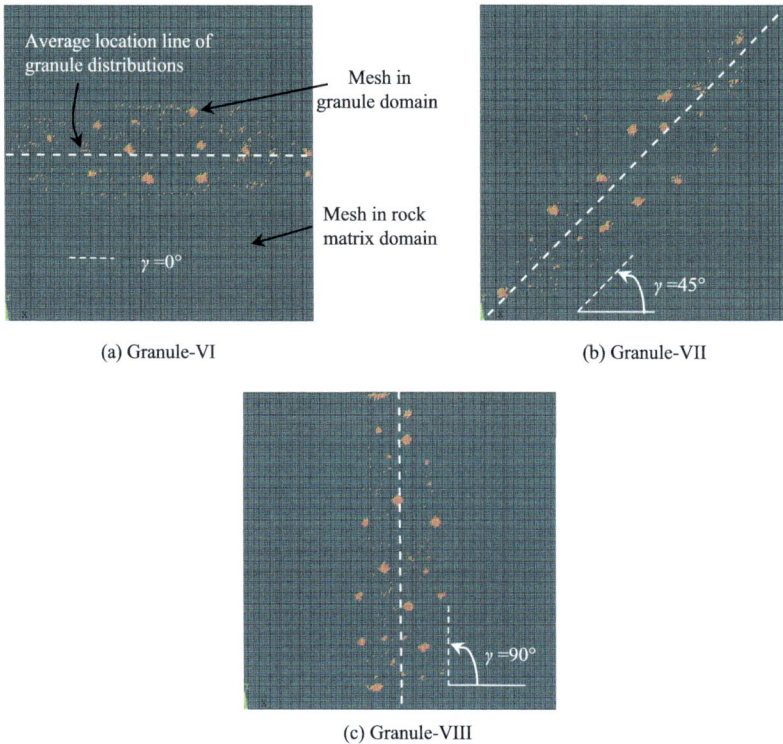

Figure 4.6. Models with different granule distributions and their meshing.

4.5   Results and discussions of dynamic propagation behaviours of hydraulic fractures in the multilayered reservoir

4.5.1   A typical example implementation of fracture propagation and pore pressure in the heterogeneous tight reservoir with beddings

A typical example (case Layer-I with deviation angle 0°) is first used to test the feasibility of the proposed method for analysing the heterogeneous reservoirs with beddings, and briefly explain the effectiveness through elaborating the results. The results of the morphology of the fracture network and pore pressure under different geomaterial properties of heterogeneous tight reservoirs embedded bedding interfaces are shown in Table 4.7:

(a) When strong geomaterial property was implemented, the hydraulic fracture did not penetrate the bedding after contacting beddings 3 and 4 closest to the water injection point. Still, it propagated along the closest bedding because the property of the bedding was greater than that of the rock matrix. The pore pressure was more significant at the water injection point and hydraulic fracture position than at other positions.

(b) When medium geomaterial property was implemented, hydraulic fractures penetrated through all six beddings, and the bedding had no significant impact on hydraulic fracture propagation, because now the bedding property was almost consistent with the rock matrix. The pore pressure along hydraulic fracture was high, and areas farther away from the fractures have low pore pressure.

(c) When weak geomaterial property was implemented, the hydraulic fracture propagated along the bedding after contacting beddings 3 and 4, and then passed through beddings 2 and 5; they also propagated along bedding 5 for a while. Finally, the hydraulic fracture stopped propagating along beddings 1 and 6. The areas of high pore pressure along hydraulic fractures and in debonding bedding were notably high.

(d) When extra-weak geomaterial property was implemented, all beddings were activated due to this weak bedding property. When the hydraulic fractures reached beddings 3 and 4, they propagated along the bedding and did not penetrate it; all beddings were activated due to the formation pressure caused by the propagation of hydraulic fractures.

Table 4.7. Results of morphology of fracture network and pore pressure (Unit: MPa) in case Layer-I (Deviation angle 0°) under different geomaterial properties of heterogeneous tight reservoirs embedded bedding interfaces.

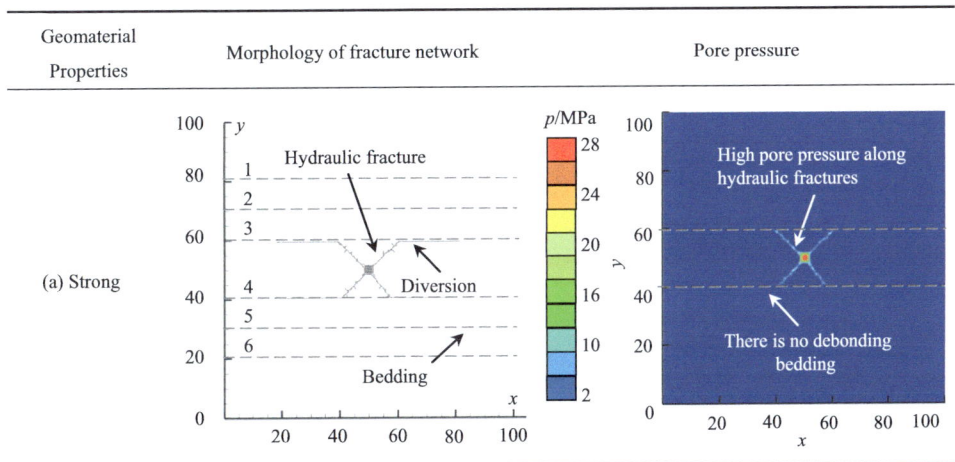

Continued

Geomaterial Properties	Morphology of fracture network	Pore pressure
(b) Medium		
(c) Weak		
(d) Extra-weak		

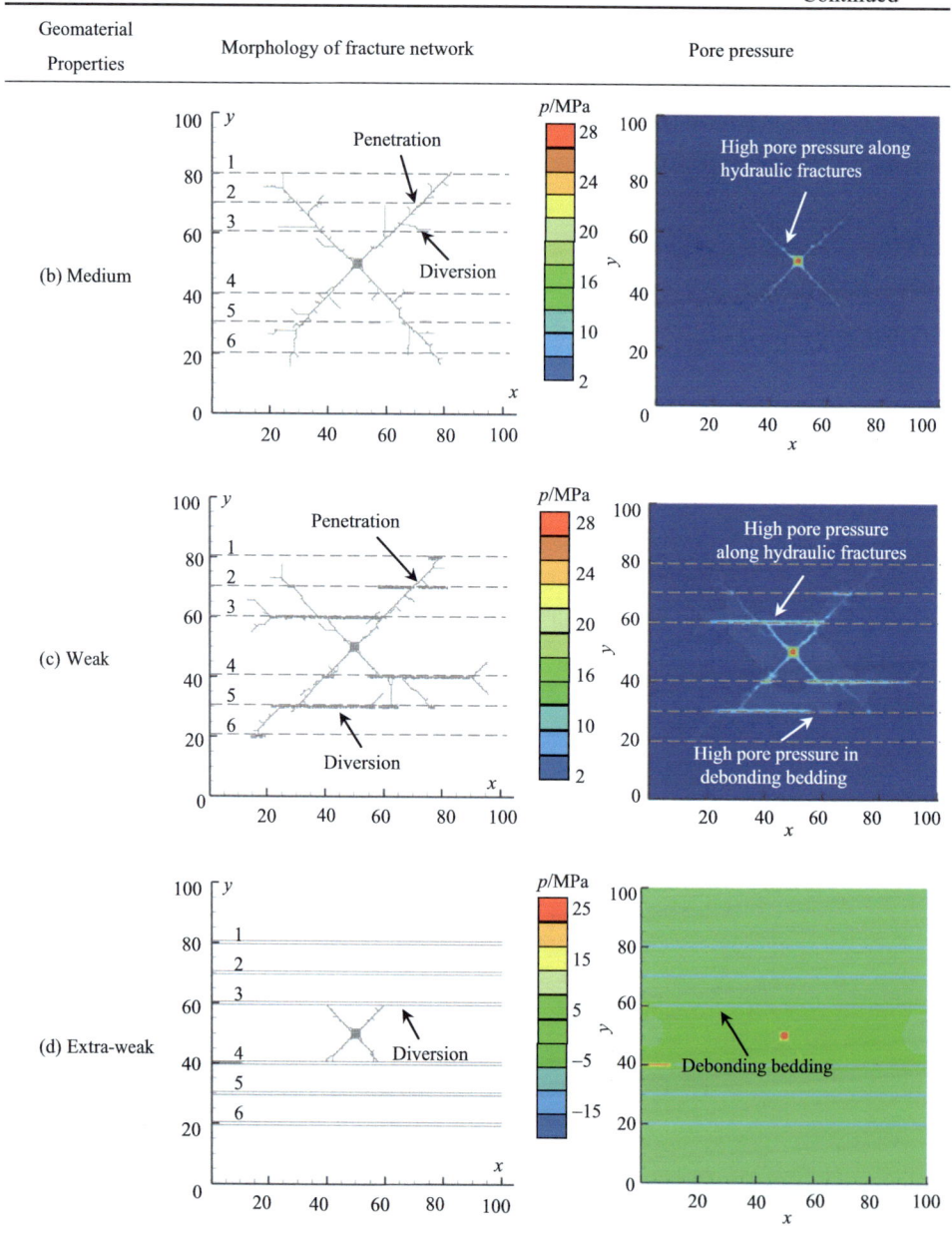

4.5.2　Influence of bedding deviation angle and geomaterial properties on hydraulic fracture propagation

To illustrate the influence of bedding deviation angle on hydraulic fracture propagation, some cases are further computed, and the representative results are selected. Figures 4.7 to 4.10 show the results of the morphology of fracture network in Layer-II (β=30°), Layer-III (β=45°), Layer-IV (β=60°), and Layer-V (β=90°) under different geomaterial properties of heterogeneous tight reservoirs embedded bedding interfaces. When the strong geomaterial property is selected for the mechanical parameters of bedding, the fracturing results of the bedding model also change with deviation angles. Figures 4.7(a), 4.8(a), 4.9(a), and 4.10(a) show the fracturing results of the bedding models with bedding deviation angles of 30°, 45°, 60°, and 90°, respectively, in stronger property of bedding. The propagation patterns of hydraulic fractures in each model vary, but the dynamic interaction behaviour between hydraulic fractures and bedding is similar. For example, after contacting bedding 3 and bedding 4

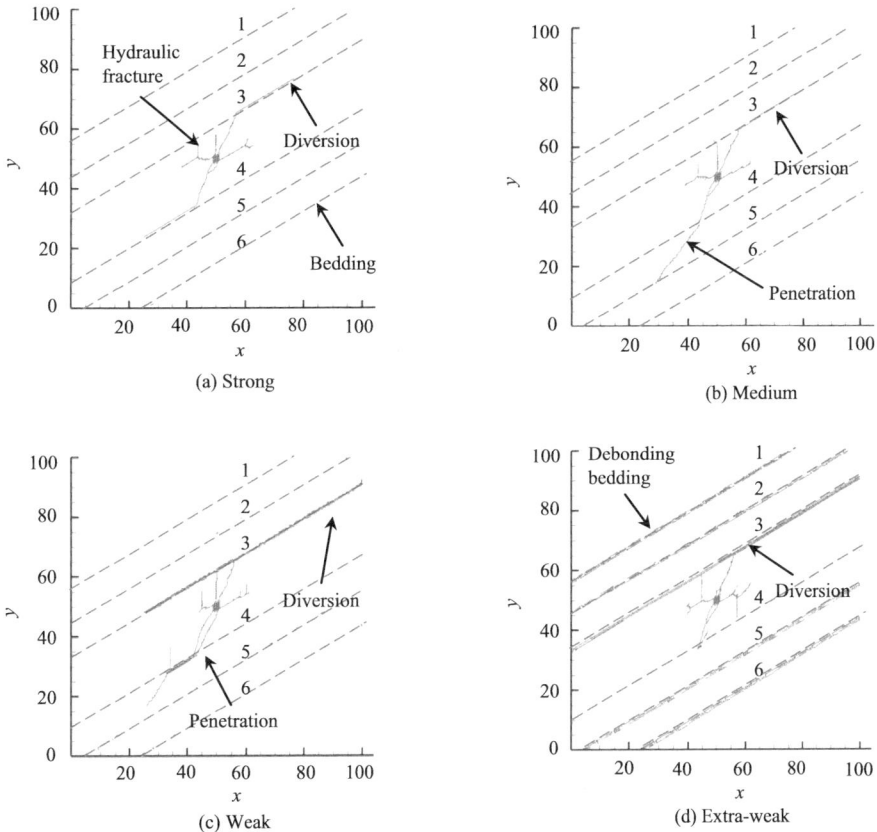

(a) Strong
(b) Medium
(c) Weak
(d) Extra-weak

Figure 4.7. Results of morphology of fracture network in case Layer-II (Deviation angle 30°) under different geomaterial properties of heterogeneous tight reservoirs embedded bedding interfaces.

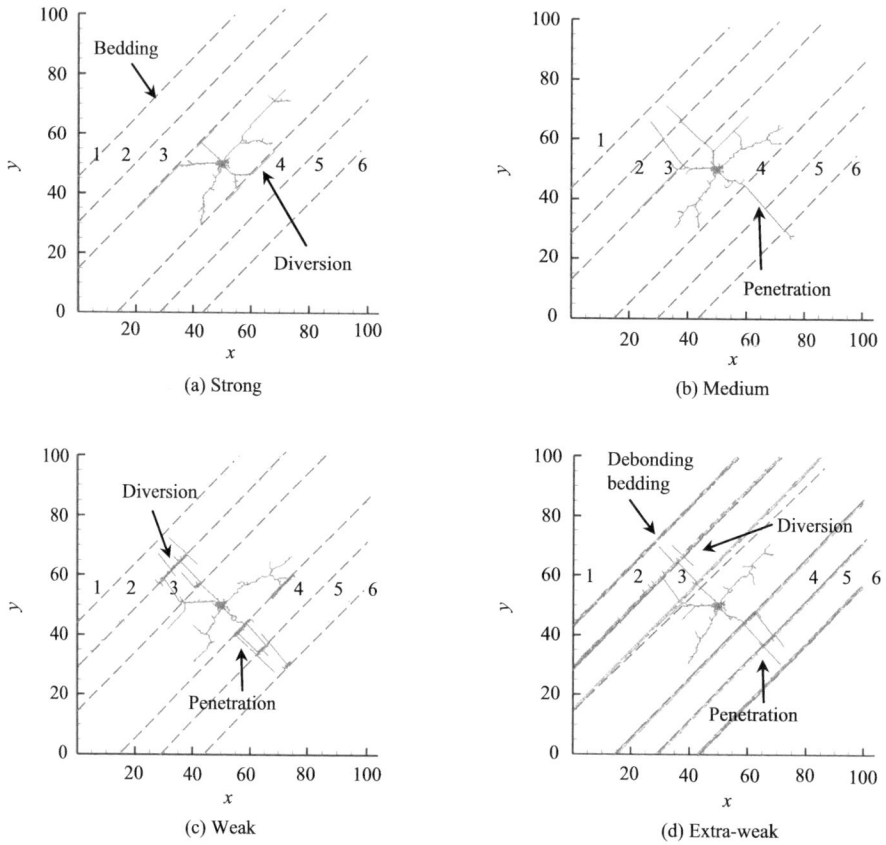

Figure 4.8. Results of morphology of fracture network in case Layer-III (Deviation angle 45°) under different geomaterial properties of heterogeneous tight reservoirs embedded bedding interfaces.

nearest to the water injection point, the hydraulic fracture did not penetrate the bedding, but propagated along the closest bedding plane because the property of bedding is greater than that of the rock matrix.

When examining medium geomaterial property of bedding, the fracturing results of the bedding model also change with alterations in deviation angle. Figures 4.7(b), 4.8(b), 4.9(b), and 4.10(b) are the fracturing results of the bedding models with bedding deviation angles of 30°, 45°, 60°, and 90°, respectively. As shown in Figure 4.7(b), in the Layer-II model, the hydraulic fracture was proximate to bedding 3 and does not propagate through it, but the lower hydraulic fracture passes through bedding 4 and ultimately stops propagating when it reaches bedding 5. As shown in Figure 4.8(b), hydraulic fractures in the Layer-III model penetrate beddings 2, 3, 4, 5, and 6. As shown in Figure 4.9(b), in the Layer-IV model, the hydraulic fracture penetrates beddings 2, 3, and 4, and deflects after passing through the bedding. As shown in Figure 4.10(b) in the Layer-V model, the vertical fracture propagation was restrained, and all beddings were activated and stretched, which form activated (or debonding) beddings.

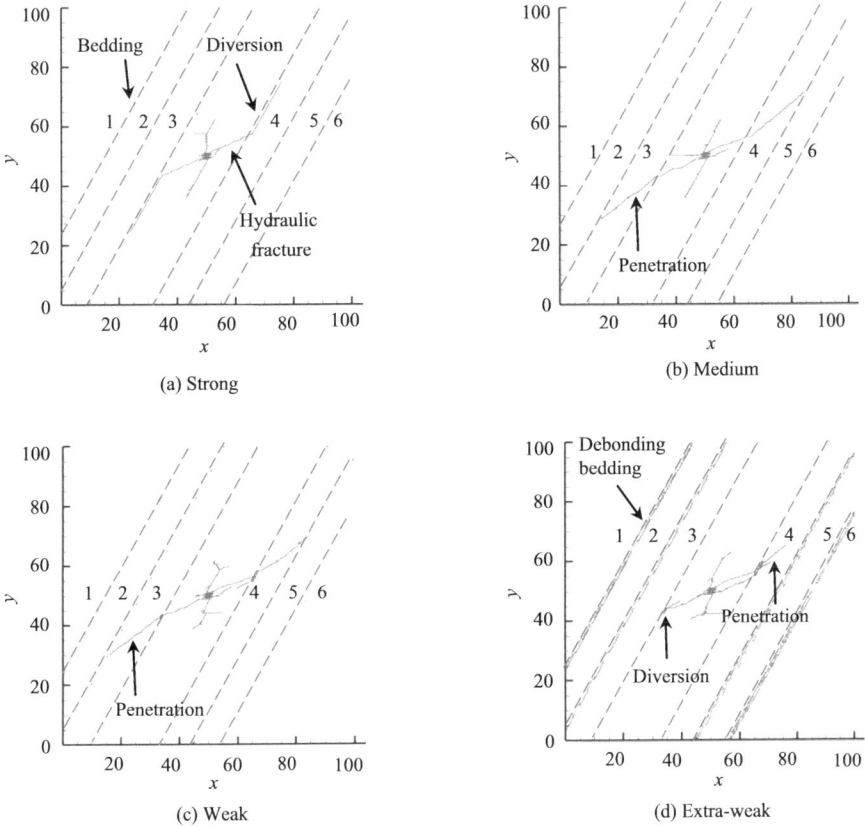

(a) Strong

(b) Medium

(c) Weak

(d) Extra-weak

Figure 4.9. Results of morphology of fracture network in case Layer-IV (Deviation angle 60°) under different geomaterial properties of heterogeneous tight reservoirs embedded bedding interfaces.

For weak geomaterial property of bedding, Figures 4.7(c), 4.8(c), 4.9(c), and 4.10(c) list the fracturing results of the bedding models with bedding deviation angles of 30°, 45°, 60°, and 90°, respectively. As shown in Figure 4.7(c) in the Layer-II model, the upper hydraulic fracture propagates along bedding 3, and the lower hydraulic fracture propagates along bedding 4 for a period before it passes through bedding 4. As shown in Figure 4.8(c) in the Layer-III model, the hydraulic fracture propagates along beddings 2, 3, 4, 5, and 6 before penetrating these beddings. As

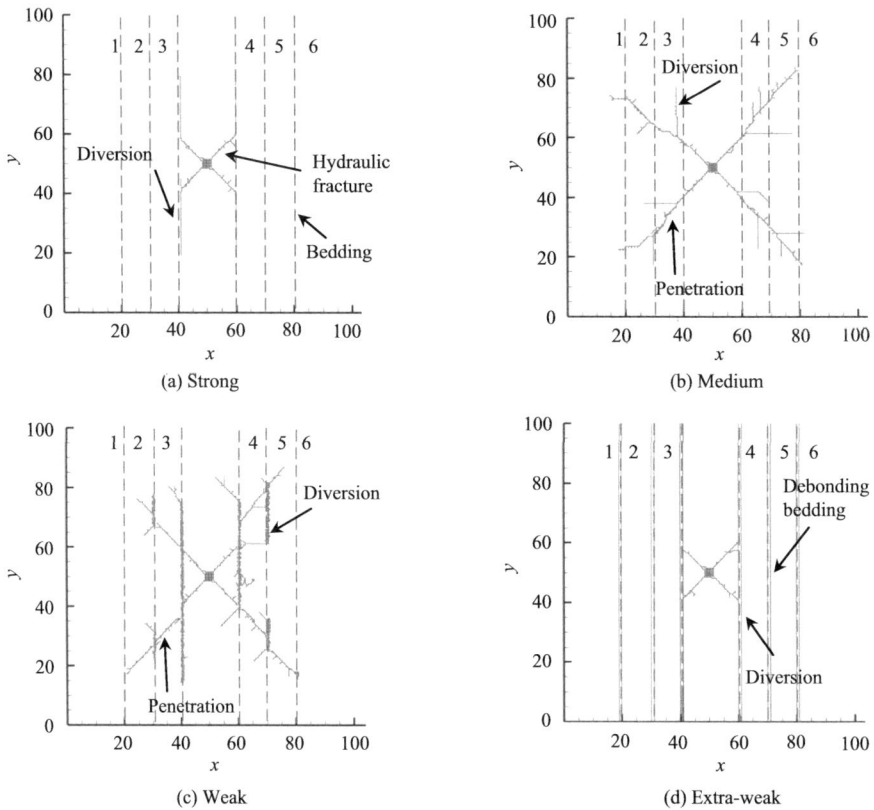

Figure 4.10. Results of morphology of fracture network in case Layer-V (Deviation angle 90°) under different geomaterial properties of heterogeneous tight reservoirs embedded bedding interfaces.

shown in Figure 4.9(c) in the Layer-IV model, the hydraulic fracture passes through beddings 3 and 4, and the deflection degree of the fracture after passing through bedding was minimal. As shown in Figure 4.10(c) in Layer-V model, the hydraulic fracture penetrates beddings 3, 4, 2, and 5, and propagates for a section along beddings 3, 4, and 5. The effects of pore pressure in Layer-II, Layer-III, Layer-IV, and Layer-V under different bedding deviation angles of heterogeneous tight reservoirs embedded bedding interfaces are shown in Figure 4.11. The pore pressure of hydraulic fractures and some activated bedding was large, and the farther away from the water injection point was, the smaller the pore pressure was. High pore pressure arises along hydraulic fractures; especially in the deviation angle 30° (Figure 4.11(a)) and 90° (Figure 4.11(d)), because there were many activated beddings, some areas with notably high pore pressure were formed around the debonding beddings.

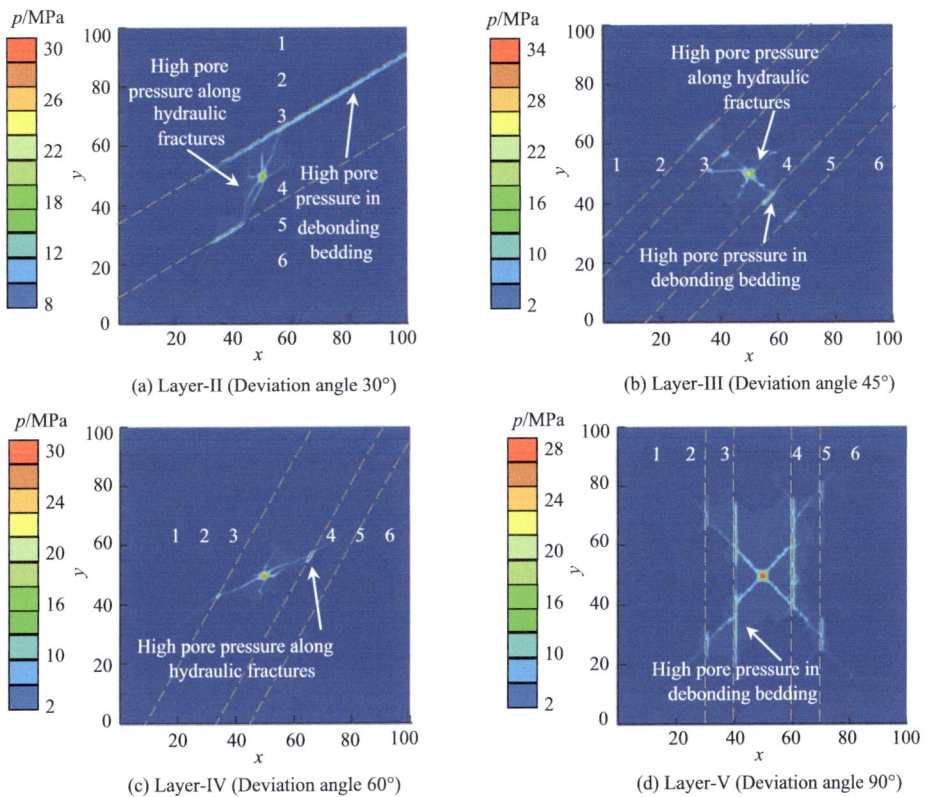

(a) Layer-II (Deviation angle 30°)

(b) Layer-III (Deviation angle 45°)

(c) Layer-IV (Deviation angle 60°)

(d) Layer-V (Deviation angle 90°)

Figure 4.11. Results of pore pressure in cases Layer-II, Layer-III, Layer-IV, and Layer-V under different bedding deviation angles of heterogeneous tight reservoirs embedded bedding interfaces.

For extra-weak geomaterial property of bedding, Figures 4.7(d), 4.8(d), 4.9(d), and 4.10(d) show the fracturing results of the bedding models with bedding deviation angles of 30°, 45°, 60°, and 90°, respectively. As shown in Figure 4.7(d) in the Layer-II model, all beddings except bedding 4 have been activated, and the hydraulic fracture propagates upward along bedding 3 and stops when bedding 4 was encountered. As shown in Figure 4.8(d) in the Layer-III model, all beddings were activated, and hydraulic fractures penetrate bedding 2, 3, 4, and 5, and they propagate downward along bedding 2. As shown in Figure 4.9(d) in the Layer-IV model, except for beddings 3 and 4, other beddings were activated, and hydraulic fractures penetrate bedding 4. As shown in Figure 4.10(d) in the Layer-I and Layer-V models, the entire bedding was activated due to weak bedding property. When the hydraulic fracture reached beddings 3 and 4, it propagates along the beddings and did not penetrate these beddings.

4.5.3 Quantitative final length and propagation states of hydrofracturing networks in heterogeneous tight reservoirs with beddings

Table 4.8 shows the final length and propagation states of hydrofracturing networks in heterogeneous tight reservoirs embedded bedding interfaces, and Figure 4.12 shows the comparison of final lengths of hydrofracturing networks. From Table 4.8 and Figure 4.12, with the weakening of bedding properties (Strong→Medium→Weak→ Extra-weak), the propagation length of hydraulic fractures will increase, because the weaker bedding is easier to be penetrated and connected by hydraulic fractures; the final length of hydraulic fractures is longer when the bedding deviation angle is 0°, 45°, and 90°, indicating that the bedding has a weak inhibition effect on the propagation of hydraulic fractures in these cases. Therefore, once the bedding deviation angles of bedding are detected, the propagation length of hydraulic fracture can be controlled in advance by designing the direction of fracturing perforation cluster and the propagation direction of initial perforation. When the bedding deviation angle was 30°, the propagation of hydraulic fractures in the horizontal direction was inhibited, whereas when the bedding deviation angle was 60°, the propagation of hydraulic fractures in the vertical direction was inhibited. When the bedding deviation angle was 45°, the propagation of hydraulic fractures along the horizontal and vertical directions was restrained to a certain extent, but the complexity of fracture network increased. Once the hydraulic fracture encountered bedding, the closer the angle between the fracture tip and bedding is to 90°, the greater the possibility of hydraulic fracture penetrating bedding was; thus, smaller angles lead to a greater degree of inhibition of hydraulic fracture propagation by bedding.

Table 4.8. Final length and propagation states of hydrofracturing fracture networks of heterogeneous tight reservoirs embedded bedding interfaces.

Cases	Strong	Medium	Weak	Extra-weak
	145.3	234.8	327.2	651.4
Layer-I	Diversion: 4	Diversion: 4	Diversion: 6	Diversion: 4
	Penetration: 0	Penetration: 4	Penetration: 4	Penetration: 0
	117.5	126.8	207.8	604.0
Layer-II	Diversion: 2	Diversion: 1	Diversion: 2	Diversion: 1
	Penetration: 0	Penetration: 1	Penetration: 1	Penetration: 0

Continued

Cases	Strong	Medium	Weak	Extra-weak
Layer-III	157.9 Diversion: 2 Penetration: 0	210.0 Diversion: 0 Penetration: 7	258.7 Diversion: 6 Penetration: 9	762.7 Diversion: 4 Penetration: 6
Layer-IV	112.4 Diversion: 2 Penetration: 0	129.9 Diversion: 0 Penetration: 2	135.6 Diversion: 0 Penetration: 2	302.5 Diversion: 2 Penetration: 1
Layer-V	145.2 Diversion: 4 Penetration: 0	234.7 Diversion: 4 Penetration: 4	327.2 Diversion: 6 Penetration: 4	651.4 Diversion: 4 Penetration: 0

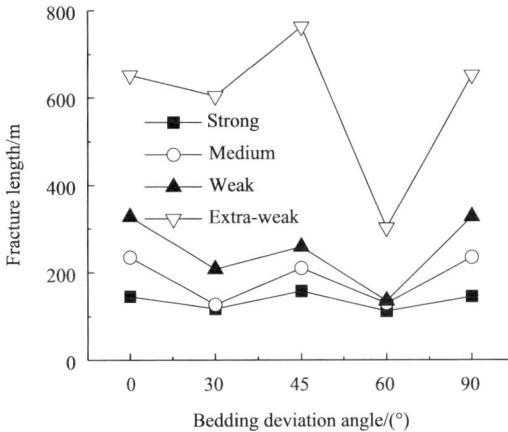

Figure 4.12. Comparison of final lengths of hydrofracturing fracture networks of heterogeneous tight reservoirs embedded bedding interfaces.

4.6   Results and discussions of dynamic propagation behaviours of hydraulic fractures in embedded multi-granule reservoir

4.6.1   A typical example implementation of fracture propagation and pore pressure in the heterogeneous tight reservoir with granules

A typical example (case Granule-VI with deviation angle 0°) is used to test the feasibility of the proposed method for analysing the heterogeneous reservoirs with granules, and briefly explain the effectiveness through elaborating the results. The results of the morphology of the fracture network and pore pressure under different geomaterial properties of heterogeneous tight reservoirs embedded granules are shown in Table 4.9:

(a) When strong granule property was implemented, five fractures propagated from the water injection point; the propagation of hydraulic fractures avoided or diverted the granules, because the granule property was greater than the property of the rock matrix. The pore pressure around the hydraulic fractures were high.

(b) When medium granule property was implemented, the existence of granules had no significant impact on the propagation of hydraulic fractures, and some granules were penetrated, because the granule property is consistent with that of the rock matrix. The fractures propagated outward uniformly in cluster form, and the area of high pore pressure presented uniform star-shaped distribution.

(c) When weak granule property was implemented, because the granule property was weaker than that of rock matrix, hydraulic fractures could directly penetrate the granules and propagate straightly; the area of high pore pressure was distributed in a belt along the straight fracture.

Table 4.9. Results of morphology of fracture network and pore pressure in case Granule-VI (Deviation angle 0°) under different geomaterial properties of heterogeneous tight reservoirs embedded granules.

Continued

Geomaterial Properties	Morphology of fracture network	Pore pressure
(b) Medium		
(c) Weak		

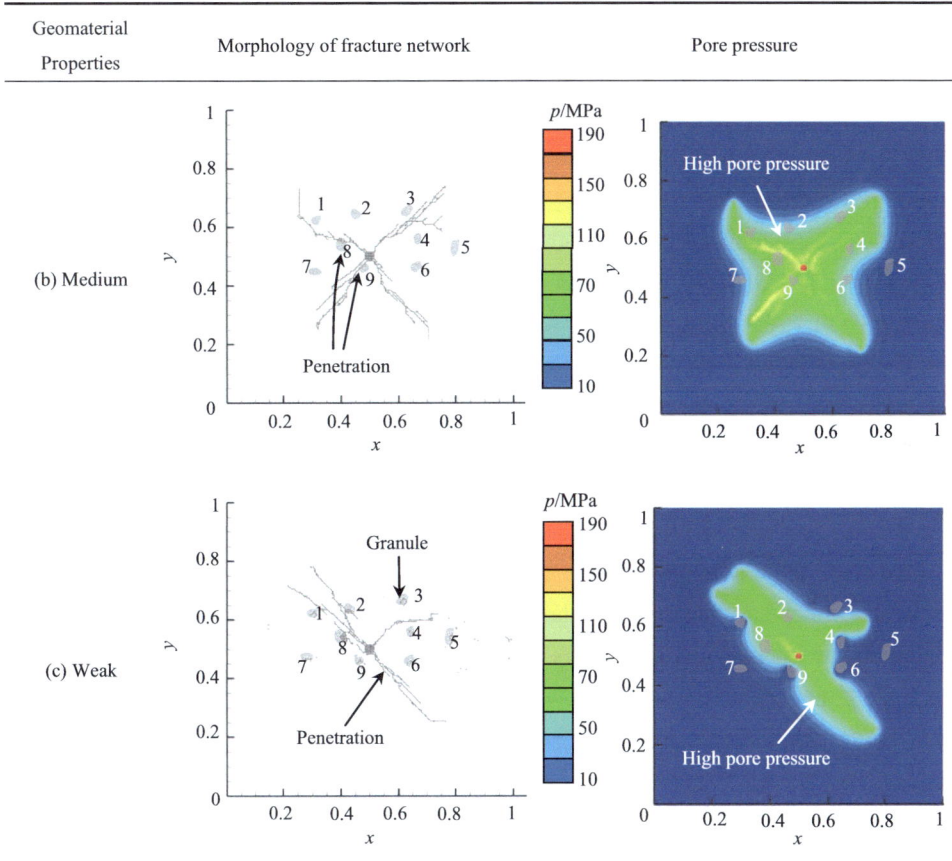

4.6.2　Influence of granule distribution and geomaterial properties on hydraulic fracture

To illustrate the influence of granule distribution on hydraulic fracture propagation, some typical and representative results are selected. Figures 4.13 and 4.14 show the morphology of fracture network results in Granule-VII (β=45°) and Granule-VIII (β=90°) under different geomaterial properties of heterogeneous tight reservoirs embedded granules. When the strong geomaterial property is selected for the mechanical parameters of granules, the fracturing results of the granule model also change with deviation angles. It shows that in Figure 4.13, the crossed hydraulic fracture formed by fracturing was easy to contact with granules distributed at 45°, thus forming more complex branch fractures. However, when the granules are distributed at

90°, the form of the fracture network was relatively simple, because the fractures and granules contact less as shown in Figure 4.14. Figures 4.13(a) and 4.14(a) show the fracturing results of the granule models with bedding deviation angles of 45° and 90°, respectively, in stronger property of bedding. Because the property of granule was greater than that of rock matrix, hydraulic fractures cannot penetrate granules. Therefore, in the process of hydraulic fracture propagation, the fractures diverted the granules and propagated around the granules. When examining medium geomaterial property of granule, the fracturing results of the granule model also change with alterations in deviation angle. Figures 4.13(b) and 4.14(b) are the fracturing results of the granule models with bedding deviation angles of 45° and 90°, respectively. With the weakening of granule properties, hydraulic fractures begin to pass through some granules when they penetrate granules. The property of granules continues to weaken. When it reaches the weak property, many granules have been penetrated, as show in Figures 4.13(c) and 4.14(c). The above fracture propagation behaviours show the influence of granule properties on the hydraulic fractures. It can be seen that this influence effect has little relationship with geometrical distribution of granules, and diversion and penetration behaviours of hydraulic fractures will occur. This shows that once the granule properties are known, the perforation angle can control the direction of fracture initiation, and the granules may change the direction of fracture propagation.

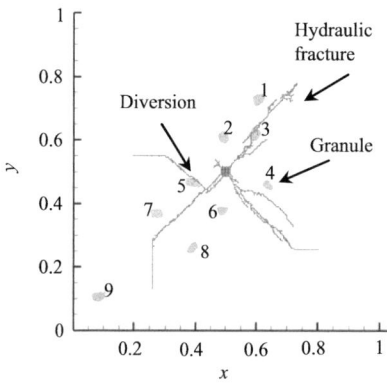

(a) Strong

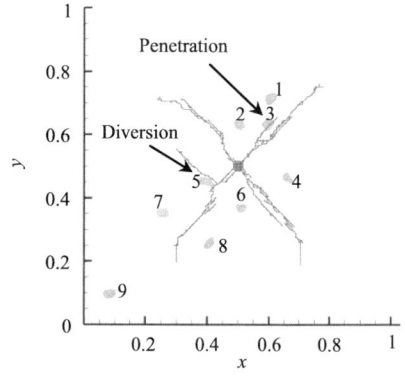

(b) Medium

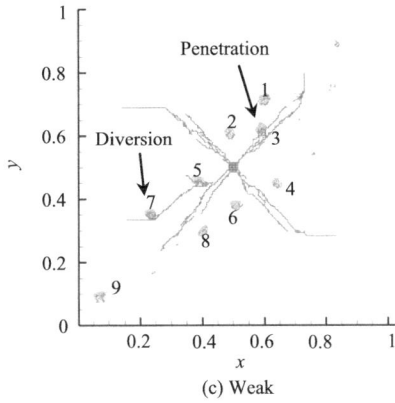

(c) Weak

Figure 4.13. Results of morphology of fracture network in case Granule-VII (Deviation angle 45°) under different geomaterial properties of heterogeneous tight reservoirs embedded granules.

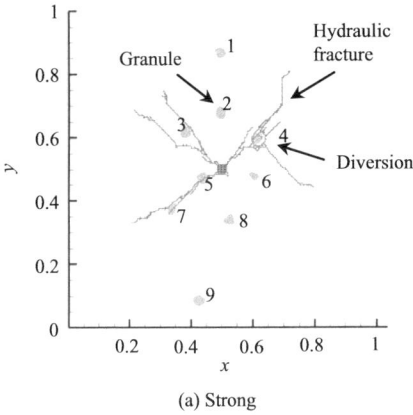

(a) Strong

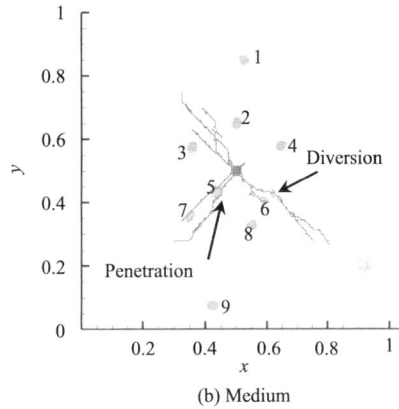

(b) Medium

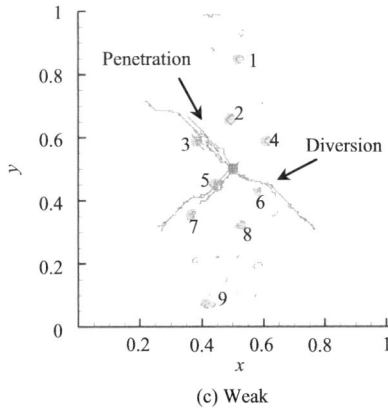

(c) Weak

Figure 4.14. Results of morphology of fracture network in case Granule-VIII (Deviation angle 90°) under different geomaterial properties of heterogeneous tight reservoirs embedded granules.

The effects of pore pressure in Granule-VII and Granule-VIII under different granule deviation angles of heterogeneous tight reservoirs embedded granules are shown in Figure 4.15. The pore pressure of hydraulic fractures and some connected granules was large, and the farther away from the water injection point was, the smaller the pore pressure was. These high pressure zones are caused by continuous water injection in hydraulic fractures, and high pressure causes some granules with weak properties to break. The pore pressure in rock matrix around the fractures was high, and some connected granules reduce the pore pressure.

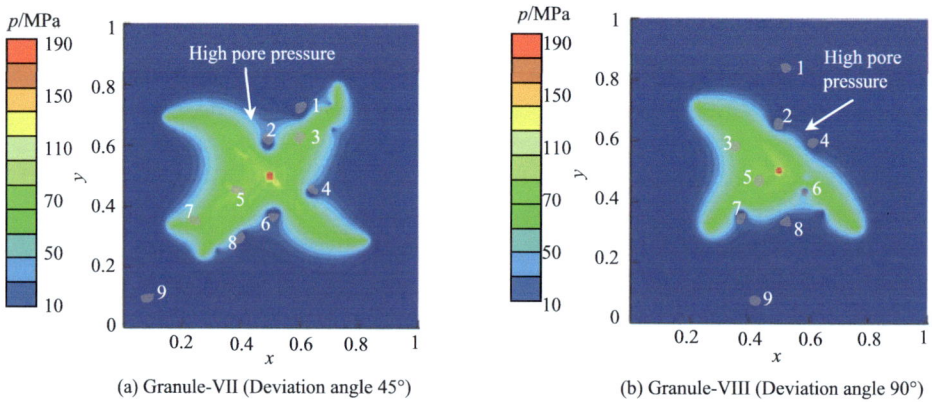

(a) Granule-VII (Deviation angle 45°)   (b) Granule-VIII (Deviation angle 90°)

Figure 4.15. Results of pore pressure in cases Granule-VII and Granule-VIII under different granules deviation angles of heterogeneous tight reservoirs embedded granules.

4.6.3   Quantitative final length and propagation states of hydrofracturing networks in heterogeneous tight reservoirs with granules

Table 4.10 shows the final length and propagation states of hydrofracturing networks of heterogeneous tight reservoirs embedded granules, and Figure 4.16 compares the final lengths of hydrofracturing networks of heterogeneous tight reservoirs embedded granules. From Table 4.10 and Figure 4.16, the deviation angle affects the final hydraulic fracture length. Under the cases conditions of various properties of granules, long hydraulic fractures were likely to occur when the granules present a 45° distribution (granule deviation angle); this may be because the *in-situ* stresses in both x and y directions in the model were set at the same value (60 MPa), and the crossed hydraulic fracture formed by fracturing was easy to contact with granules distributed at

45°, thus forming more complex branch fractures. Under the condition that the properties of granules become weak (Strong→Medium→Weak), the granules were easier to penetrate by hydraulic fractures, so relatively long hydraulic fractures were formed. On the contrary, under the condition of strong properties of granules, granules block the penetration of fractures, thus forming relatively short fractures. However, there were also some exceptional cases, such as the strong property case at 0° (granule deviation angle) and the medium property case at 90°, which reflects the influence of the randomness of granules distribution on the fracture growth.

Table 4.10. Final length and propagation states of hydrofracturing fracture networks of heterogeneous tight reservoirs embedded granules.

Cases	High	Medium	Weak
Granule-VI	2.535	2.685	2.417
	Bypass: 2	Bypass: 0	Bypass: 0
	Penetration: 0	Penetration: 2	Penetration: 2
Granule-VII	2.395	2.622	3.449
	Bypass: 2	Bypass: 1	Bypass: 1
	Penetration: 0	Penetration: 1	Penetration: 2
Granule-VIII	2.088	1.888	2.180
	Bypass: 4	Bypass: 1	Bypass: 1
	Penetration: 0	Penetration: 2	Penetration: 2

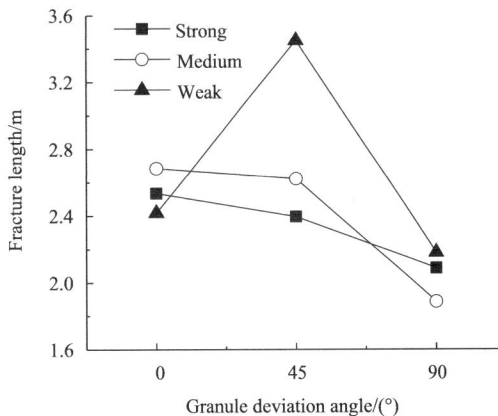

Figure 4.16. Comparison of final lengths of hydrofracturing fracture networks of heterogeneous tight reservoirs embedded granules.

4.7　Conclusions

The main conclusions are as follows:

(1) A heterogeneous continuum-discontinuum computation method was proposed for analysing dynamic diversion and penetration behaviours of hydraulic fractures contacting multi-layers and granules in tight reservoir rock. The geometrical region division and parameter assignment for multi-materials were implemented in modeling the heterogeneous medium, and consequently some representative bedding and granule models were established. To obtain the reliable numerical results, the FE, DE, and FV methods were combined to simulate the fluid-driven fracture propagation in bedding and granule models considering hydraulic-mechanical coupling.

(2) To investigate the diversion and penetration behaviours of hydraulic fractures contacting multi-layers, the bedding models with bedding deviation angles of 0°, 30°, 45°, 60°, and 90° were established and analysed. Once the hydraulic fracture encountered bedding, the closer the angle between the fracture tip and bedding is to 90°, the greater the possibility of hydraulic fracture penetrating bedding was; thus, smaller angles lead to a greater degree of inhibition of hydraulic fracture propagation by bedding. When the geomaterial properties of bedding are extremely strong or weak, the hydraulic fracture could not easily penetrate the bedding: strong bedding properties will hinder the penetration of hydraulic fractures; whereas when the bedding properties was weaker than that of the rock matrix, the hydraulic fracture could easily propagate along the weak bedding plane.

(3) The granule models with deviation angles of 0°, 45°, and 90° were established and analysed. Different granule distribution patterns not only affected the fractures in contact with granules but also significantly impacted the overall number of fractures and the direction of fracture propagation. Once the granule property was strong, the hydraulic fracture did not easily penetrate the granule; when the granule property was weaker than that of the rock matrix, the hydraulic fracture easily penetrated the granule. With the change in granule property, the number of hydraulic fractures and the specific deflection of fractures also changed.

This study used the numerical method to study the behaviour of bedding and granules in the reservoir to hydraulic fractures. This model is two-dimensional, whereas the fracture morphology in the actual engineering project is three-dimensional and spatial. In future studies, it is necessary to extend the numerical method in this study to three dimensions, and use the three-dimensional heterogeneous model to discuss the spatial propagation behaviour of hydraulic fractures.

References

Al-Bazali, T.M., Zhang, J., Wolfe, C., Chenevert, M.E., Sharmat, M.M. (2009), "Wellbore instability of directional wells in laminated and naturally fractured shales. Journal of Porous Media, Vol. 12 No. 2, pp. 119–130.

Alfano, M., Furgiuele, F., Leonardi, A., Maletta, C., Paulino, G.H. (2009), "Model fracture of adhesive joints using tailored cohesive zone models", International Journal of Fracture, Vol. 157 No. 1, pp. 193–204.

Almia, S., Masoa, G.D., Toader, R. (2014), "Quasi-static crack growth in hydraulic fracture", Nonlinear Analysis-real World Applications, Vol. 109, pp. 301–318.

Chen, D., Pan, Z., Ye, Z. (2015), "Dependence of gas shale fracture permeability on effective stress and reservoir pressure: Model match and insights", Fuel, Vol. 139, pp. 383–392.

Cho, J., Kim, H., Jeon, S., Min, K. (2012), "Deformation and strength anisotropy of Asan gneiss, Boryeong shale, and Yeoncheon schist", International Journal of Rock Mechanics and Mining Sciences, Vol. 50, pp. 158–169.

Chong, Z., Karekal, S., Li, X., Hou, P., Yang, G., Liang, S. (2017a), "Numerical investigation of hydraulic fracturing in transversely isotropic shale reservoirs based on the discrete element method", Journal of Natural Gas Science and Engineering, Vol. 46, pp. 398–420.

Chong, Z., Li, X., Hou, P., Wu, P., Zhang, J., Chen, T., Liang, S. (2017b), "Numerical investigation of bedding plane parameters of transversely isotropic shale", Rock Mechanics and Rock Engineering, Vol. 50 No. 5, pp. 1183–1204.

Daneshy, A.A. (1974), "Hydraulic fracture propagation in the presence of planes of weakness", SPE European Spring Meeting, Amsterdam, Netherlands, SEP-4852.

Guo, T., Zhang, S., Qu, Z., Zhou, T., Xiao, Y., Gao, J. (2014), "Experimental study of hydraulic fracturing for shale by stimulated reservoir volume", Fuel, Vol. 128 No. 14, pp. 373–380.

Guo, Y., Yang, C., Wang, L., Xu, F. (2018), "Study on the influence of bedding density on hydraulic fracturing in shale", Arabian Journal for Science and Engineering, Vol. 43 No. 11, pp. 6493–6508.

He, Z., Tian, S., Li, G., Wang, H., Shen, Z., Xu, Z. (2015), "The pressurization effect of jet fracturing using supercritical carbon dioxide", Journal of Natural Gas Science and Engineering, Vol. 27, pp. 842–851.

Hofmann, H., Babadagli, T., Yoon, J.S., Zang, A., Zimmermann, G. (2015), "A grain based modeling study of mineralogical factors affecting strength, elastic behavior and micro fracture development during compression tests in granites", Engineering Fracture Mechanics, Vol. 147, pp. 261–275.

Hou, B., Zeng, C., Chen, D., Fan, M., Chen, M. (2017), "Prediction of wellbore stability in conglomerate formation using discrete element method", Arabian Journal for Science and Engineering, Vol. 42 No. 4, pp. 1609–1619.

Hou, Z.K., Yang, C.H., Wang, L., Liu, P.J., Li, Z. (2016), "Hydraulic fracture propagation of shale horizontal well by large-scale true triaxial physical simulation test", Rock and Soil Mechanics, Vol. 37 No. 2, pp. 407–414.

Hu, J., Zhang, C. (2017), "Fractured horizontal well productivity prediction in tight oil reservoirs", Journal of Petroleum Science and Engineering, Vol. 151, pp. 159–168.

Ju, Y., Liu, P., Chen, J., Yang, Y., Ranjith, P.G. (2016), "CDEM-based analysis of the 3D initiation and propagation of hydrofracturing cracks in heterogeneous glutenites", Journal of Natural Gas Science and Engineering, Vol. 35, pp. 614–623.

Ju, Y., Chen, J., Wang, Y., Gao, F., Xie, H. (2018), "Numerical analysis of hydrofracturing behaviors and mechanisms of heterogeneous reservoir glutenite, using the continuum-based discrete element method while considering hydromechanical coupling and leak-off effects", Journal of Geophysical Research: Solid Earth, Vol. 123 No. 5, pp. 3621–3644.

Lee, H., Ong, S.H., Azeemuddin, M., Goodman, H. (2012), "A wellbore stability model for formations with anisotropic rock strengths", Journal of Petroleum Science and Engineering, Vol. 96, pp. 109–119.

Li, C.F., Zhang, Z.N. (2019), "Modeling shale with consideration of bedding plane by cohesive finite element method", Theoretical and Applied Genetics, Vol. 9 No. 6, pp. 397–402.

Li, L., Meng, Q., Wang, S., Li, G., Tang, C. (2013), "A numerical investigation of the hydraulic fracturing behaviour of conglomerate in Glutenite formation", Acta Geotechnica, Vol. 8 No. 6, pp. 597–618.

Li, M., Guo, Y., Wang, H., Li, Z., Hu, Y. (2020), "Effects of mineral composition on the fracture propagation of tight sandstones in the Zizhou area, east Ordos Basin, China", Journal of Natural Gas Science and Engineering, Vol. 78, 103334.

Lin, C., He, J., Li, X., Wang, X., Zheng, B. (2017), "An experimental investigation into the effects of the anisotropy of shale on hydraulic fracture propagation", Rock Mechanics and Rock Engineering, Vol. 50, pp. 543–554.

Ma, X., Zou, Y., Li, N., Chen, M., Zhang, Y., Liu, Z. (2017), "Experimental study on the mechanism of hydraulic fracture growth in a glutenite reservoir", Journal of Structural Geology, Vol. 97, pp. 37–47.

Men, X.X., Tang, C.A., Ma, T.H. (2013), "Numerical simulation on influence of rock mass parameters on fracture propagation during hydraulic fracturing", Journal of Northeastern University, Vol. 5, pp. 700–703.

Niandou, H., Shao, J.F., Henry, J.P., Fourmaintraux, D. (1997), "Laboratory investigation of the mechanical behaviour of Tournemire shale", International Journal of Rock Mechanics and Mining Sciences, Vol. 34 No. 1, pp. 3–16.

Norman, L., Jessen, F.W. (1963), "The effects of existing fractures in rocks on the extension of hydraulic fractures", Journal of Petroleum Technology, Vol. 15 No. 2, pp. 203–209.

Okland, D., Cook, J.M. (1998), "Bedding-related borehole instability in high-angle wells", SPE/ISRM Rock Mechanics in Petroleum Engineering, Trondheim, Norway, SPE-47285.

Ong, S.H., Roegiers, J.C. (1993), "Influence of anisotropies in borehole stability", International

Journal of Rock Mechanics & Mining Sciences & Geomechanics Abstracts, Pergamon, Vol. 30 No. 7, pp. 1069–1075.

Potyondy, D.O., Cundall, P.A. (2004), "A bonded-particle model for rock",International Journal of Rock Mechanics and Mining Sciences, Vol. 41 No. 8, pp. 1329–1364.

Rui, Z., Metz, P.A., Chen, G., Zhou, X., Wang, X. (2012), "Regressions allow development of compressor cost estimation models", Oil & Gas Journal, Vol. 110 No. 1, pp. 110–115.

Rui, Z., Guo, T., Feng, Q., Qu, Z., Qi, N., Gonget, F. (2018), "Influence of gravel on the propagation pattern of hydraulic fracture in the glutenite reservoir", Journal of Petroleum Science and Engineering, Vol. 165, pp. 627–639.

Shao, S., Wasantha, P.L.P., Ranjith, P.G., Chen, B.K. (2014), "Effect of cooling rate on the mechanics behavior of heated Strathbogie granite with different grain sizes", International Journal of Rock Mechanics and Mining Sciences, Vol. 70 No. 9, pp. 381–387.

Shen, B., Siren, T., Rinne, M. (2015), "Modelling fracture propagation in anisotropic rock mass", Rock Mechanics and Rock Engineering, Vol. 48 No. 3, pp. 1067–1081.

Shimizu, H., Murata, S., Ishida, T. (2011), "The distinct element analysis for hydraulic fracturing in hard rock considering fluid viscosity and particle size distribution", International Journal of Rock Mechanics and Mining Sciences, Vol. 48 No. 5, pp. 712–727.

Suarez-Rivera, R., Burghardt, J., Stanchits, S., Edelman, E., Surdi, A. (2013), "Understanding the effect of rock fabric on fracture complexity for improving completion design and well performance", International Petroleum Technology Conference, Beijing, China, IPTC-17018-MS.

Tan, X., Konietzky, H., Frühwirt, T., Dan, D. (2015), "Brazilian tests on transversely isotropic rocks: Laboratory testing and numerical simulations", Rock Mechanics and Rock Engineering, Vol. 48, pp. 1341–1351.

Tian, W.L., Yang, S.Q., Xie, L.X., Wang, Z.L. (2018), "Cracking behavior of three types granite with different grain size containing two non-coplanar fissures under uniaxial compression", Archives of Civil and Mechanical Engineering, Vol. 18 No. 4, pp. 1580–1596.

Wang, H., Liao, X.W., Zhao, X.L. (2014), "The progress of reservoir stimulation simulation technology in unconventional oil and gas reservoir", Special Oil & Gas Reservoir, Vol. 21, pp. 8–15.

Wang, Y.L., Zhang, X. (2022), "Dual bilinear cohesive zone model-based fluid-driven propagation of multiscale tensile and shear fractures in tight reservoir", Engineering Computations, Vol. 39 No. 10, pp. 3416–3441.

Wang, Y., Li, C.H., Hu, Y.Z. (2018), "Experimental investigation on the fracture behaviour of black shale by acoustic emission monitoring and CT image analysis during uniaxial compression", Geophysical Journal International, Vol. 213 No. 1, pp. 660–675.

Wang, Y.L., Wang, J., Li, L.C. (2022), "Dynamic propagation behaviors of hydraulic fracturenetworks considering hydro-mechanical coupling effects in tight oil and gasreservoirs: A multi-thread parallel computation method", Computers and Geotechnics, Vol. 152, 105016.

Zangeneh, N., Eberhardt, E., Bustin, R.M. (2015), "Investigation of the influence of natural fractures and *in-situ* stress on hydraulic fracture propagation using a distinct-element approach", Canadian Geotechnical Journal, Vol. 52 No. 7, pp. 926–946.

Zhang, Y., He, J., Li, X., Lin, C. (2019), "Experimental study on the supercritical CO_2 fracturing of shale considering anisotropic effects", Journal of Petroleum Science and Engineering, Vol. 173, pp. 932–940.

Zou, C.N., Zhu, R.K., Wu, S.T. (2012), "Types, characteristics, genesis and prospects of conventional and unconventional hydrocarbon accumulations: taking tight oil and tight gas in China as an instance", Acta Petrolei Sinice, Vol. 33, pp. 173–187.

Zou, Y., Ma, X., Zhang, S., Zhou, T., Li, H. (2016a), "Numerical investigation into the influence of bedding plane on hydraulic fracture network propagation in shale formations", Rock Mechanics and Rock Engineering, Vol. 49, pp. 3597–3614.

Zou, Y., Zhang, S., Zhou, T., Zhou, X., Guo, T. (2016b), "Experimental investigation into hydraulic fracture network propagation in gas shales using CT scanning technology", Rock Mechanics and Rock Engineering, Vol. 49 No. 1, pp. 1–13.

Zou, Y., Ma, X., Zhou, T., Li, M., Chen, M., Li, S., Zhang, Y, Li, H. (2017), "Hydraulic fracture growth in a layered formation based on fracturing experiments and discrete element modeling", Rock Mechanics and Rock Engineering, Vol. 50, pp. 2381–2395.

Chapter 5 Dynamic propagation and intersection of hydraulic fractures and pre-existing natural fractures involving the sensitivity factors

5.1 Introduction

Unconventional tight oil and gas reservoirs consist of highly heterogeneous and discontinuous rock masses (Cai *et al.*, 2014; Gale *et al.*, 2014; Liang *et al.*, 2014; Raterman *et al.*, 2018; Zeng *et al.*, 2016), in which the hydraulic fractures in the fractured reservoir will intersect with the pre-existing natural fractures and propagate dynamically during the hydrofracturing process (Kresse *et al.*, 2011; Mi *et al.*, 2018; Yan *et al.*, 2018). The properties of natural fractures have a substantial impact on the morphology of the fracture network and subsequent gas production (Fall *et al.*, 2015; Garland *et al.*, 2012; Pitman *et al.*, 2001; Teufel and Clark, 1984). A schematic of the dynamic propagation and intersection of hydraulic fractures and pre-existing natural fractures is shown in Figure 5.1. Various types of natural fractures meet and connect with hydraulic fractures, which change the geometric propagation form of the hydraulic fractures. If these natural fractures are simplified and their characteristics are classified, the geometric shapes of these natural fractures are determined by the orientation (direction), spacing, length, and persistence (Craig *et al.*, 2019; Hannes *et al.*, 2014; Li, 2020). In addition, if it is a more complex natural fracture of an underground reservoir, it also includes the geometric nonlinear and nonplanar complex characteristics of the natural fractures. There are many factors to control the pre-existing natural fractures, as shown above, so the fracture network in a rock mass is different from that in a homogeneous reservoir. If the representative properties of natural fractures cannot be accurately extracted and appropriate numerical models are not established to describe the propagation and intersection behaviours of these natural fractures, it is difficult to determine the fracturing scheme for natural fractured reservoirs and to obtain the expected fracture network and gas production.

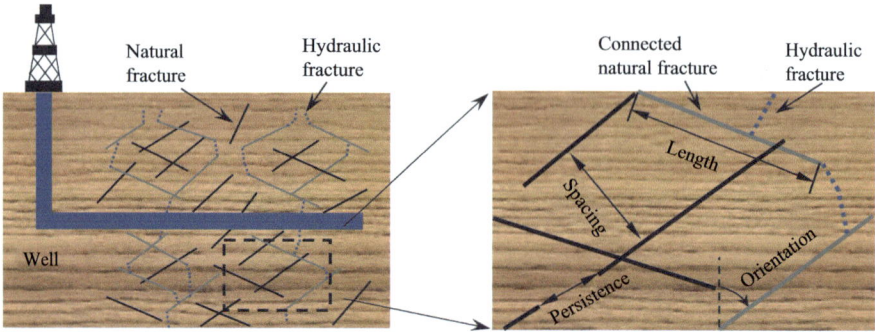

Figure 5.1. Schematic simulation of dynamic propagation and intersection of hydraulic fractures and pre-existing natural fractures.

A hydraulic fracture network with a random distribution of natural fractures is more conducive to generating complex fracture networks than that with a regular distribution; however, the random distribution of natural fractures often leads to the reorientation of fractures and a sudden decrease in branch fractures (Li, 2020). Some studies have analysed the sensitivity factors of natural fractures with single geometric properties to study and detect the influence of the geometry of natural fractures on the fracture network. Generally, hydraulic fractures propagate along the direction of maximum horizontal *in-situ* stress and readily form a single fracture with a large stress difference. However, owing to the existence of highly differentiated geometrical natural fractures, the fracturing process and fracture morphology become more complex. Subsequently, some studies began to focus on the sensitive geometric factors of natural fractures that affect the dynamic evolution of the fracture network. For the orientation of pre-existing natural fractures, which is a critical factor, the small orientation angle of the natural fracture causes hydraulic fractures to propagate along natural fractures (Suo *et al.*, 2020). As the orientation increases, the stress concentration is mainly concentrated at the tip of the obtuse angle side of the natural fractures, and the induced fracture offset will become larger (Li *et al.*, 2020); the propagation morphology of hydraulic fractures changes from open-type to cross-type, and the length of the hydraulic fracture increases. When the angle between the natural fracture and the hydraulic fracture is 90°, the adsorbed gas in the rock matrix is easily desorbed, which can effectively supplement the formation pressure; hence, this angle is considered to be preferable (Wang *et al.*, 2020). In addition, the length of the natural fractures is also considered to be a significant factor affecting the propagation, arrest, and bifurcation behaviours of the fracture network. The length of the natural fracture may disturb the fracture propagation path; the longer the natural fracture is, the easier it is to deflect the fracture network path (Craig *et al.*, 2019). When the orientation angle of natural fractures remains the same, the density and length of the natural fractures are positively correlated with the complexity of the fracture network (Hou *et al.*, 2016). The length of the natural fracture affects the behaviour of the hydraulic fracture meeting or bypassing it (Song *et al.*, 2020; Wan *et al.*, 2018). Some studies have reported the quantitative length of natural fractures; for example, when the natural fracture length is less than 10 m, the hydraulic fracture will bypass the natural fracture; otherwise, when the natural fracture length is greater than 20 m, the natural fracture will prevent the propagation of hydraulic fractures (Rahman and Rahman, 2013).

Some experiments have been conducted to study the dynamic propagation and intersection of hydraulic fractures and pre-existing natural fractures (Hu and Ghassemi, 2021; Jian and Xue, 2011; Wang *et al.*, 2018). However, owing to the limitations of the experimental specimen size and observation techniques, it is difficult to directly observe the interaction evolution process between hydraulic fractures and natural fractures, and the fracture evolution behaviour with time is even more difficult to derive. Therefore, numerical methods and models that consider the effects of natural fractures have been developed, such as the finite element method (FEM) (Wang *et al.*, 2019), discrete element method (DEM) (Yao *et al.*, 2021), extended finite element method (XFEM) (Zheng *et al.*, 2020), displacement discontinuity method (DDM) (Zhang *et al.*, 2016), and the cohesive zone method (CZM) (Ru *et al.*, 2020). The discrete fracture network (DFN) was developed to effectively simulate natural fractures in tight fractured reservoirs (Li, 2020; Katsaga *et al.*, 2015; Wang *et al.*, 2018; Yao *et al.*, 2020; Yuan *et al.*, 2018; Zhao *et al.*, 2019), which can flexibly control the two- and three-dimensional geometries of natural fractures. Therefore, this model was regarded as a potential technique for simulating natural fractures. The authors of this study used the DFN model to simulate the natural fractures and the interaction behaviours in tight reservoirs to analyse the behaviours of the supercritical CO_2 fracturing of horizontal wells and microseismic modelling (Wang *et al.*, 2018, 2019). Based on previous research, this study further developed the DFN models according to the geometric sensitivity factors of natural fractures. Further, the effects of natural fractures with different sensitive factors on fracturing behaviours and effects (growth, intersection, and distribution of hydraulic fractures, fracturing fluid flowback, and gas production) were studied.

The remainder of this chapter proceeds as follows. Section 2 introduces the combined finite element (FE)-discrete element (DE) method for hydrofracturing, considering hydro-mechanical (HM) coupling (fluid-solid coupling), and presents the relevant governing equations and DFN models. Section 3 presents the geometric and finite element models and typical pre-existing natural fractures involving the sensitivity factors of orientation, spacing, length, and persistence. Section 4 presents the results and discussion of the sensitivity factors of the natural fracture, quantitative length and volume of the fracture network, and gas production of the enhanced permeability fractured reservoir. Finally, Section 5 summarises the main conclusions of this study.

5.2   Combined finite element-discrete element method for hydrofracturing in fractured reservoirs

5.2.1   Geomechanical equations in hydrofracturing and gas production

This study involves the HM coupling behaviours in the hydrofracturing process in porous rock media, and the proppant transport in the fractures is considered (Appendix A). The mechanical governing equation for the solid stress field is as follows:

$$\boldsymbol{L}^{\mathrm{T}}(\boldsymbol{\sigma}^{\mathrm{e}} - \alpha \boldsymbol{m} p_{\mathrm{s}}) + \rho_{\mathrm{b}} \boldsymbol{g} = 0 , \tag{5.1}$$

where $\boldsymbol{L}$ is the spatial differential operator; $\boldsymbol{\sigma}^{\mathrm{e}}$ is the effective stress tensor; α is Biot's coefficient; $\boldsymbol{m}$ is the identity tensor; p_{s} is the pore fluid pressure in the rock formation; ρ_{b} is the wet bulk density; and $\boldsymbol{g}$ is the gravity vector. The basic physical parameters and meanings of the governing equations are listed in Table 5.1.

The governing equations for liquid seepage and fracture fluid flow are as follows:

$$\mathrm{div}\left[\frac{k}{\mu_{\mathrm{l}}}(\nabla p_{\mathrm{l}} - \rho_{\mathrm{l}} g)\right] = \left(\frac{\phi}{K_{\mathrm{l}}} + \frac{\alpha - \phi}{K_{\mathrm{s}}}\right)\frac{\partial p_{\mathrm{l}}}{\partial t} + \alpha \frac{\partial \varepsilon_{v}}{\partial t} , \tag{5.2}$$

$$\frac{\partial}{\partial x}\left[\frac{k^{\mathrm{fr}}}{\mu_{n}}(\nabla p_{n} - \rho_{\mathrm{fn}} g)\right] = S^{\mathrm{fr}}\frac{\mathrm{d} p_{n}}{\mathrm{d} t} + \alpha\left(\Delta \dot{e}_{\varepsilon}\right) , \tag{5.3}$$

where k is the intrinsic permeability of the porous media; μ_{l} is the viscosity of the pore liquid; p_{l} is the pore liquid pressure; ρ_{l} is the density of the pore liquid; ϕ is the porosity of the porous media; K_{l} is the bulk stiffness of the pore liquid; K_{s} is the bulk stiffness of the solid grains; ε_{v} is the volumetric strain of the porous media; k^{fr} is the intrinsic permeability of the fractured region; μ_n is the viscosity of the fracturing fluid; p_n is the fracturing fluid pressure; ρ_{fn} is the density of the fracture fluid; S^{fr} is the storage coefficient (which is effectively a measure of the compressibility of the fractured region when a fluid is present); and $\Delta \dot{e}_\varepsilon$ is the aperture strain rate.

Table 5.1. Basic physical parameters.

Parameter	Value
Depth of horizontal well /m	4000
Horizontal *in-situ* stress in x direction S_h /MPa	40
Vertical *in-situ* stress in y direction S_v /MPa	60
Fluid injection rate Q /(m³/s)	0.125
Pore pressure p_s/MPa	30
Biot's coefficient α	0.8
Young's modulus E/GPa	31
Poisson's ratio v	0.2
Porosity ϕ	0.05
Tensile strength σ_t /MPa	1.0
Fracture energy G_f /(N•m)	50
Leak-off coefficient C_I /(m³/s$^{1/2}$)	0.1×10^{-6}
Leak-off coefficient C_II /(m³/s$^{1/2}$)	0.1×10^{-6}
Density ρ_b /(kg/m³)	2.615×10^{3}
Permeability k/nD	50
Gravity g/(m/s²)	9.81
Liquid density of the pore fluid ρ_g /(kg/m³)	1×10^{3}
Liquid density of the fracturing fluid ρ_{fn} /(kg/m³)	1×10^{3}

The governing equations of the gas seepage and gas network for gas recovery and production are as follows:

$$\mathrm{div}\left[\frac{k(p_{\mathrm{g}})}{\mu_{\mathrm{g}}}\nabla p_{\mathrm{g}} - \rho_{\mathrm{g}}g\right] = \left[\phi\frac{\partial\rho_{\mathrm{g}}}{\partial p_{\mathrm{g}}} + (\rho_{\mathrm{g}}-q)\frac{\partial\phi}{\partial p_{\mathrm{g}}} + (1-\phi)\frac{\partial q}{\partial p_{\mathrm{g}}}\right]\frac{\partial p_{\mathrm{g}}}{\partial t}, \qquad (5.4)$$

$$\frac{\partial}{\partial x}\left[\frac{K^{\mathrm{fr}}}{\mu_{\mathrm{g}}}(\nabla p_{\mathrm{g}} - \rho_{\mathrm{g}}g)\right] = \phi\left(C_{\mathrm{g}} - \frac{\rho_{\mathrm{g}}}{Z}\frac{\partial\rho_{\mathrm{g}}}{\partial Z}\right)\frac{\partial\rho_{\mathrm{g}}}{\partial t}, \qquad (5.5)$$

where μ_g is the viscosity of the pore gas; p_g is the pore gas pressure; ρ_g is the density of the pore gas; q is the mass of adsorbed gas per unit volume; C_g is the gas compressibility; Z is the gas compressibility factor.

5.2.2 Leak-off of fracturing fluid

The Carter leak-off model and bulk flow rate for considering the proppant were utilised. The model assumes an initial volume loss V_{sp} per unit area over a spurt time, t_{sp}, followed by a two-term leak-off coefficient C. The formulas for the leak-off of the fracturing fluid are given as follows:

$$t - t_{exp} < t_{sp}, \quad q_1 = \frac{V_{sp}}{t_{sp}}, \tag{5.6a}$$

$$t - t_{exp} \geqslant t_{sp}, \quad q_1 = \frac{C}{\sqrt{t - t_{sp}}}, \tag{5.6b}$$

with

$$C = \frac{2C_{\mathrm{I}} C_{\mathrm{II}}}{C_{\mathrm{I}} + \sqrt{C_{\mathrm{I}}^2 - 4C_{\mathrm{II}}^2}}, \quad C_{\mathrm{I}} = \left[\frac{k_f \phi_f \Delta p}{2\mu_f}\right]^{0.5}, \quad C_{\mathrm{II}} = \left[\frac{k_r \phi_r c_{\mathrm{T}}}{\mu_r}\right]^{0.5} \Delta p, \tag{5.7}$$

where t is the current time, t_{exp} is the time at which a fracture surface is exposed for leak-off, and q_1 is the one-dimensional (1D) normal leak-off velocity.

5.2.3 Discrete fracture network model

In the 2D DFN model, fracture H is represented as a line segment connecting two points in the rock matrix (Seifollahi *et al.*, 2014):

$$H \equiv \left[P(x_1, y_1), Q(x_2, y_2)\right], \tag{5.8}$$

where P and Q are the endpoints of the fracture. The position of the DFN natural fracture can be represented by four parameters, $w = (w_1, w_2, w_3, w_4)$, where (w_1, w_2) are the coordinates of the centre, and w_3 and w_4 are the orientation and the length of the fracture, respectively.

Based on the 2D DFN model, the rock-matrix-embedded natural fracture can be discretized by 2D finite elements (linear triangular elements) and one-dimensional (1D) fracture elements (linear elements) (Karimi-Fard and Firoozabadi, 2001), as shown in Figure 5.2. The discrete DFN model can be applied to any fractured medium by:

$$\int_\Omega f(x,y)\,\mathrm{d}\Omega = \int_{\Omega_m} f(x,y)\,\mathrm{d}\Omega_m + \int_{\Omega_f} f(x,y)\,\mathrm{d}\Omega_f$$
$$= \int_{\Omega_m} f(x,y)\,\mathrm{d}\Omega_m + e \times \int_{\overline{\Omega}_f} f(x,y)\,\mathrm{d}\overline{\Omega}_f \quad, \tag{5.9}$$

where $f(x,y)$ represents the flow equations of the rock matrix and fracture media, as shown in Eqs. (5.2) and (5.3); Ω represents the global domain containing the rock matrix Ω_m and fracture Ω_f; and $\overline{\Omega}_f$ represents the fracture part of the domain as a 1D domain, and the fracture width is represented by e.

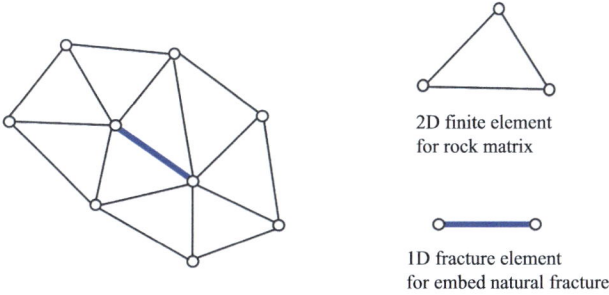

Figure 5.2. Numerical discretization of rock matrix embedded natural fracture.

5.2.4   Numerical discretization

The above governing equations are discretized using the conventional FEM as follows:

$$\begin{bmatrix} M & 0 & 0 \\ 0 & 0 & 0 \\ 0 & 0 & 0 \end{bmatrix}\begin{Bmatrix} \ddot{u} \\ \ddot{p}_s \\ \ddot{p}_n \end{Bmatrix} + \begin{bmatrix} Q_n^T & 0 & 0 \\ 0 & S_s & 0 \\ 0 & 0 & S_n \end{bmatrix}\begin{Bmatrix} \dot{u} \\ \dot{p}_s \\ \dot{p}_n \end{Bmatrix} + \begin{bmatrix} K & 0 & 0 \\ 0 & H_s & 0 \\ 0 & 0 & H_n \end{bmatrix}\begin{Bmatrix} u \\ p_s \\ p_n \end{Bmatrix} = \begin{Bmatrix} f_u \\ f_s \\ f_n \end{Bmatrix} \tag{5.10}$$

using shape function spatial gradient matrices for the structure, seepage, and network fields:

$$B_u = L_u N_u, \quad B_s = L_s N_s, \quad B_n = L_n N_n, \tag{5.11}$$

where L_u, L_s, and L_n are the gradient operators; and N_u, N_s, and N_n are the shape function matrices.

5.3　Numerical models of fractured reservoir embedded discrete fracture networks

5.3.1　Geometrical and finite element models

The geometrical model of hydrofracturing in a fractured reservoir with two sets of DFN pre-existing natural fractures is shown in Figure 5.3, with a side length of 1000 m and height of 600 m. There is a single initial perforation cluster in the horizontal well. The domain of natural fractures around the cluster is local and possesses two side lengths of 400 m and 200 m, avoiding redundant computation caused by excessive existing fractures. The FE model and mesh refinement accompanying the fracture propagation process are shown in Figure 5.4. To obtain higher computational accuracy, the initial refined finite element meshes around the domain of natural fractures were used, as shown in Figure 5.4 (a). To enhance the modelling efficiency and computational reliability, the initial FE meshes are adaptively refined in the fracture propagation process (Appendix B), so that the stress solutions near the fracture tips and fracture propagation paths can be guaranteed in comparison to conventional approaches, as shown in Figure 5.4 (a), in which the computational parameters for mesh refinement and coarsening are listed in Table 5.2. Using the proposed FE model, multistage fracturing was implemented, and Table 5.3 lists the duration of the fracturing and gas production stages, including the initial *in-situ* balance of the numerical model, hydrofracturing, flowback, and gas production.

Table 5.2. Computational parameters for mesh refinement and coarsening.

Parameter	Value
Initial intensive region in x direction /m	$400 \leqslant x \leqslant 600$
Initial intensive region in y direction /m	$100 \leqslant y \leqslant 500$
Small detail size	4
Fracture mesh size factor	0.25
Mesh density factor	1
Mesh density	0.125
Bubble size	3
Coarsening frequency	10
Coarsening density factor	1
Coarsening density	0.5

Continued

Parameter	Value
Coarsening threshold factor	0.9
Coarsening threshold	0.45
Non-coarsening zone factor	30
Non-coarsening zone	15
Max coarsening zone factor	15
Max coarsening zone	7.5

Table 5.3. Duration of fracturing and gas production stages.

Stage	Duration/s	Total time/s
Initial balance	2	2
Hydrofracturing	1500	1502
Flowback	300	1802
Gas production	32,400,000	32,401,802

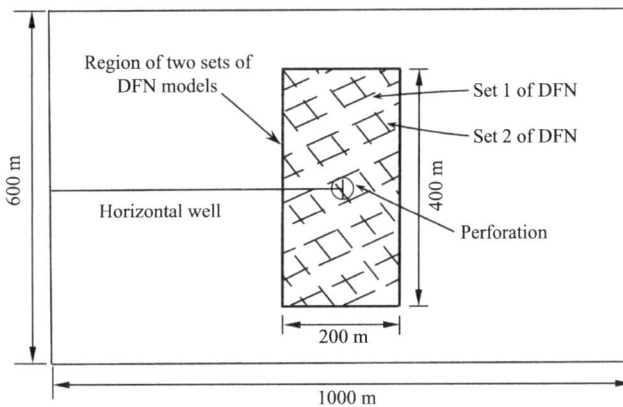

Figure 5.3. Geometrical model of hydrofracturing in a fractured reservoir with two sets of DFN models.

5.3.2   Cases study for typical pre-existing natural fractures

The DFN can be generated as pre-existing fractures, which can help to concentrate on the effects of natural pre-existing fractures during hydrofracturing fracture propagation. Each pre-existing natural fracture set is created using the properties of orientation, spacing, length, and persistence, which can define the position of the DFN natural fracture, as introduced in Section 2.3. The values of the above four properties of natural fractures in the 13 numerical models are provided in Table 5.4, focusing on the relatively large-scale natural fractures. The 13 numerical cases are divided into four groups to investigate the effects of each sensitive factor on the hydraulic fracture propagation separately. Case I is used as the target model, and the other cases (the sensitivity parameters of variation are denoted by an underline in Table 5.4) are analysed by comparing its results:

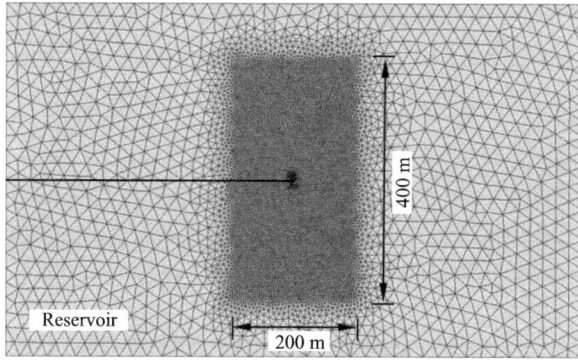

(a) Initial FE mesh.

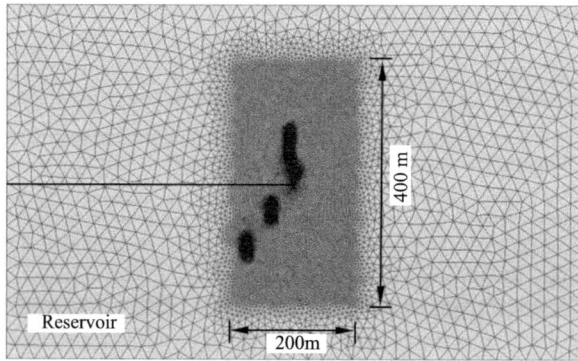

(b) Mesh refinement in fracture propagation process.

Figure 5.4. FE model and mesh refinement accompanying the fracture propagation process.

Table 5.4. Pre-existing fracture sets of DFN models: orientation, spacing, length, and persistence.

DFN	State	Cases												
		I	II	III	IV	V	VI	VII	VIII	IX	X	XI	XII	XIII
Set 1	Orientation/(°)	60	30	90	120	60	60	60	60	60	60	60	60	60
	Spacing/m	30	30	30	30	20	40	50	30	30	30	30	30	30
	Length/m	50	50	50	50	50	50	50	40	60	70	50	50	50
	Persistence/m	20	20	20	20	20	20	20	20	20	20	10	30	40
Set 2	Orientation/(°)	135	105	165	195	135	135	135	135	135	135	135	135	135
	Spacing/m	25	25	25	25	15	35	45	25	25	25	25	25	25
	Length/m	35	35	35	35	35	35	35	25	45	55	35	35	35
	Persistence/m	25	25	25	25	25	25	25	25	25	25	15	35	45

(1) **Orientation:** This sensitive property is given in degrees measured clockwise from the vertical (direction of maximum *in-situ* stress); Cases I, II, III, and IV are used to investigate the effects of orientation on hydraulic fracture propagation.

(2) **Spacing:** This sensitive property is the perpendicular distance between adjacent fractures; Cases I, V, VI, and VII are used to investigate the effects of spacing on hydraulic fracture propagation.

(3) **Length:** This sensitive property is the distance along the fracture tips; Cases I, VIII, IX, and X are used to investigate the effects of length on hydraulic fracture propagation.

(4) **Persistence:** This sensitive property is the longitudinal distance between the ends of adjacent fractures. Cases I, XI, XII, and XIII are used to investigate the effects of persistence on hydraulic fracture propagation.

For the above-mentioned cases, the natural fracture morphologies are shown in Figure 5.5, where it can be seen that the crossed natural fracture network is formed via two sets of DFN natural fractures, which can be used to observe the propagation and intersection behaviours of fracturing fractures when they encounter these natural fractures.

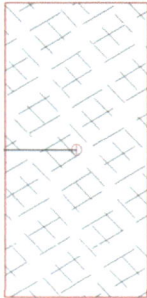

Set 1
Orientation: 60
Spacing: 30
Length: 50
Persistence: 20

Set 2
Orientation: 135
Spacing: 25
Length: 35
Persistence: 25

(a) Case I

Set 1
Orientation: 30
Spacing: 30
Length: 50
Persistence: 20

Set 2
Orientation: 105
Spacing: 25
Length: 35
Persistence: 25

(b) Case II

Set 1
Orientation: 90
Spacing: 30
Length: 50
Persistence: 20

Set 2
Orientation: 165
Spacing: 25
Length: 35
Persistence: 25

(c) Case III

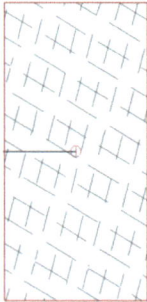

Set 1
Orientation: 120
Spacing: 30
Length: 50
Persistence: 20

Set 2
Orientation: 195
Spacing: 25
Length: 35
Persistence: 25

(d) Case IV

Set 1
Orientation: 60
Spacing: 20
Length: 50
Persistence: 20

Set 2
Orientation: 135
Spacing: 15 .
Length: 35
Persistence: 25

(e) Case V

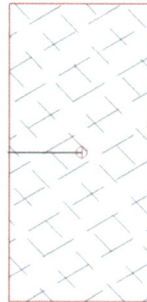

Set 1
Orientation: 60
Spacing: 40
Length: 50
Persistence: 20

Set 2
Orientation: 135
Spacing: 35
Length: 35
Persistence: 25

(f) Case VI

Set 1
Orientation: 60
Spacing: 50
Length: 50
Persistence: 20

Set 2
Orientation: 135
Spacing: 45
Length: 35
Persistence: 25

(g) Case VII

Set 1
Orientation: 60
Spacing: 30
Length: 40
Persistence: 20

Set 2
Orientation: 135
Spacing: 25
Length: 25
Persistence: 25

(h) Case VIII

Set 1
Orientation: 60
Spacing: 30
Length: 60
Persistence: 20

Set 2
Orientation: 135
Spacing: 25
Length: 45
Persistence: 25

(i) Case IX

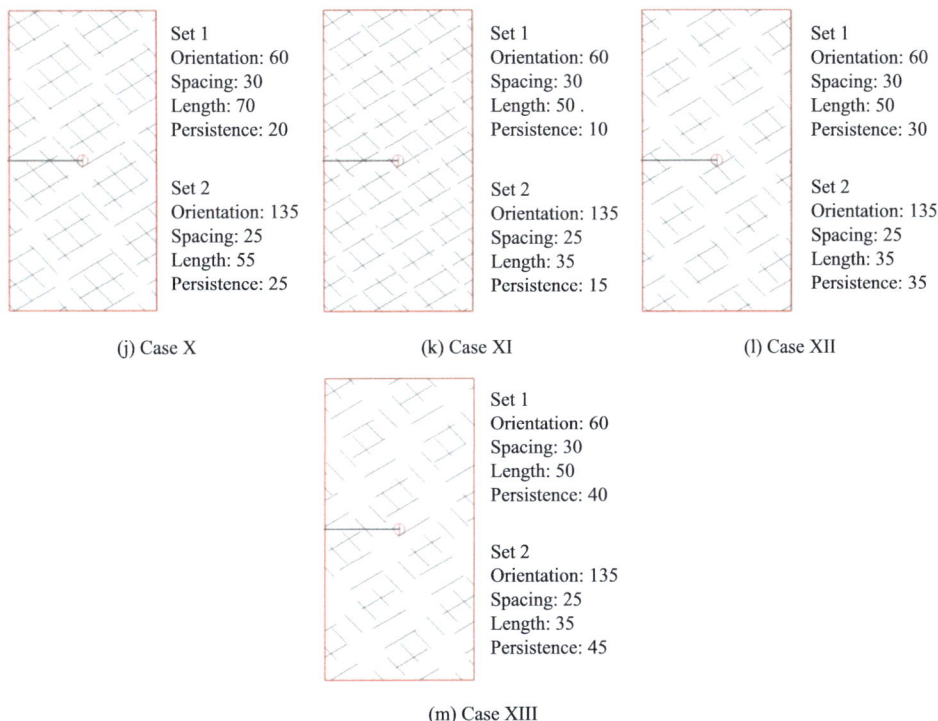

Set 1
Orientation: 60
Spacing: 30
Length: 70
Persistence: 20

Set 2
Orientation: 135
Spacing: 25
Length: 55
Persistence: 25

(j) Case X

Set 1
Orientation: 60
Spacing: 30
Length: 50 .
Persistence: 10

Set 2
Orientation: 135
Spacing: 25
Length: 35
Persistence: 15

(k) Case XI

Set 1
Orientation: 60
Spacing: 30
Length: 50
Persistence: 30

Set 2
Orientation: 135
Spacing: 25
Length: 35
Persistence: 35

(l) Case XII

Set 1
Orientation: 60
Spacing: 30
Length: 50
Persistence: 40

Set 2
Orientation: 135
Spacing: 25
Length: 35
Persistence: 45

(m) Case XIII

Figure 5.5. Cases for the sensitivity factors: orientation, spacing, length, and persistence.

5.4　Results and discussion

In this section, the numerical cases are computed using the proposed models with the program package ELFEN TGR (Rockfield Software Ltd., 2016). The following section discusses the fracture propagation and intersection behaviours under various natural fracture conditions, and analyses the results of quantitative fracture length, volume, and gas production.

5.4.1　Sensitivity factors of pre-existing natural fractures

The results of stress, dynamic propagation, and the intersection of hydraulic fractures and pre-existing natural fractures are shown in Figure 5.6. The hydraulic fractures propagate along the y-direction where the maximum *in-situ* stress (S_y = 60 MPa) is located. Figure 5.6 (a) provides the results of the sensitivity analysis of Case I used as comparison targets. Through comparative analysis, the following will introduce and discuss the fracture propagation pattern affected by changes in the properties of natural fractures.

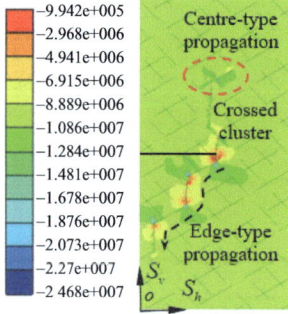

(a) Case I

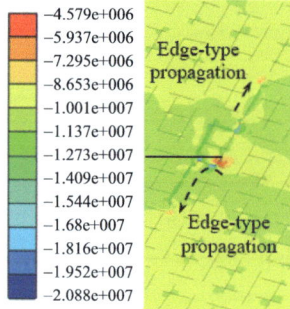

(b) Case II

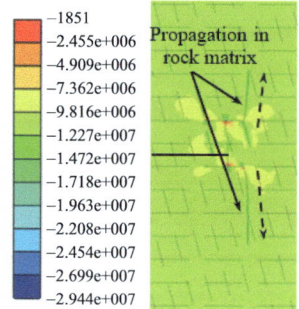

(c) Case III

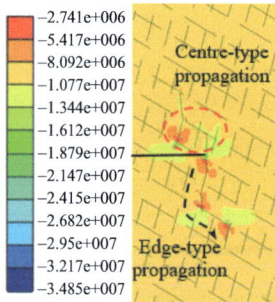

(d) Case IV

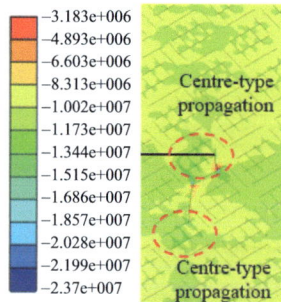

(e) Case V

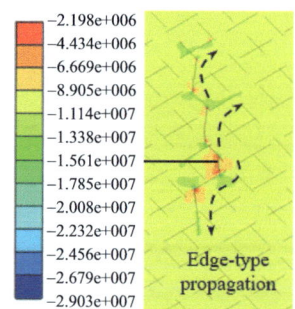

(f) Case VI

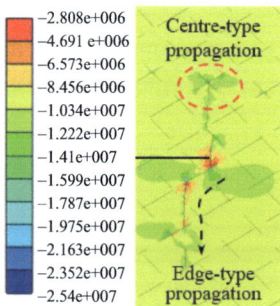

(g) Case VII

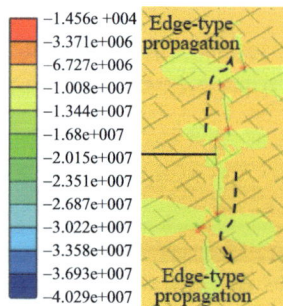

(h) Case VIII

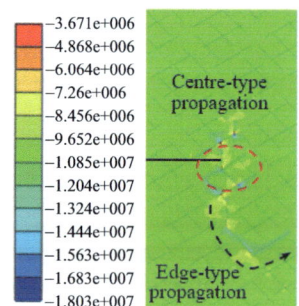

(i) Case IX

(j) Case X   (k) Case XI   (l) Case XII

(m) Case XIII

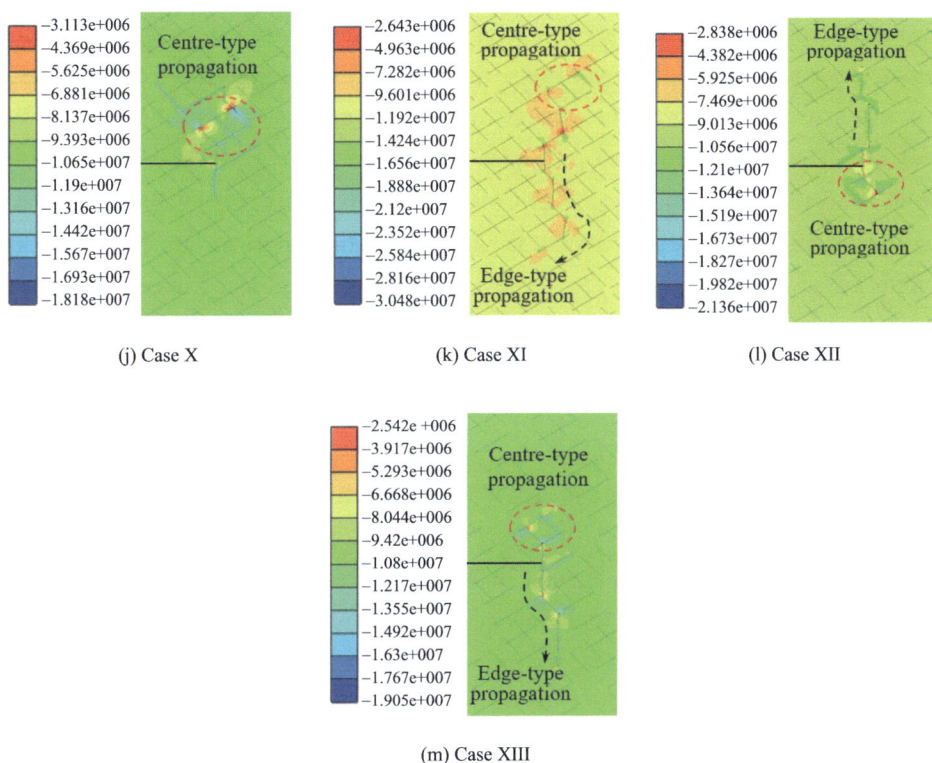

Figure 5.6. Results of stress (first principal stress, Pa), dynamic propagation and intersection of hydraulic fractures and pre-existing natural fractures.

5.4.1.1   Orientation of inclined natural fractures

The results of the orientation of the initial pre-existing inclined natural fractures for the sensitivity analyses of Cases II, III, and IV are shown in Figures 5.6 (b)-(d). The orientation angle of the natural fractures varied between these cases by 30°, with one set of natural fractures at 30°, 60°, 90°, and 120°; the other set of natural fractures at 105°, 135°, 165°, and 195°. There are two typical types of fracture network morphologies when hydraulic fractures intersect with natural fractures:

(1) Centre-type propagation is the result of the intersection of hydraulic fractures and crossed clusters of natural fractures.

(2) Edge-type propagation is the result of the intersection of hydraulic fractures and the edge of the natural fractures.

As shown in Figure 5.6 (a), hydraulic fractures will normally propagate along the edges of natural fractures; once the hydraulic fracture contacts the crossed cluster of the natural fractures, the cluster is activated. With the adjustment of the orientation angles, the fracture may propagate along the edges of the natural fractures (Figure 5.6 (b)) and may not encounter the natural fracture (Figure 5.6 (c)). In these cases, the hydraulic fractures are prone to propagate straight and form a single crack. Notably, the 60° and 120° orientation angles in Cases I (Figure 5.6 (a)) and IV (Figure 5.6 (d)) are equally effective, which possess the same inclination angle of 60° clockwise and anticlockwise compared with the vertical direction (Direction of maximum *in-situ* stress); finally, the centre-type and edge-type propagations are formed.

5.4.1.2   Spacing of adjacent natural fractures

The results of fracture spacing for the sensitivity analyses of Cases V, VI, and VII are provided in Figures 5.6 (e)-(g). When the spacing of adjacent natural fractures is relatively narrow and the density of natural fractures is large (Case V; fracture spacing of 20 m), the centre-type propagation of hydraulic fractures may arise, as shown in Figure 5.6 (e). As the spacing increases and the density of natural fractures decreases (Cases I, VI, and VII; 30 m, 40 m, and 50 m of fracture spacing), the fracture begins to propagate from the centre to the edge of the cluster of natural fractures, as shown in Figures 5.6 (a), (f), and (g). This demonstrates that once the spacing of adjacent natural fractures is exceptionally small in the reservoirs, many initial fracture networks with a small cross density will be formed, which will increase the probability that it will connect with hydraulic fractures. Once the fractures meet, the centre-type propagation of fractures is readily formed. When there is no natural fracture in the reservoir or its spacing is tiny, a single fracture is typically formed.

5.4.1.3   Length of straight natural fractures

The results of the fracture length sensitivity analyses of Cases VIII, IX, and X are shown in Figures 5.6 (h)-(j). When the length of straight natural fractures and density of natural fractures are relatively large (Case X; fracture spacing of 20 m), the centre-type propagation of hydraulic fractures may increase, as demonstrated in Figure 5.6 (j). As the length decreases and the density of natural fractures decreases (Cases I, VIII, and IX; 40 m, 50 m, and 60 m of fracture length), the fracture begins to propagate from the edge to the centre of the cluster of natural fractures, as shown in Figures 5.6 (a), (h), and (i). Similar to the sensitivity relationship of fracture length, once the natural fracture length increases, many original fracture networks with small cross density will be formed, which will increase the probability of meeting with hydraulic fractures. Once meeting, a centre-type propagation will readily. On the contrary, if the length of the straight natural fractures is too short, it will greatly reduce the probability of meeting with hydraulic fractures and form a single fracture.

5.4.1.4   Persistence of adjacent natural fractures

The results of the sensitivity analyses of fracture persistence for of Cases XI, XII, and XIII are provided in Figures 5.6 (k)-(m). When the persistence between adjacent natural fractures is small (Case XI; 10 m of fracture persistence), the hydraulic fractures may propagate from the initial perforation to meet the natural fracture clusters more rapidly, and centre-type propagation occurs, as demonstrated in Figure 5.6 (k). With the increase in fracture persistence (Cases I, XII, and XIII; 20 m, 30 m, and 40 m of fracture persistence), the spacing between the natural fracture clusters becomes larger, and the fracture easily propagates along the edge of the natural fractures or through the intervals between the natural fractures, as shown in Figures 5.6 (a), (l), and (m).

5.4.2   Quantitative length and volume of fracture networks

To investigate the influences of the sensitive factors of natural fractures on the fracture network, the quantitative results of the length and volume of the fracture networks may be derived and discussed in detail as follows.

5.4.2.1   Fracture length

The evolution of the length of the hydraulic and connected DFN natural fractures during hydrofracturing is demonstrated in Figure 5.7. Figure 5.7 (a) shows the total fracture length considering the sensitivity factor of the orientation of natural fractures. The orientation angles of natural fractures under these conditions are different; however, the number of natural fractures remains almost the same (including the spacing, length, and persistence of natural fractures), so there is no significant difference in each curve. When the orientation angle of natural fractures is 30°, all the hydraulic fractures extend along the edges of the natural fractures (no activation of natural fracture clusters), and the fracture length was the shortest in the hydrofracturing process. Figure 5.7 (b) provides the total fracture length considering the sensitivity factor of the spacing of the natural fracture. According to the analysis of the results in Figure 5.6 (b), hydraulic fractures are most likely to activate and connect natural fracture clusters when the natural fracture spacing is large (20 m and 15 m for two sets of natural fractures) in Case V, forming the longest natural fracture length in the hydrofracturing process. Figure 5.7 (c) shows the total fracture length considering the sensitivity factor of the length of the natural fracture. Hydraulic fractures are most likely to activate and connect natural fracture clusters when the length of the natural fracture is short (70 m and 55 m for two sets of natural fractures) in Case X, forming the longest natural fracture length in the hydrofracturing process. Figure 5.7 (c) displays the total fracture length considering the sensitivity factor of the persistence of natural fractures. Hydraulic fractures are most likely to activate and connect natural fracture clusters when the persistence of natural fracture is short (10 m and 15 m for two sets of natural fractures) in Case XI, forming the longest natural fracture length in the hydrofracturing process.

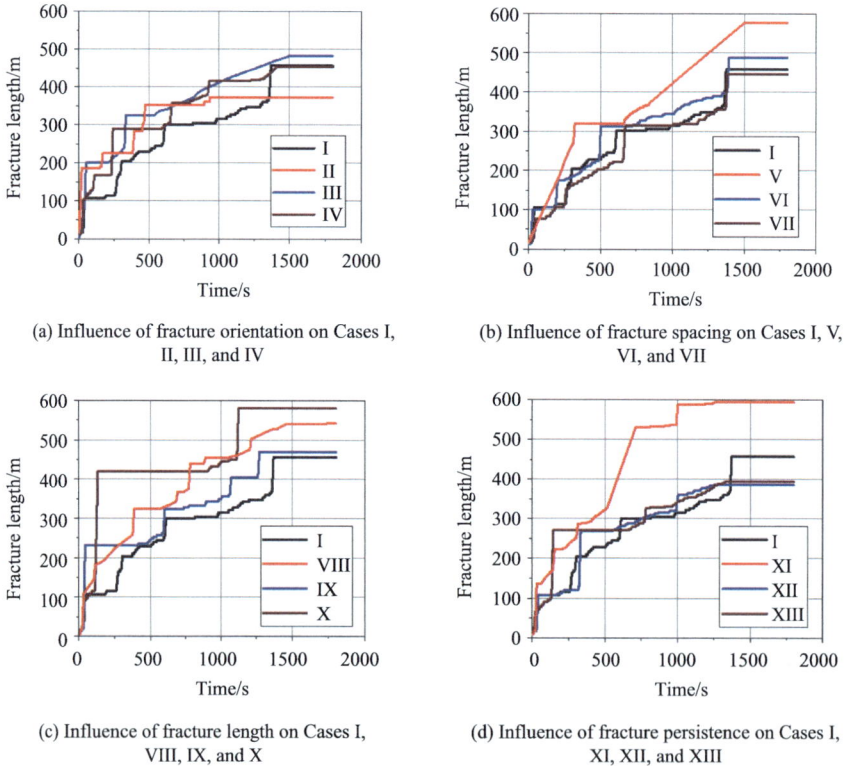

(a) Influence of fracture orientation on Cases I, II, III, and IV

(b) Influence of fracture spacing on Cases I, V, VI, and VII

(c) Influence of fracture length on Cases I, VIII, IX, and X

(d) Influence of fracture persistence on Cases I, XI, XII, and XIII

Figure 5.7. Evolution for fracture length involving the sensitivity factors: orientation, spacing, length, and persistence.

Table 5.5 lists the final length of the hydraulic fractures and connected DFN natural fractures of each case, in which the results with the longest total fracture length in each case is underlined. For a more intuitive display and analysis, the comparison of fracture length in different cases is shown in Figure 5.8. One of the factors causing the substantial fracture lengths is the high density of the initial natural fractures to increase their chances of being activated by the hydraulic fracturing. Furthermore, in the first group when analysing the orientation effects of the natural fracture, the initial fracture length is almost the same in each case, and the final fracture in Case III is the longest. In the analysis of the spacing, length, and persistence of natural fracture, the longest lengths of final fractures in Cases V, X, and XI are primarily constituted by the connected natural fractures and relatively shorter hydraulic fractures.

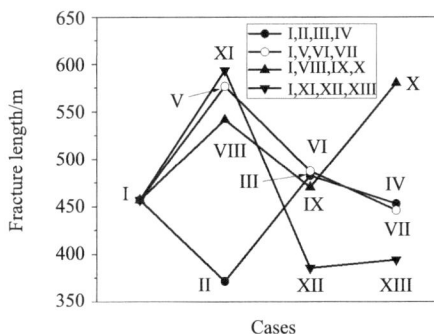

Figure 5.8. Comparison of fracture length in different cases.

Table 5.5. Final length of hydraulic fractures and connected DFN natural fractures.

Sensitivity factor	Cases	Length of hydraulic fracture/m	Length of hydraulic and connected DFN natural fractures/m	Initial DFN natural fractures/m
Benchmark	I	167	457	3698
Orientation	II	66	371	3731
	III	227	482	3965
	IV	128	453	3740
Spacing	V	81	576	5889
	VI	182	487	2750
	VII	191	446	2160
Length	VIII	217	541	3375
	IX	95	470	4036
	X	85	580	4248
Persistence	XI	133	593	4479
	XII	130	385	3269
	XIII	138	394	2836

5.4.2.2   Fracture volume

In all cases, the injection volume of the fracturing fluid is the same (same injection rate and duration time), which may flow into the hydraulic and natural fractures and leak into the rock matrix. Therefore, the evolution process of the fracture volume is different in various cases. The evolution of the fracture volume with respect to the sensitivity factors is shown in Figure 5.9. A long fracture network during the fracturing stage generally leads to a large fracture volume. After backflow, the fracture volume decreases sharply owing to the loss of fracturing fluid and the lack of fluid support.

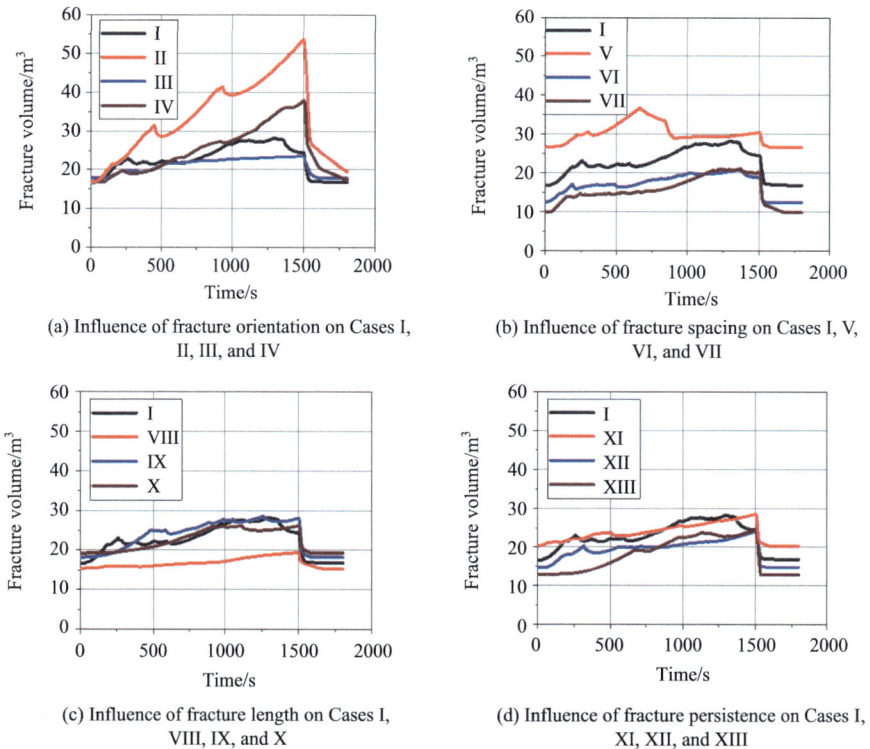

(a) Influence of fracture orientation on Cases I, II, III, and IV

(b) Influence of fracture spacing on Cases I, V, VI, and VII

(c) Influence of fracture length on Cases I, VIII, IX, and X

(d) Influence of fracture persistence on Cases I, XI, XII, and XIII

Figure 5.9. Evolution for fracture volume involving the sensitivity factors: orientation, spacing, length, and persistence.

Table 5.6 lists the final volume of hydraulic fractures or the accompanying connected DFN natural fractures at the end of fracturing (1502 s) and the end of flowback (1802 s) of each case, respectively. For the case in each group with longest fracture length listed in Table 5.5, the corresponding values of the fracture volumes are also underlined in Table 5.6. The underlined data for volume exhibit larger values, which varied considerably from the result of reservoirs without natural fractures. Because there is no natural fracture connection and fracturing fluid input, the volume of hydraulic fractures is often inversely proportional to their length.

Table 5.6. Final volume of hydraulic fractures and connected DFN natural fractures.

Sensitivity factor	Cases	Volume of hydraulic fracture (m^3) at end of fracturing (1502 s)	Volume of hydraulic and connected DFN natural fractures (m^3) at end of fracturing (1502 s)	Volume of hydraulic fracture (m^3) at end of flowback (1802 s)	Volume of hydraulic and connected DFN natural fractures (m^3) at end of flowback (1802 s)
Benchmark	I	21.34	24.47	1.76	16.86
Orientation	II	9.69	53.48	1.17	19.36
	III	18.07	23.50	1.18	17.92
	IV	17.19	37.91	1.07	17.41
Spacing	V	5.69	30.38	0.53	26.62
	VI	17.80	18.75	1.19	12.43
	VII	30.05	20.02	1.30	9.78
Length	VIII	20.99	19.45	1.27	15.29
	IX	10.33	28.10	0.52	18.28
	X	7.95	26.10	0.40	19.33
Persistence	XI	15.97	28.48	0.84	20.25
	XII	29.18	24.09	0.73	14.78
	XIII	27.59	24.58	0.83	12.87

In order to investigate the relationship between the propagation length and volume in hydraulic fractures, Figure 5.10 demonstrates the relationship of the fracture length and volume of the hydraulic fractures at end of fracturing (1502 s) and the end of flowback (1802 s). Figure 5.10 (a) indicates that the evolution trend of the fracture length and fracture volume is almost the same at the end of fracturing, which indicates that the fracture length and volume are positively correlated. Figure 5.10 (b) demonstrates that at the end of flowback, due to the flowback of fracturing fluid, the fracture loses support and gradually closes, and the fracture volume drastically decreases. As established by the above results, the length and volume of the fracture network are positively correlated, excluding the natural fractures that store the fracturing fluid and the leak-off of fluid in the hydraulic fractures.

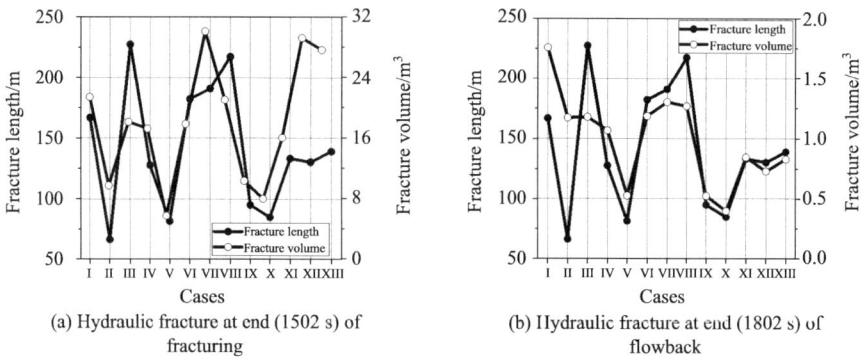

(a) Hydraulic fracture at end (1502 s) of
fracturing

(b) Hydraulic fracture at end (1802 s) of
flowback

Figure 5.10. Relationship of fracture length and volume.

5.4.3　Gas production in enhanced permeability fractured reservoirs

Gas production is affected by the length of the natural fractures, and relatively long fractures with high hydraulic fracturing conductivity are required to obtain high gas production (Ren and Lau, 2020; Tayong *et al.*, 2019; Yuan *et al.*, 2018). Gas reservoirs with a high natural density and shorter primary fractures are superior (Han *et al.*, 2013). There is a discernible linear relationship between the cumulative gas production and fracture permeability (Hu *et al.*, 2020). These models provide gas production in enhanced permeability fractured reservoirs.

　　Figure 5.11 displays the gas pressure in each case at the initial gas production stage after the hydrofracturing process. The gas pressure near the hydraulic and connected natural fractures is lower than that of other domains in the global reservoir, which forms low-pressure seepage channels that allow the gas flow from the rock matrix into the fracture networks. Through these seepage channels, the gas products evolve over time, as shown in Figure 5.12. Through the comparison results of each group, it can be seen that under the condition of the formation of a local centre-type fracture network and relatively developed fracture length, the gas in the matrix around the fracture network easily migrates into the fracture through the network, increasing the continuous gas production. The centre-type fracture network is formed in Case I in Figure 5.6 (a), Case V in Figure 5.6 (e), Case X in Figure 5.6 (j), and Case XI in Figure 5.6 (k), resulting in substantial gas production. Figure 5.12 (a) indicates that the gas production of Case I is the largest for the sensitive factor of orientation, which indicates that when the orientation of the natural fracture is close to the direction of the maximum *in-situ* stress (60° and 135° in two sets of natural fractures), the hydraulic fracture easily propagates to form the seepage channels. Figure 5.12 (b) demonstrates that for the sensitive factor of spacing, Case V has the largest gas production, which indicates that when the fracture spacing is small (20 m and 15 m in two sets of natural fractures), natural fractures form crossed clusters, and hydraulic fractures readily intersect with these crossed clusters to form complex fracture networks in the process of propagation, which also increases the gas production. Figure 5.12 (c) shows that for the sensitive factor of fracture length, Case X has the largest gas production, indicating that when the length of the natural fractures is large (50 m and 55 m in two sets of natural fractures), the fractures easily intersect and connect to form crossed clusters, and hydraulic fractures can easily intersect with these crossed clusters to form a complex fracture network in the process of propagation, which also allows the gas production to become larger. Figure 5.12 (d) demonstrates that for the sensitive factor of persistence, Case I provides the largest values because of the narrow persistence (50 m and 55 m in two sets of natural fractures).

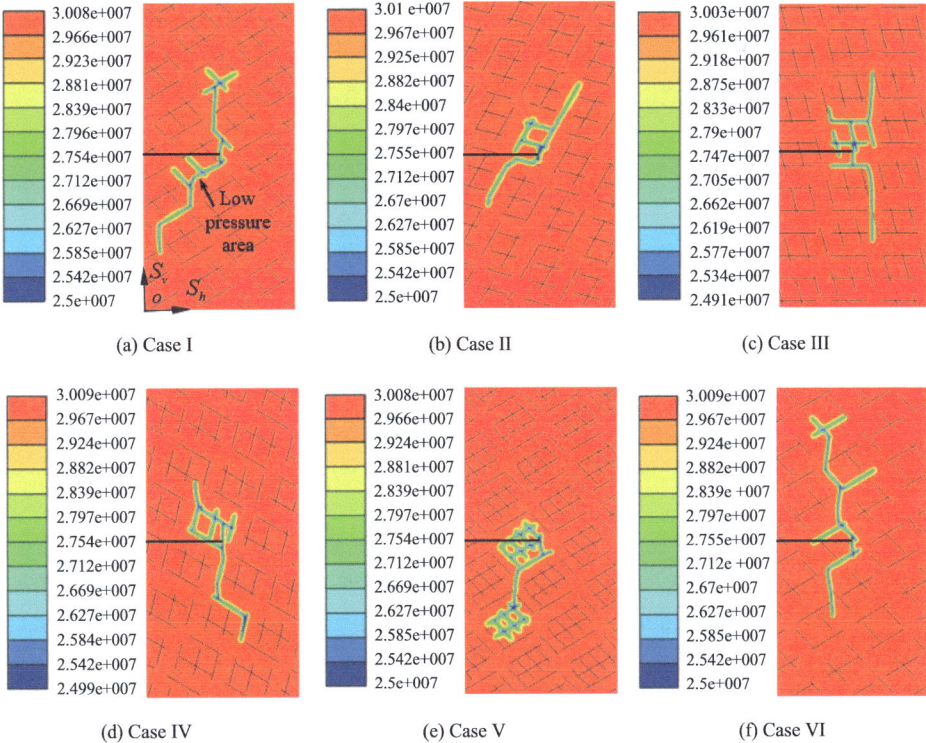

(a) Case I   (b) Case II   (c) Case III

(d) Case IV   (e) Case V   (f) Case VI

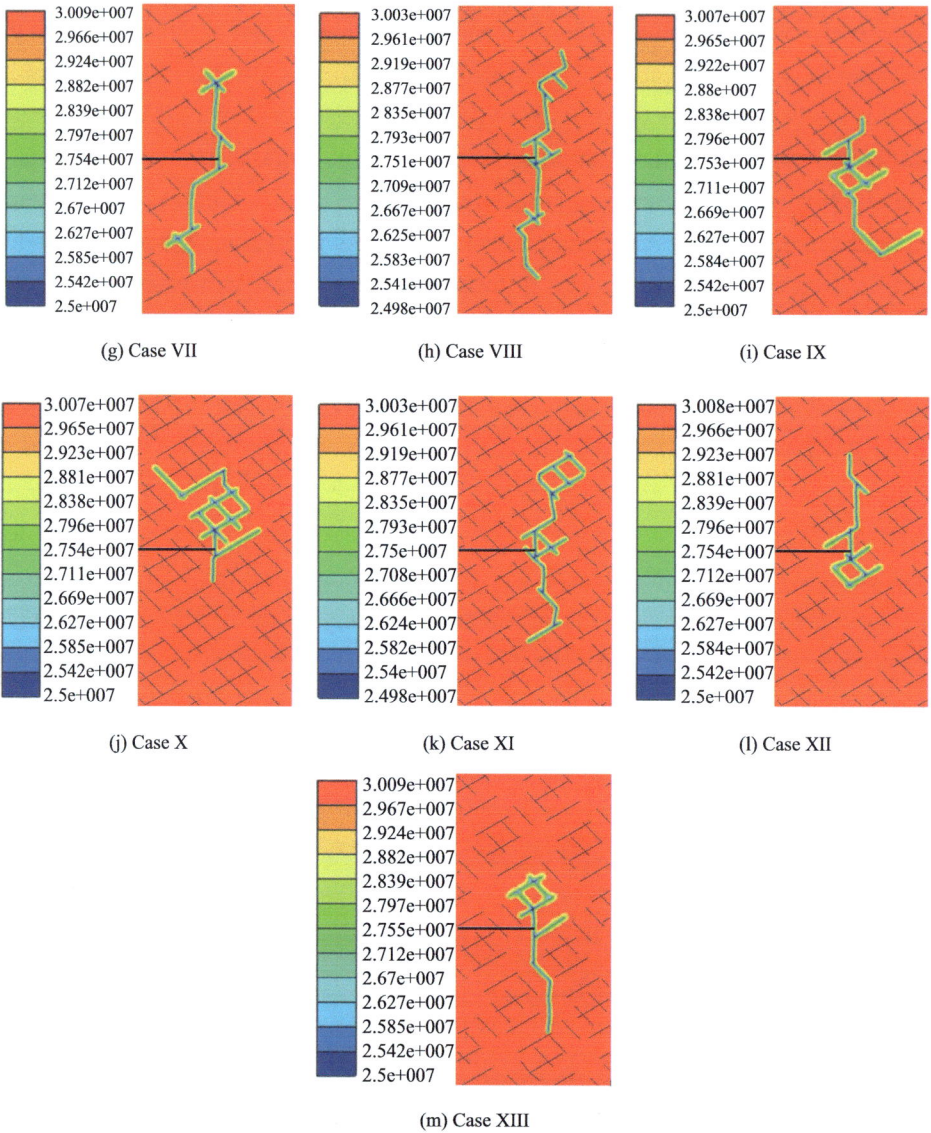

(g) Case VII

(h) Case VIII

(i) Case IX

(j) Case X

(k) Case XI

(l) Case XII

(m) Case XIII

Figure 5.11. Pressure of reservoir and fracture networks at the initial gas production stage.

Figure 5.13 compares the gas production volume in all cases, in which the values of Cases V, VIII, X, and XI are significantly larger than the others in the same group. Combined with the phenomena of Figure 5.12, the results of Cases V, X, and XI indicate that longer fracture networks may induce more gas production volume. Figure 5.13 displays two extreme cases: Cases II and X. As demonstrated in Figure 5.11 (b), the edge-type fracture network is formed and the total conceded fracture length (371

m) is short, which is unfavourable for gas flow, with the final gas production volume of 4.43×10^6 m^3. In Figure 5.11 (j), the centre-type fracture network is formed that is predominantly concentrated together, which is exceptionally conducive to gas flow and production, and the final gas production is 8.69×10^6 m^3.

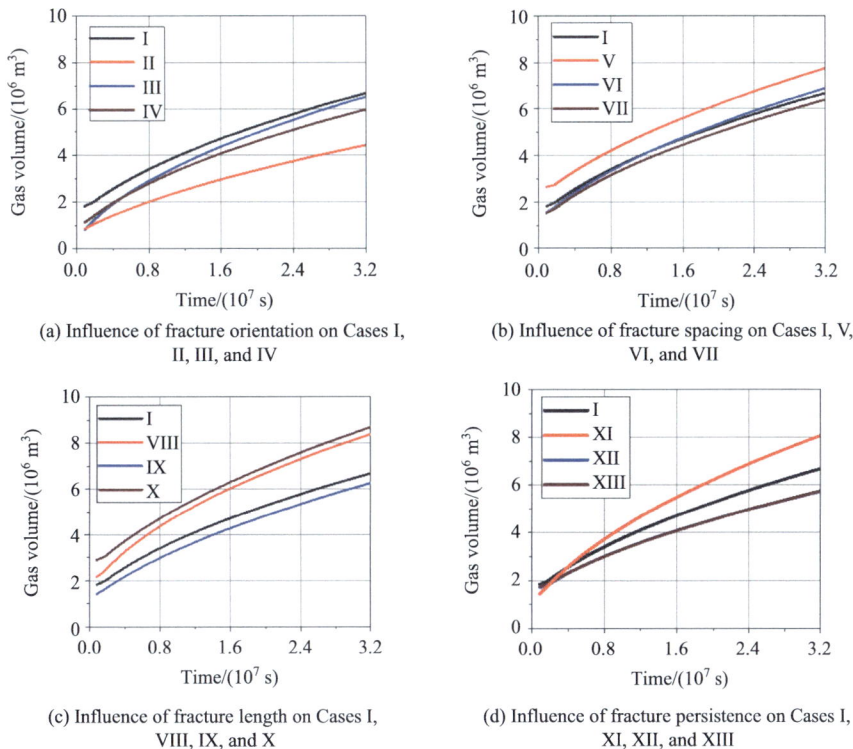

(a) Influence of fracture orientation on Cases I, II, III, and IV

(b) Influence of fracture spacing on Cases I, V, VI, and VII

(c) Influence of fracture length on Cases I, VIII, IX, and X

(d) Influence of fracture persistence on Cases I, XI, XII, and XIII

Figure 5.12. Evolution for gas production volume involving the sensitivity factors: orientation, spacing, length, and persistence.

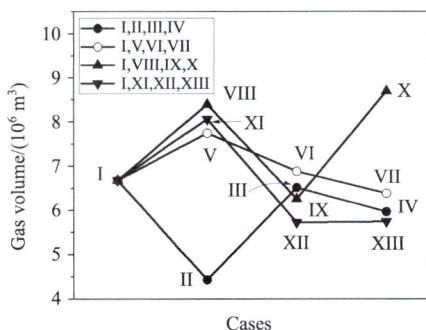

Figure 5.13. Comparison of gas production volume in different cases.

To investigate the relationship between the fracture length and gas production volume, the numerical cases were reordered based on the magnitude of the induced fracture length, as listed in Table 5.7, which lists the final length of the hydraulic fractures and connected DFN natural fractures and gas production volume by sequence. To quantitatively obtain the relationship, the fitting curve is derived via the least squares method, as shown in Figure 5.14, and the equation of the curve is as follows:

$$V = 2 \times L^2 - 1.29 \times 10^4 \times L \tag{5.12}$$

$$R^2 = 0.94 \tag{5.13}$$

where V is the volume of gas production volume (m^3); L is the fracture length (m); R^2 and is the R-square of the correlation coefficient. Figure 5.14 provides the basic relationship between the length of the fracture network and gas production volume in a fractured reservoir, which can be used to evaluate gas exploitation. However, this relationship is only suitable for the basic mechanical properties and natural fracture distribution studied in this investigation. For more general situations, the gas volume is not only related to fractures but also involves other factors, such as the permeability, heterogeneity of the reservoir, and gas production duration.

Table 5.7. Final length of hydraulic fractures and connected DFN natural fractures and Gas production volume.

Cases	Length of hydraulic and connected DFN natural fractures/m	Gas production volume /(10^6 m^3)
II	371	4.43
XII	385	5.72
XIII	394	5.74
VII	446	6.38
IV	453	5.97
I	457	6.68
IX	470	6.26
III	482	6.52
VI	487	6.88
VIII	541	8.39
V	576	7.74
X	580	8.69
XI	593	8.06

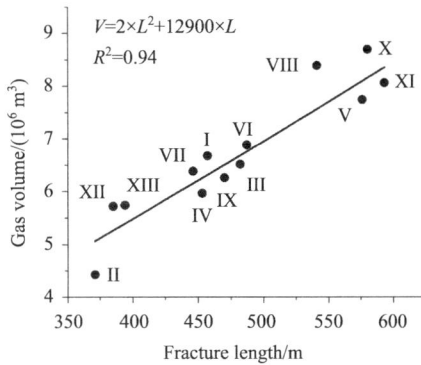

Figure 5.14. Relationship of fracture length and gas production volume.

5.5   Conclusions

In this study, the numerical models and analysis for the dynamic propagation and intersection of hydraulic fractures and pre-existing natural fractures involving the sensitivity factors (orientation, spacing, length, and persistence) are implemented. The models considered significant effects in hydrofracturing, including HM coupling and fluid leak-off. Under the same conditions, the natural fracture models considering different sensitive factors were simulated, and the effects of these factors on the growth and distribution of hydraulic fractures, fracturing fluid flowback, and gas production were studied. In addition, the evolution of some quantitative targets, such as the length and volume of the fractures and gas production, enabled us to analyse the influence of natural fractures. The main conclusions are as follows:

(1) Numerical models for the hydrofracturing of horizontal wells with pre-existing natural fractures considering sensitive factors (orientation, spacing, length, and persistence) were proposed. DFN models have been used to describe pre-existing natural fractures in naturally fractured reservoirs. To investigate the sensitive factors, typical numerical cases were carefully designed and established, and the results of the dynamic propagation and intersection of hydraulic fractures and pre-existing natural fractures were derived.

(2) Two typical types of fracture network morphologies were detected when the hydraulic fractures intersect the natural fractures: centre-type (intersection of hydraulic fractures and crossed cluster of the natural fractures) and edge-type (intersection of hydraulic fractures and the edge of the natural fractures). The propagation direction of

the hydraulic fractures depends on the orientation angle between the maximum *in-situ* stress and the natural fractures. When the natural fractures are relatively dense, that is, small fracture spacing, large fracture length, or small fracture persistence of natural fractures, especially when they are connected in crossed clusters, the hydraulic fractures easily intersect with the clusters and form a complex centre-type fracture network.

(3) Quantitative results of the length and volume of fracture networks combined with hydraulic and natural fractures were derived and discussed. In the hydrofracturing process, it is easier to form a relatively substantial length and volume of the fracture network by the centre-type propagation of hydraulic and natural fractures, and understanding the properties (orientation, spacing, length, and persistence) of natural fractures is key to accurately evaluating the propagation type and obtaining an accurate quantitative length and volume of the fracture network. At the end of the flowback, owing to the flowback process of the fracturing fluid, the fracture network loses support and gradually closes, and the fracture volume drastically decreases.

(4) The influence of pre-existing natural fractures on gas production in the enhanced permeability fractured reservoirs was analysed. Under the condition of formation of a local centre-type fracture network and relatively developed fracture length, the gas in the matrix around the fracture network easily migrates into the fracture through the network, enhancing the continuous gas production. A basic functional relationship between the length of the fracture network and gas production volume in a fractured reservoir was established, which can be used to evaluate gas exploitation.

This study used numerical models to simulate the hydrofracturing of tight reservoirs with natural fractures considering the sensitivity factors. Focusing on the influence mechanisms of natural fractures on the dynamic propagation of hydraulic fractures, the length, spacing, and persistence of fractures are relatively large, simulating large-scale natural fractures in the reservoirs. In fact, the sizes and distribution of natural fractures vary considerably in the stratum; therefore, it may be necessary to further study the properties of different scales and the randomness of natural fractures on hydrofracturing networks. Relevant research is in progress and will be reported in the future.

References

Cai, B., Ding, Y., Cui, Z., Yang, Z., Shen, H. (2014), "Hydraulic fracturing technology design in natural fracture reservoirs", Advanced Materials Research, Vol. 3226 No. 941–944, pp. 2521–2524.

Craig, K., Evensen, D., Van, D. (2019), "How distance influences dislike: responses to proposed fracking in fermanagh, Northern Ireland", Moravian Geographical Reports, Vol. 27 No. 2, pp. 92–107.

ELFEN TGR user and theory manual. (2016), Rockfield Software Ltd., United Kingdom, 2016.

Fall, A., Eichhubl, P., Bodnar, R.J., Laubach, S.E., Davis, J.S. (2015), "Natural hydraulic fracturing of tight-gas sandstone reservoirs, Piceance Basin, Colorado", Geological Society of America Bulletin, Vol. 127 No. 1–2, pp. 61–75.

Gale, J.F., Laubach, S.E., Olson, J.E., Eichhubl, P., Fall, A. (2014), "Natural fractures in shale: a review and new observations", AAPG Bulletin, Vol. 98 No. 11, pp. 2165–2216.

Garland, J., Neilson, J., Laubach, S.E., Whidden, K.J. (2012), "Advances in carbonate exploration and reservoir analysis", Geological Society, London, Special Publications, Vol. 370 No. 1, pp. 1–15.

Han, X.L., Zhou, F.J., Xiong, C.M., Liu, X.F., Yang, X.Y. (2013), "A new method of optimization design of hydraulic fracture parameters in low porosity and fractured gas reservoir", Applied Mechanics and Materials, Vol. 2617 No. 380–384, pp. 1656–1659.

Hannes, H., Tayfun, B., Günter, Z. (2014), "Numerical simulation of complex fracture network development by hydraulic fracturing in naturally fractured ultratight formations", Journal of Energy Resources Technology, Vol. 136 No. 4.

Hou, B., Chen, M., Cheng, W., Diao, C. (2016), "Investigation of hydraulic fracture networks in shale gas reservoirs with random fractures", Arabian Journal for Science and Engineering, Vol. 41 No. 7, pp. 2681–2691.

Hu, B., Wang, J., Ma, Z. (2020), "A fractal discrete fracture network based model for gas production from fractured shale reservoirs", Energies, Vol. 13 No. 7, 1857.

Hu, L., Ghassemi, A. (2021), "Laboratory-scale investigation of the slippage of a natural fracture resulting from an approaching hydraulic fracture", Rock Mechanics and Rock Engineering, Vol. 54 No. 3, pp. 1–12.

Jian, Z., Xue, C. (2011), "Experimental Investigation of fracture interaction between natural fractures and hydraulic fracture in naturally fractured reservoirs", SPE EUROPEC/EAGE Annual Conference and Exhibition, Vienna, Austria, SPE-142890-MS.

Karimi-Fard, M., Firoozabadi, A. (2001), "Numerical simulation of water injection in 2D fractured media using discrete-fracture model", SPE Annual Technical Conference and Exhibition, New Orleans, Louisiana, USA, SPE-71615-MS.

Katsaga, T., Riahi, A., DeGagne, D. O., Valley, B., Damjanac, B. (2015), "Hydraulic fracturing operations in mining: conceptual approach and DFN modeling example", Mining Technology, Vol. 124 No. 4, pp. 255–266.

Kresse, O., Cohen, C., Weng, X., Wu, R., Gu, H. (2011), "Numerical modeling of hydraulic fracturing in naturally fractured formations", 45th U.S. Rock Mechanics / Geomechanics Symposium, San Francisco, California, ARMA-11-363.

Li, Y. (2020), "Simulation of the interactions between multiple hydraulic fractures and natural fracture network based on discrete element method numerical modeling", Energy Science and Engineering, Vol. 8 No. 8, pp. 2922–2937.

Li, Z., Li, X., Yu, J., Cao, W., Liu, Z., Wang, M., Liu, Z., Wang, X. (2020), "Influence of existing natural fractures and beddings on the formation of fracture network during hydraulic fracturing based on the extended finite element method", Geomechanics and Geophysics for Geo-Energy and Geo-Resources, Vol. 6 No. 4, pp. 1286–1303.

Liang, C., Jiang, Z., Zhang, C., Guo, L., Yang, Y., Li, J. (2014), "The shale characteristics and shale gas exploration prospects of the Lower Silurian Longmaxi shale, Sichuan Basin, South China", Journal of Natural Gas Science and Engineering, Vol. 21, pp. 636–648.

Mi, L., Jiang, H., Mou, S., Li, J., Pei, Y., Liu, C. (2018), "Numerical simulation study of shale gas reservoir with stress-dependent fracture conductivity using multiscale discrete fracture network model", Particulate Science and Technology, Vol. 36 No. 2, pp. 202–211.

Pitman, J.K., Price, L.C., Lefever, J.A. (2001), "Diagenesis and fracture development in the Bakken Formation, Williston Basin: implications for reservoir quality in the middle member", US Department of the Interior, US Geological Survey.

Rahman, M.M., Rahman, S.S. (2013), "Fully coupled finite-element-based numerical model for investigation of interaction between an induced and a preexisting fracture in naturally fractured poroelastic reservoirs: fracture diversion, arrest, and breakout", International Journal Geomechanics, Vol. 13 No. 4, pp. 390–401.

Raterman, K.T., Farrell, H.E., Mora, O.S., Janssen, A.L., Gomez, G.A., Busetti, S., McEwen, J., Friehauf, K., Rutherford, J., Reid, R., Jin, G., Roy, B., Warren, M. (2018), "Sampling a stimulated rock volume: an eagle ford example", SPE Reservoir Evaluation and Engineering, Vol. 21 No. 4, pp. 927–941.

Ren, W., Lau, H.C. (2020), "Analytical modeling and probabilistic evaluation of gas production from a hydraulically fractured shale reservoir using a quad-linear flow model", Journal of Petroleum Science and Engineering, Vol. 184, pp. 106516.

Ru, Z., Hu, J., Madni, A., An, K. (2020), "A study on the optimal conditions for formation of complex fracture networks in fractured reservoirs", Journal of Structural Geology, Vol. 135, 104039.

Seifollahi, S., Dowd, P.A., Xu, C., Fadakar, A.Y. (2014), "A spatial clustering approach for stochastic fracture network modelling", Rock Mechanics and Rock Engineering, Vol. 47 No. 4, pp. 1225–1235.

Song, Y., Lu, W., He, C., Bai, E. (2020), "Numerical simulation of the influence of natural fractures on hydraulic fracture propagation", Geofluids, Vol. 2020, 8878548.

Suo, Y., Chen, Z., Rahman, S., Yan, H. (2020), "Numerical simulation of mixed-mode hydraulic fracture propagation and interaction with different types of natural fractures in shale gas reservoirs", Environmental Earth Sciences, Vol. 79 No. 7, pp. 153–167.

Tayong A, R., Alhubail, M.M., Maulianda, B., Barati, R. (2019), "Simulating yield stress variation along hydraulic fracture face enhances polymer cleanup modeling in tight gas reservoirs",

Journal of Natural Gas Science and Engineering, Vol. 65, pp. 32–44.

Teufel, L., Clark, J. (1984), "Hydraulic fracture propagation in layered rock: experimental studies of fracture containment", Society of Petroleum Engineers Journal, Vol. 24 No. 1, pp. 19–32.

Wan, L., Chen, M., Hou, B., Kao, J., Zhang, K., Fu, W. (2018), "Experimental investigation of the effect of natural fracture size on hydraulic fracture propagation in 3D", Journal of Structural Geology, Vol. 116 No. 1, pp. 1–11.

Wang, L., Chen, W., Tan, X., Tan, X., Yang, J., Yang, J., Zhang, X. (2020), "Numerical investigation on the stability of deforming fractured rocks using discrete fracture networks: a case study of underground excavation", Bulletin of Engineering Geology and the Environment, Vol. 79 No.1, pp. 133–151.

Wang, Q., Hu, Y., Zhao, J. (2019), "Effect of natural fractures on stress evolution of unconventional reservoirs using a finite element method and a fracture continuum method", Geofluids, Vol. 2019, pp. 1–16.

Wang, W., Olson, J.E., Prodanović, M., Schultz, R.A. (2018), "Interaction between cemented natural fractures and hydraulic fractures assessed by experiments and numerical simulations", Journal of Petroleum Science and Engineering, Vol. 167, pp. 506–516.

Wang, Y. (2020), "Adaptive finite element–discrete element analysis for stratal movement and microseismic behaviours induced by multistage propagation of three-dimensional multiple hydraulic fractures", Engineering Computations, Vol. 38 No. 5, pp. 1350–1371.

Wang, Y., Ju, Y., Yang, Y. (2018), "Adaptive finite element-discrete element analysis for microseismic modelling of hydraulic fracture propagation of perforation in horizontal well considering pre-existing fractures", Shock and Vibration, Vol. 2018, 2748408.

Wang, Y., Ju, Y., Chen, J., Song, J. (2019), "Adaptive finite element-discrete element analysis for the multistage supercritical CO_2 fracturing of horizontal wells in tight reservoirs considering pre-existing fractures and thermal-hydro-mechanical coupling", Journal of Natural Gas Science and Engineering, Vol. 61, pp. 251–269.

Yan, B., Mi, L., Wang, Y., Tang, H., An, C., Killough, J.E. (2018), "Multi-porosity multi-physics compositional simulation for gas storage and transport in highly heterogeneous shales", Journal of Petroleum Science and Engineering, Vol. 160, pp. 498–509.

Yao, W., Mostafa, S., Yang, Z., Xu, G. (2020), "Role of natural fractures characteristics on the performance of hydraulic fracturing for deep energy extraction using discrete fracture network (DFN)", Engineering Fracture Mechanics, Vol. 230, 106962.

Yao, Y., Guo, Z., Zeng, J., Li, D., Lu, J., Liang, D., Jiang, M. (2021), "Discrete element analysis of hydraulic fracturing of methane hydrate-bearing sediments", Energy & Fuels, Vol. 35 No. 8, pp. 6644–6657.

Yuan, Y., Yan, W., Chen, F., Li, J., Xiao, Q., Huang, X. (2018), "Numerical simulation for shale gas flow in complex fracture system of fractured horizontal well," International Journal of Nonlinear Sciences and Numerical Simulation, Vol. 19 No. 3–4, pp. 367–377.

Zeng, L., Lyu, W., Li, J., Zhu, L., Weng, J., Yue, F., Zu, K. (2016), "Natural fractures and their influence on shale gas enrichment in Sichuan Basin, China", Journal of Natural Gas Science and

Engineering, Vol. 30, pp. 1–9.

Zhang, Z., Li, X., Cowen, E. (2016), "Numerical Study on the Formation of Shear Fracture Network", Energies, Vol. 9 No. 4, pp. 299.

Zhao, Y., Li, N., Zhang, L., Zhang, R. (2019), "Productivity analysis of a fractured horizontal well in a shale gas reservoir based on discrete fracture network model", Journal of Hydrodynamics, Vol. 31 No. 3, pp. 552–561.

Zheng, H., Pu, C., Sun, C. (2020), "Study on the interaction between hydraulic fracture and natural fracture based on extended finite element method", Engineering Fracture Mechanics, Vol. 230, 106981.

Appendix

Appendix A. Proppant transport model

The 1D proppant transport equation used in this study for an idealised parallel plate flow for proppant component i is

$$\frac{\partial(w_{\mathrm{eff}}C_i)}{\partial t} + \frac{\partial(q_{\mathrm{fr}}C_i)}{\partial x} = 0 \qquad i = 1, 2, \cdots, N_c, \tag{A.5.1}$$

The flow rate for the fluid phase in the fracture can be replaced by a mixed flow rate, defined as

$$q_{\mathrm{fr}} = -K_{\mathrm{b}}(C_p)\left(\frac{\partial P_{\mathrm{f}}}{\partial x} - \rho_b n_{\mathrm{f}} g\right), \tag{A.5.2}$$

using the parameters defined in the previous researches.

Appendix B. Fracture propagation criteria and mesh refinement

In the model used in this study, the domain around the fracture tip was damaged due to stress concentration. The slope H of the stress-strain curve and Young's modulus $\tilde{E}$ in the damage stage are given by:

$$H = \frac{\sigma_t^2 C_l}{2G_{\mathrm{f}}}, \tag{B.5.1}$$

$$\tilde{E} = E(1-d), \tag{B.5.2}$$

where d is a damage variable. Once the damage variable d is zero, the fracture begins. Furthermore, this study uses the mesh refinement and coarsening techniques to ensure the accuracy of the solutions around the fracture tip and the accuracy of fracture propagation.

Chapter 6   Unstable propagation of multiple hydraulic fractures and stress shadow effects in multilayered reservoirs

6.1   Introduction

With the increasing demand for energy and the fluctuation of oil price, oil and gas energy in tight reservoirs attracts more and more attention (Huang *et al.*, 2020). Simultaneous hydrofracturing of multiple perforation clusters in vertical wells for forming fracturing fracture network is widespreadly applied in oil and gas energy exploitation (Wang, 2016; Li *et al.*, 2018), that is confronted with some challenges issues in common multilayered reservoirs containing the reservoir and interlayer strata, such as the stress shadow effects between the multiple hydraulic fractures (He *et al.*, 2017; Taghichian *et al.*, 2014), the propagation behaviours of fractures and energy exploitation induced microseisms in the plastic reservoir strata and the bedded interfaces between reservoir and interlayer strata (Zhang and Jeffrey, 2008; Papanastasiou, 1997), and the influences of *in-situ stress* on fracture deflection pattern (Zhou *et al.*, 2008). If the unstable propagation mechanisms of multiple hydraulic fractures system (Bažant *et al.*, 2014; Bažant and Cedolin, 2010) in vertical wells controlled by above unsolved issues are comprehended, which may help to optimize the fracturing effects and improve energy exploitation rate.

     The fractures produced by multiple perforation clusters change the initial *in-situ* stress field, and each fracture will form a local disturbed area of *in-situ* stress (Latham *et al.*, 2013). The disturbed stress areas induced by fracture propagation will affect the stress field around the adjacent fracture, forming the superposition areas or shadow areas of the stress field (Yoon *et al.*, 2015; Kresse *et al.*, 2013). The range of stress superposition area directly affects the deflection degree of fracture propagation. The decisive factor that affects the range of stress shadow area is the distance between multiple fractures, that is, the space between initial perforation clusters (Cheng, 2012; Guo *et al.*, 2015a, b). The perforation cluster with larger space will form a very small stress superposition area; once the space becomes smaller, the stress superposition area will become larger, and the stress on both sides of the fracture will not be equal to induce fracture deflection and unstable propagation. Quantitative analysis and optimization of the relationship between stress shadow effects and fracture deflection are crucial in simultaneous hydrofracturing (Peirce and Bunger, 2015).

The brittleness and plasticity of the reservoir are the basic properties of the rock mass, which may have significant influences on the fracturing fracability. At present, the study of rock fracturing mainly focuses on brittle reservoirs, but in recent years, exploration and experimental research have found that there are a large number of plastic shale reservoirs (Sone and Zoback, 2011; Masoudian et al., 2018), which show different physical and mechanical characteristics from brittle reservoirs. Brittle reservoir has a good fracturing property and is easy to form complex network (Bai, 2016; Zhang et al., 2016). However, the influence of the plastic property of rock mass on the hydrofracturing behaviour is still not clear (Papanastasiou, 1997). It is needed to make clear the fracture propagation behaviour characteristics of rock mass in different brittle or plastic reservoirs. In the process of oil and gas energy exploitation, it is found that it is difficult to avoid microseismic events induced by hydrofracturing in brittle reservoirs. It is not clear whether the similar microseismic behaviours will be induced in fracturing activities in plastic reservoir (Fischer et al., 2008). The similarities and differences between microseismic distribution and magnitude and brittle reservoir also need to be comprehended.

For multilayered reservoir, the factors that affect the fracture propagation behaviours include the in-situ stress states and fracture passing through the bedded interfaces between reservoir and interlayer strata (Cai, 2008; Zhang and Jeffrey, 2008). The ratio of in-situ stress will affect the fracture deflection direction, and the deflection will sometimes occur when the fractures meet the bedded interfaces. In practical engineering, the above two factors often exist simultaneously (Gudmundsson et al., 2010; Zhang and Jeffrey, 2008), therefore, it is necessary to make clear the influence of each factor and investigate the coupling behaviours of these two factors.

For conquering above challenging issues, some specific methods and models have been developed for oil and gas energy exploitation. However, the theoretical models are difficult to deal with such complex fracturing behaviours because of simplified treatment (Yan et al., 2010), and the field detections of hydrofracturing are challenging to accurately monitor due to the limitations of current technical means (Xiao et al., 2016), further, the physical experiments for propagation behaviours of multiple fractures in laboratory cannot achieve engineering-scale simulation (Ishida et al., 2012;

He *et al.*, 2020). Hence, numerical methods, such as finite element method (Chen, 2012), discrete element method (Deng *et al.*, 2014), extended finite element method (Lecampion, 2009; Gordeliy and Peirce, 2020), phase field method (Miehe and Mauthe, 2016; Wang *et al.*, 2020), boundary element method (Zhou *et al.*, 2020), are introduced as an alternative to experimental investigation techniques. In this study, the engineering-scale models for simultaneous hydrofracturing in vertical wells are established, and the combined finite element (FE)-discrete element (DE) method (Munjiza *et al.*, 1995; Wang, 2020, 2021) is used, comprehensive implementing of the advantages of the finite element method for computing continuous medium and the discrete element method for the simulating of fractures propagation. In these models, the hydro-mechanical (HM) coupling and leak-off effects (Barati and Liang, 2014) for simulating the flow behaviours of fracturing fluids in porous media are fully considered. Meanwhile, the mesh refinement strategy is adopted to improve the accuracy of stress solutions and reliability of fracture path of fracturing fracture networks; based on these fracturing fracture networks, energy production of oil and gas are computed and evaluated quantitatively, which technologies are used to simulate the whole process of hydraulic fracturing and energy recoveries.

This paper is organized as follows. Firstly, the numerical methods for HM coupling and bedded interfaces containing the geomechanical equations, characterization technology for multilayered reservoirs bedded interfaces, and local remeshing and microseismicity analysis, are introduced in Section 2. The numerical models of multilayered reservoirs are introduced in Section 3. Numerical results and discussion are analysed in Section 4. Finally, some conclusions are summarized in Section 5.

6.2   Numerical methods for hydro-mechanical coupling and bedded interfaces

6.2.1   Geomechanical equations

The solid and fluid (liquid seepage in porous media and fracturing fluid flow in fracture network) fields fully couple in the computation process of hydraulic fracturing. The effects of matrix poroelastic deformation are considered by treating the rock as porous medium and the mechanical governing equation of solid fields is given as (Wang, 2020):

$$L^{\mathrm{T}}\left(\sigma^{\mathrm{e}} - \alpha m p_{\mathrm{s}}\right) + \rho_{\mathrm{b}} g = \rho_{\mathrm{b}} \ddot{u}, \tag{6.1}$$

where σ^{e} is the effective stress tensor, with the expression $\left\{ \sigma_x \quad \sigma_y \quad \tau_{xy} \right\}^{\mathrm{T}}$; m is $\left\{ 1 \quad 1 \quad 0 \right\}^{\mathrm{T}}$; α is Biot's coefficient for porous medium; g is the gravity vector; p_{s} is the pore fluid pressure; and ρ_{b} is the density; L is the spatial differential operator, with the expression for two-dimensional plane problem

$$L^{\mathrm{T}} = \begin{bmatrix} \dfrac{\partial}{\partial x} & 0 & \dfrac{\partial}{\partial y} \\[3mm] 0 & \dfrac{\partial}{\partial y} & \dfrac{\partial}{\partial x} \end{bmatrix},$$

The liquid seepage equation that describes the porous flow in a rock matrix is:

$$\mathrm{div}\left[\frac{k}{\mu_{\mathrm{l}}}(\nabla p_{\mathrm{l}} - \rho_{\mathrm{l}} g)\right] = \left(\frac{\phi}{K_{\mathrm{l}}} + \frac{\alpha - \phi}{K_{\mathrm{s}}}\right)\frac{\partial p_{\mathrm{l}}}{\partial t} + \alpha \frac{\partial \varepsilon_v}{\partial t}, \tag{6.2}$$

where k is the intrinsic permeability of the rock formation; μ_{l} is the viscosity of the pore liquid; p_{l} is the pore liquid pressure; ρ_{l} is the density of the pore liquid; ϕ is the porosity of the rock formation; K_{l} is the bulk stiffness of the pore liquid; K_{s} is the bulk stiffness of the solid grains; ε_v is the volumetric strain; and t is the time.

The fluid flow equation that describes the fluid in the fracture network is:

$$\frac{\partial}{\partial x}\left[\frac{k^{\mathrm{fr}}}{\mu_n}(\nabla p_n - \rho_{\mathrm{fn}} g)\right] = S^{\mathrm{fr}}\frac{\mathrm{d}\, p_n}{\mathrm{d}\, t} + \alpha\left(\Delta \dot{e}_\varepsilon\right), \tag{6.3}$$

where k^{fr} is the intrinsic permeability of the fractured region (according to the parallel plate theory, the intrinsic permeability k^{fr} of a fractured region is related to the fracture aperture e by $k^{\mathrm{fr}} = \dfrac{e^2}{12}$); μ_n is the viscosity coefficient of the fracturing fluid; p_n is the fracturing fluid pressure; ρ_{fn} is the density of the fracture fluid; S^{fr} is the storage coefficient; and $\Delta \dot{e}_\varepsilon$ is the aperture strain rate. There is a term $\alpha\left(\Delta \dot{e}_\varepsilon\right)$ at the right end of the above fluid flow equation for describing the fluid in the fracture network, which considers the influence of solid deformation with time.

The leak-off of fracturing fluids will cause fluid exchange between fractures and porous rock media, and the leak-off simulation techniques consider the behaviours of the leakage and flow behaviours of fracturing fluid from hydraulic fractures into rock mass matrix. The above geomechanical equations can be discretizated and solved by conventional finite element method, and then the FE nodes will be separated once the fracture criteria are meet. The fracture criteria for fracture initiation and propagation used are based on the damage evolution as defined in Appendix A, in which the fracture will initiate when the damage reach fully damaged state (damage variable becomes the value of 1) and it will generate two new fracture surfaces. In this study, the above well-developed leak-off simulation techniques and fracture criteria are introduced.

The governing geomechanical equations (1)-(3) can be discretized using the shape functions in conventional FE method as follows:

$$B_{\mathrm{u}} = L_{\mathrm{u}} N_{\mathrm{u}}, \quad B_{\mathrm{s}} = L_{\mathrm{s}} N_{\mathrm{s}}, \quad B_{\mathrm{n}} = L_{\mathrm{n}} N_{\mathrm{n}}, \tag{6.4}$$

where N_{u}, N_{s}, and N_{n} are shape functions for the structure, seepage, and network fields, respectively; B_{u}, B_{s}, and B_{n} are the shape function spatial gradient matrices; L_{u}, L_{s}, and L_{n} are the gradient operators.

Further, the corresponding numerical discretized equations can be derived in the following matrix form:

$$\begin{bmatrix} M & 0 & 0 \\ 0 & 0 & 0 \\ 0 & 0 & 0 \end{bmatrix} \begin{Bmatrix} \ddot{u} \\ \ddot{p}_s \\ \ddot{p}_n \end{Bmatrix} + \begin{bmatrix} 0 & 0 & 0 \\ 0 & S_s & 0 \\ Q_n^{\mathrm{T}} & 0 & S_n \end{bmatrix} \begin{Bmatrix} \dot{u} \\ \dot{p}_s \\ \dot{p}_n \end{Bmatrix} + \begin{bmatrix} K & 0 & 0 \\ 0 & H_s & 0 \\ 0 & 0 & H_n \end{bmatrix} \begin{Bmatrix} u \\ p_s \\ p_n \end{Bmatrix} = \begin{Bmatrix} f_{\mathrm{u}} \\ f_s \\ f_n \end{Bmatrix} \tag{6.5}$$

using the matrices and vectors defined as follows:

$$M = \int_{\Omega_u} (N_u)^{\mathrm{T}} \rho_{\mathrm{b}} N_u \, \mathrm{d}\Omega_u, \quad K = \int_{\Omega_u} (\nabla N_u)^{\mathrm{T}} D \nabla N_u \, \mathrm{d}\Omega_u,$$

$$Q_n = \int_{\Omega_n} (\nabla N_u)^{\mathrm{T}} m N_n \, \mathrm{d}\Omega_n, \quad S_n = \int_{\Omega_n} (N_n)^{\mathrm{T}} S^{\mathrm{fr}} N_n \, \mathrm{d}\Omega_n,$$

$$H_n = \int_{\Omega_n} (N_n)^{\mathrm{T}} \frac{e^2}{12\mu_n} \nabla N_n \, \mathrm{d}\Omega_n, \quad f_u = \int_{\Omega_u} (N_u)^{\mathrm{T}} \rho_B b \, \mathrm{d}\Omega_u + \int_{\Gamma_u} (N_u)^{\mathrm{T}} \bar{t} \, \mathrm{d}\Gamma,$$

$$f_n = -\int_{\Omega_n} (N_n)^{\mathrm{T}} \nabla^{\mathrm{T}} \left(\frac{e^2}{12\mu_n} \rho_{\mathrm{fn}} b \right) \mathrm{d}\Omega_n + \int_{\Gamma_n} (N_n)^{\mathrm{T}} q \, \mathrm{d}\Gamma$$

where D is the material stiffness matrix; Γ_u is the boundary region of solid field; Γ_n is the boundary region of fracture network; $\bar{t}$ is the external traction load of fluid pressure along the exposed fracture surface; and q is the fracturing fluid flux.

6.2.2 Characterization technology for bedded interfaces in multilayered reservoirs

For simulating the behaviours of hydraulic fractures passing through the bedded interfaces between the reservoir and interlayer strata, contact conditions are required to govern the behaviour of bedded interfaces, in which the characterization technology and frictional contact setting technique for interfaces are utilized. The adhesion laws with friction are utilized for normal and tangential contacts as follows:

$$f_n = \alpha_n g_n ,\tag{6.6}$$

$$f_t = \alpha_t g_t .\tag{6.7}$$

where f_n and f_t are normal and tangential frictional forces, respectively; α_n and α_t are normal and tangential friction coefficients, respectively; g_n and g_t are normal and tangential contact stresses, respectively. Actually, the displacement-version finite element method is used in this study, and the contact and tangential forces are evaluated by the derivative solutions of displacements .

The details of the basic techniques and the definitions of the physical parameters can be found in the literature (Rockfield Software Ltd., 2016), which has been effectively used to simulate the behaviours of the fracture propagation through the interfaces between multiple layers. Table 6.1 provides the parameters of the contact settings for bedded interfaces between reservoir stratum and interlayer stratum.

Table 6.1. Contact settings between reservoir stratum and interlayer stratum.

Parameters	Values
Friction coefficient	0.1
Cohesion coefficient	0
Contact type	Node-edge
Zone factor	2.5
Field factor	0.75
Normal penalty factor	0.01
Normal penalty	3.863×10^8
Tangential penalty factor	0.0001
Tangential penalty	38630
Damping type	Velocity/Momentum-dependent
Damping value	0.1

6.2.3   Local remeshing and microseismicity analysis

Because fracture propagation through the elements is undesirable, remeshing is required to refine the elements around the fracture tips to detect the propagation direction and the length of the two-dimensional fracture (Wang, 2020). In general, the engineering-scale reservoirs are many hundreds of meters and the local remeshing strategy improve the computation accuracy and efficiency. Table 6.2 demonstrates the computational parameters for mesh refinement in the proposed models. Once the stress solutions are obtained, the computed differential stress $\Delta\boldsymbol{\sigma}^{e}$ from the pre- and post-events can be calculated over a single time step. Furthermore, the moment tensors can be computed if $\boldsymbol{M}$ is rotated into its principle direction using the unit eigenvectors. The damaged events (Mode I, tensile failure) and contact slip events (Mode II and Mode III, shear failure) are distinguished, using the microseismicity analysis technique by evaluating moment tensors, and the Gutenberg-Richter law (Gutenberg and Richter, 1956) can be used in hydraulic fracturing to give an indication of fault reactivation by determining the maximum magnitude of the expected seismic events, as introduced in Appendix B.

Table 6.2. Computational parameters for mesh refinement.

Parameters	Values
Small detail size/m	2.5
Fracture mesh size factor	0.75
Mesh density factor	1
Mesh density	1.875
Bubble size	3

6.3   Numerical models of multilayered reservoirs

Figure 6.1 shows the two-dimensional geometrical models of initial perforation cluster in vertical well. This model is a simulation engineering-scale case for multilayered reservoirs, with a model length l of 1000 m. The reservoir is in the middle reservoir layer with thickness h_r of 420 m; the upper and lower layers in model are interlayers with thickness h_i of 90 m. There is a well in the middle of the model and initial multiple perforations along the vertical well, in which numbering sequences of perforations are 1st, 2nd, 3rd, 4th, and 5th from bottom to top sides.

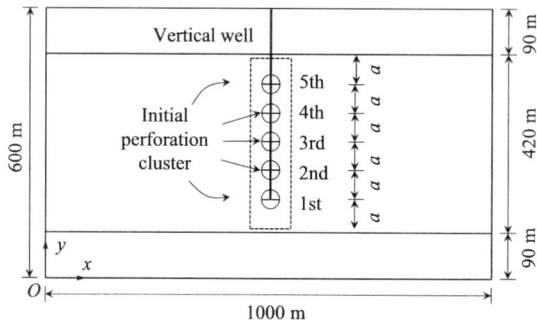

Figure 6.1. Geometrical models of initial perforation cluster in vertical well.

Table 6.3 lists the numerical cases of initial multiple perforation clusters for hydraulic fracturing in reservoir stratum. In this study, there are five cases for simulating different number of initial perforation clusters, e.g., 1~5 perforations. These perforation clusters are located at the equidistant points of the reservoir, in which the cluster space a is computed as follows:

$$a = \frac{h_r}{n+1} \ (\text{m}) \tag{6.8}$$

where n is the number of the initial perforation clusters.

The perforation spaces under five conditions is listed in Table 6.3. We set these perforation spaces to study the stress shadow effects between fractures and the mutual interference between fractures. The definitions and values of the physical parameters in above governing equations are presented in Table 6.4. Table 6.5 lists the differentiated physical parameters for reservoir and interlayer strata. The model uses different physical properties for reservoir and interlayer. In order to study the influence of reservoir properties, especially brittleness and plasticity on fracturing, the reservoir is divided into brittle-enhanced, medium, plastic-enhanced reservoir strata.

Table 6.3. Numerical cases of initial multiple perforation clusters for hydraulic fracturing in reservoir stratum.

Cases	Perforation clusters in reservoir stratum	
	Cluster number n	Cluster space a/m
I	1	210
II	2	140
III	3	105
IV	4	84
V	5	70

Table 6.4. Basic physical parameters.

Parameters	Values
Depth of horizontal well /m	2300
Fluid injection rate Q/(m^3/s)	0.1
Fluid injection volume in one fracturing stage/m^3	2000
Fluid injection duration in one fracturing stage/s	20000
Leak-off coefficient C/(m^3/s$^{1/2}$)	1.0×10^{-15}
Pore pressure p_s /MPa	25
Gravity g/(m/s^2)	9.81
Porosity ϕ	0.05
Biot's coefficient α	0.75
Dynamic viscosity coefficient of the pore fluid μ_g /(Pa·s)	1.00×10^{-3}
Dynamic viscosity coefficient of the fracturing fluid μ_n /(Pa·s)	1.67×10^{-3}
Liquid density of the pore fluid ρ_g /(kg/m^3)	1×10^3
Liquid density of the fracturing fluid ρ_{fn} /(kg/m^3)	1×10^3
Bulk modulus of the pore fluid K_g /MPa	2050
Bulk modulus of the fracturing fluid K_f^{fr} /MPa	2000

Table 6.5. Differentiated physical parameters for reservoir and interlayer strata.

Parameters	Rock lithology type			
	Interlayer stratum	Brittle-enhanced reservoir stratum	Medium reservoir stratum	Plastic-enhanced reservoir stratum
Young's modulus E /GPa	8.62	38.63	8.62	3.14
Poisson's ratio v	0.28	0.23	0.28	0.35
Tensile strength σ_t/MPa	5.8	1.962	5.8	1.65
Fracture energy G_f/(N·m)	160	50	160	50
Permeability k/nD	1×10^{-16}	4.90×10^{-21}	4.90×10^{-21}	4.90×10^{-21}
Biot's coefficient α	0.75	1.0	1.0	1.0
Density ρ_b /(kg/m^3)	2.6×10^3	2.615×10^3	2.615×10^3	2.615×10^3
Cohesion C /MPa	13	25	25	25
Friction angle ϕ_f /(°)	38	45	45	45

In this study, we introduced the combined FE-DE method (Munjiza *et al.*, 1995; Wang, 2020), comprehensive implementing of the advantages of the finite element method for computing continuous medium and the discrete element method for the simulating of fractures propagation. Figure 6.2 shows the FE models and initial meshes, and the bedded interfaces are set between the reservoir and interlayer in the models. In order to improve the accuracy of solutions around the initial perforation, dense meshes are only used in the local domain I around perforations. In the outer domain II, the sparse meshes are used. The domain of detailed mesh for numerical cases with multiple perforations in hydraulic fracturing is $450 \leqslant x \leqslant 550$ and $100 \leqslant y \leqslant 500$, and the element size in this domain is 2.5 m. Table 6.6 lists the *in-situ* stress conditions of reservoir. In order to study the influence of different *in-situ* stress ratio on fracture propagation, three typical *in-situ* stress states are set up in this study, i.e., $S_h : S_v = 1.0 : 1.2$, $1.0 : 1.3$, and $1.0 : 1.5$. In fact, the fracturing behaviours of reservoirs with different depths is simulated by theses *in-situ* stress conditions, in which the differences between the *in-situ* stresses are $\Delta S = 7$, 10, and 17 MPa, respectively. It should be noted that the symbol S_h is the horizontal *in-situ* stress, rather than the maximum or minimum one, and S_v is the vertical *in-situ* stress. This study is different from the conventional studies mainly focusing on the propagation of fracturing fractures along the principal *in-situ* stress direction (Olson, 2008; Guo *et al.*, 2015a,b). The two-dimensional section of this model is not exactly parallel to the maximum or minimum horizontal *in-situ* stress, but parallel to the section of the initial perforation, which can help to well observe the behaviours of fracture propagation from the perforation; besides, the *in-situ* stress in the vertical direction of the model is larger than that in the horizontal direction, which can help to well observe the behaviours of fractures propagating along the vertical direction and then penetrating the horizontal bedded interlayer.

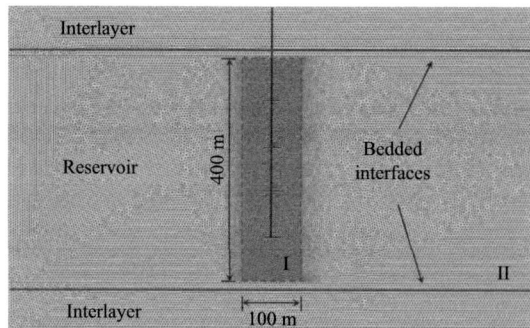

Figure 6.2. FE models and initial meshes (increasingly dense FE meshes for domains II and I).

Table 6.6. *In-situ* stress conditions of tight reservoir.

States	$S_h : S_v$	*In-situ* stress difference ΔS/MPa	Horizontal *in-situ* stress in x direction S_h /MPa	Vertical *in-situ* stress in y direction S_v /MPa
A	1.0 : 1.2	7	34	41
B	1.0 : 1.3	10	34	44
C	1.0 : 1.5	17	34	51

6.4   Results and discussion

The representative cases were executed applying package ELFEN TGR (Rockfield Software Ltd., 2016), and using the well-established models and aforementioned mesh refinement strategy.　The above program package has been verified and tested by some examples, and has shown good results in the previous research, therefore, it is introduced in this study into the application (Wang, 2000).

6.4.1   Dynamic unstable propagation of multiple hydraulic fractures

To explain the reliability of fracture propagation and stress field solution in multiple layer reservoirs, Figure 6.3 shows the illustration of remeshing in dynamic propagation of fracturing fracture in three- and four-layer reservoirs, in Case IV of four multiple perforation clusters under *in-situ* stress $S_h : S_v = 1.0:1.2$. It can be seen from the figure that with the fracture propagation, smaller size of triangular element is used around the fracture, so a dense mesh is formed to ensure the accuracy of the solutions

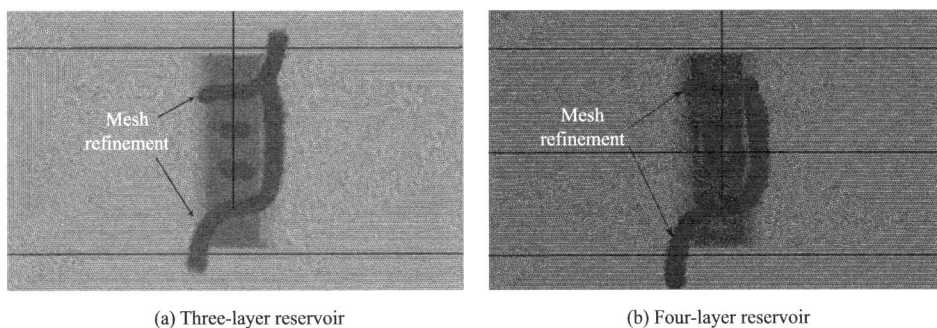

(a) Three-layer reservoir　　　　　　　　(b) Four-layer reservoir

Figure 6.3. Illustration of remeshing in dynamic propagation of fracturing fracture, in Case IV of four multiple perforation clusters.

around the fracture tips and the reality of fracture propagation path. Other domains that are not concerned about apply relatively lager element to establish sparse meshes, which can improve the efficiency of entire computation.

In order to detect the dynamic propagation behaviours of hydraulic fractures, Table 6.7 lists the final morphology of fracturing fracture network and first principal stress under different *in-situ* stress conditions in brittle-enhanced reservoir stratum. Evidently, the stress field appears typical discontinuous phenomenon around the interfaces of reservoir and interlayer. The initial perforation fracture is along the horizontal direction, and the fractures deflect gradually towards the direction of maximum vertical *in-situ* stress S_v. For single perforation Case I, when the ratio of *in-situ* stress is very small as $S_h : S_v = 1.0 : 1.2$, the slight fracture deflection arises to form symmetrical morphology of fracturing fracture network; once the ratio is increased as $S_h : S_v = 1.0 : 1.3$ even reaching $S_h : S_v = 1.0 : 1.5$, the deflection is intensified, that is to say, it soon deflects towards the direction of S_v, and almost propagates straight (incidence angle $\theta = 90^\circ$); the multiple perforations cases have the coincident deflection behaviours, e.g., in Case IV, the incidence angle $\theta = 90^\circ$ arises as the *in-situ* stress ratio becoming larger ($S_h : S_v = 1.0 : 1.3$ or $1.0 : 1.5$). For multiple perforations Cases II-V, the fracture deflection causes the intersection of multiple fractures, and the fractures continue to propagate and connect together, i.e., Cases II with $S_h : S_v = 1.0 : 1.2$. Analogously, the propagation of multiple fractures also results in the shielding effects between fractures, inducing that some fractures cannot propagate, but mainly grow to form a main fracture and several minor fractures, i.e., Cases III with $S_h : S_v = 1.0 : 1.2$. In the cases of the maximum *in-situ* stress ratio as $S_h : S_v = 1.0 : 1.5$, the fracture deflects quickly, and the behaviours of multiple fracture shielding are most likely to occur. The above-mentioned fracture intersection and shielding effects due to the unstable propagation behaviours of parallel fractures (Bažant et al., 2014; Bažant and Cedolin, 2010; Cheng, 2012), which become tough issues in hydraulic fracturing engineering affecting the optimization of fracturing effects.

Table 6.7. Final morphology of fracturing fracture network and first principal stress (Pa) under different *in-situ* stress conditions in brittle-enhanced reservoir stratum.

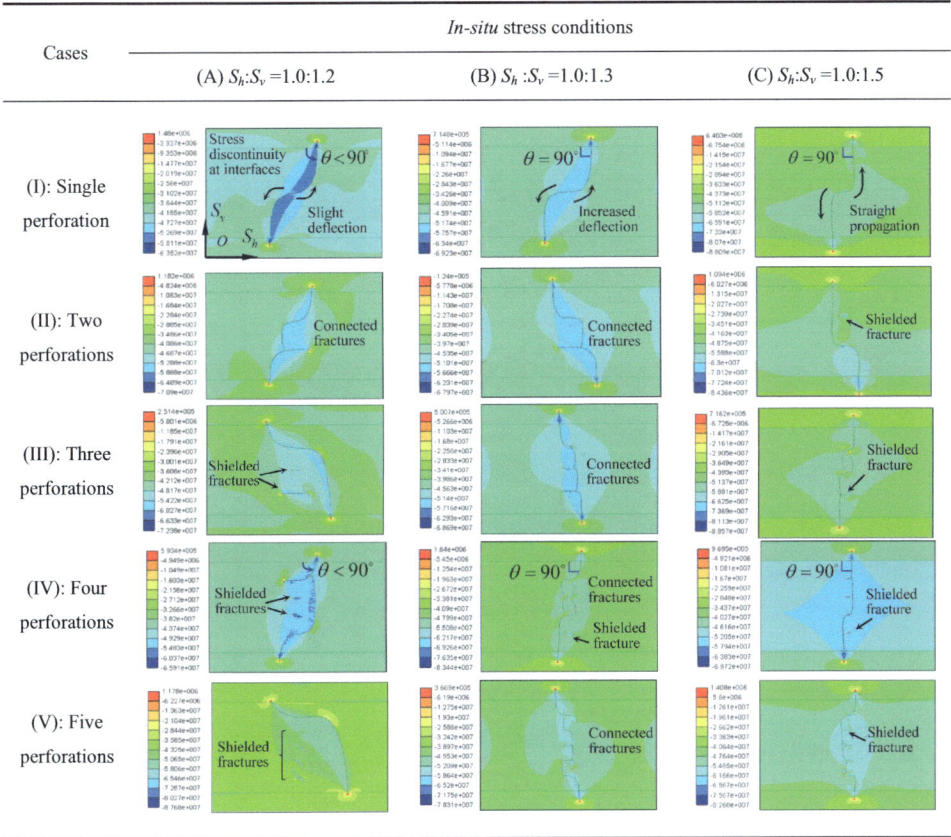

Cases	*In-situ* stress conditions		
	(A) S_h:S_v =1.0:1.2	(B) S_h :S_v =1.0:1.3	(C) S_h:S_v =1.0:1.5
(I): Single perforation			
(II): Two perforations			
(III): Three perforations			
(IV): Four perforations			
(V): Five perforations			

Table 6.8 lists the final morphology of fracturing fracture network and first principal stress for hydraulic fracturing of vertical well of *in-situ* stress $S_h : S_v$ =1.0:1.2 under different physical attributes for reservoir and interlayer strata. With the plasticity increase of reservoir, the fracture in plastic-enhanced reservoir propagates more slowly, compared with medium reservoir. Under the same fracturing fluids injection, the fracture in plastic-enhanced reservoir shortens and the fracture network does not develop. Plasticity property becomes a crucial factor that is not conducive to fracture propagation.

Table 6.8. Final morphology of fracturing fracture network and first principal stress (Pa) for hydraulic fracturing of vertical well of *in-situ* stress $S_h : S_v = 1.0 : 1.2$ under different physical attributes for reservoir and interlayer strata.

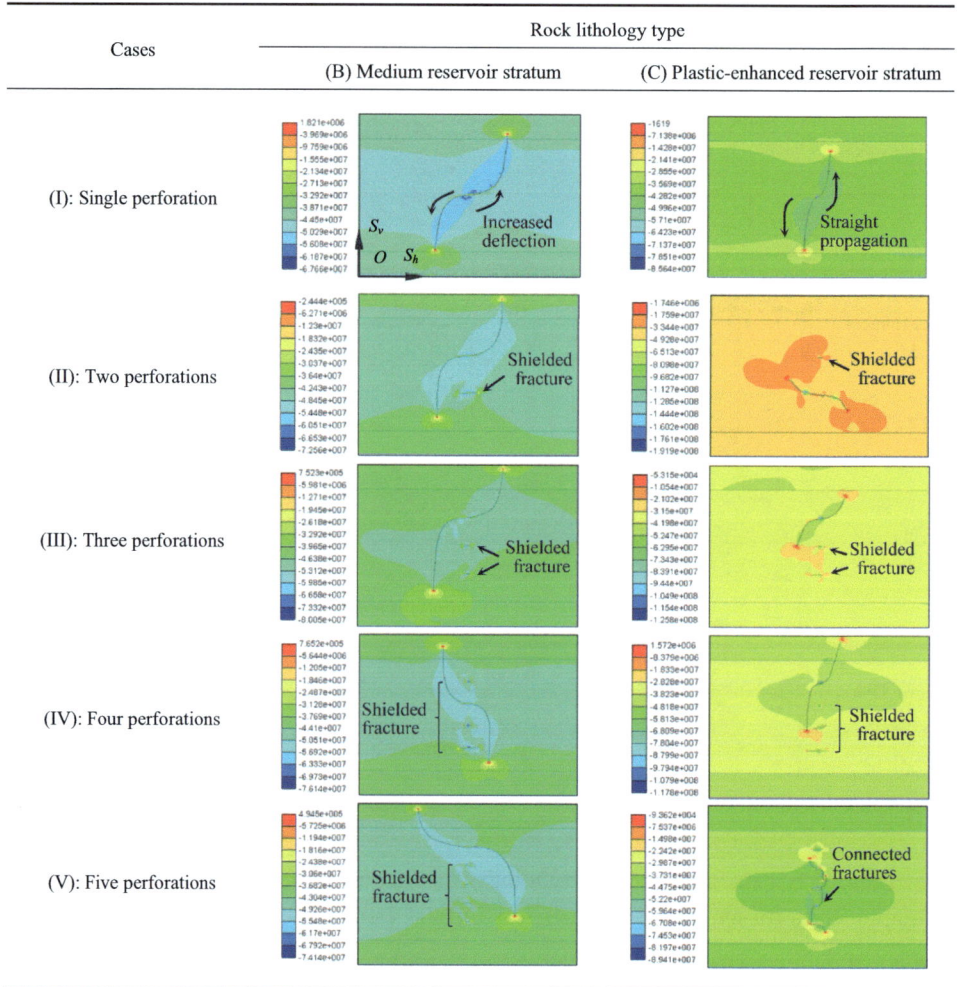

Cases	Rock lithology type	
	(B) Medium reservoir stratum	(C) Plastic-enhanced reservoir stratum
(I): Single perforation		
(II): Two perforations		
(III): Three perforations		
(IV): Four perforations		
(V): Five perforations		

6.4.2　Evolution of stress field and injected fluid volume

The reasons for fracture propagation are the injection and driving of the fracturing fluid, which causes the change of the *in-situ* stress field. Therefore, it is needed to study the evolution of the stress field and fluid volume. The superposition and attenuation of the shear stress field caused by fractures propagation can reflect the mutual interference between fractures. Table 6.9 lists the evolution of the shear stress for selected three stages in simultaneous fracturing using four perforation clusters, at three typical moments $t = 1452$ s, 3310 s, and 20002 s, respectively. The areas circled by the dotted lines in these figures are the main change domains of the *in-situ* stress field changed by the fracture tips propagation, and the positive and negative values of the shear stress at these places marked by the signs "+" or "−", respectively. In the first stage at $t = 1452$ s as shown in Table 6.9 (a), it can be seen that there are different positive and negative shear stress domains on both sides of the fracture tips. The positive and negative domains of shear stress fields between adjacent fractures are superposed to form stress shadows, which will shield propagation of the adjacent fractures. In the fracture network, the stress fields induced by the 2nd and 3rd perforating fractures in the inner domains are affected by the superposition of the 1st and 4th fractures on the outer side, and the areas circled by the dotted lines for the change of the *in-situ* stress field are very small; while the stress fields induced by the 2nd and 3rd perforating fractures are very large, because it is not superposed by others. In the latter two stages at $t = 3310$ s and 20002 s as shown in Tables 6.9 (b) and 6.9 (c), with the propagation of fracture network, two adjacent fractures intersect, the phenomenon of stress field superposition and mutual offset will arise. It can be seen that when the stress fields at the fracture tips of the 1st and 4th fractures on the outer side are not offset and larger stress fields are formed, the fracture is easier to propagate to form larger length. Actually, the initial *in-situ* stress field changes caused by fractures propagation and the superposition of stress shadows between them make the fractures deflect. Table 6.10 provides the evolution of the shear stress for selected three stages in simultaneous fracturing using four perforation clusters in four-layer reservoir. It can be seen that the initial stress field changes due to the existence of the increased interlayer, which affects the fracture propagation and the formation of different fracture morphologies. The gas production migration was used once the above fracturing network was formed (Wang, 2021). Figure 6.4 shows the quantitative evolution of gas production of the multiple fractures (the 1st, 2nd, 3rd, and 4th fractures, respectively) in three- and four-layer reservoirs. The fracturing fractures under four-layer case are more likely to form higher gas production than the three-layer case, indicating that the gas production is positively related to the initial discontinuity of the reservoirs.

Table 6.9. Evolution of the shear stress for selected three stages in simultaneous fracturing using four perforation clusters in three-layer reservoir.

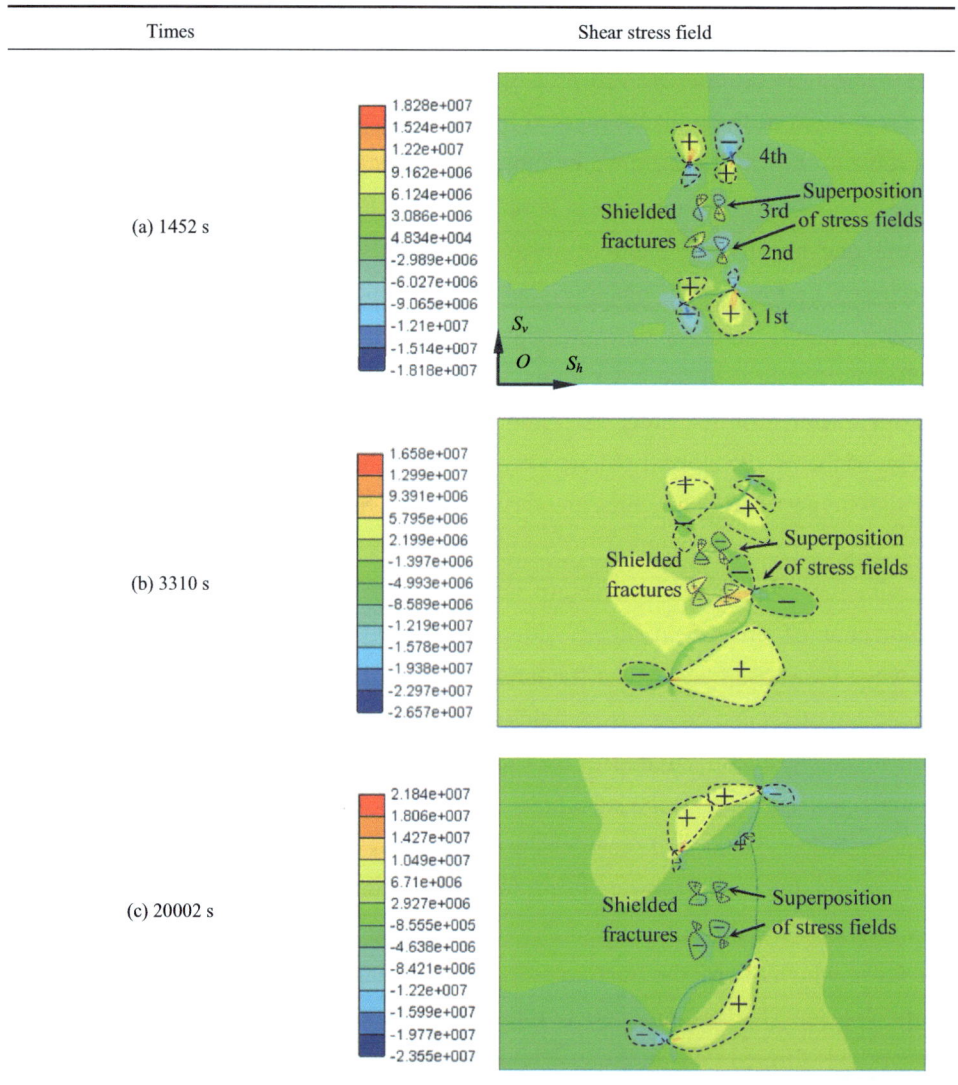

Times	Shear stress field
(a) 1452 s	
(b) 3310 s	
(c) 20002 s	

Table 6.11 lists the evolution of fluid volume of hydraulic fracturing of brittle-enhanced reservoir stratum under different *in-situ* stress conditions. For single perforation in Case I, fracture volume increases almost linearly with time accompanying the injection of fracturing fluid. For three perforations in Case III, a main single fracture develops and more fracturing fluid is injected; other minor fractures propagate slightly and little fracturing fluid is injected. This is due to the

influence of stress shadow on some fractures as discussed above, resulting in the shielding effect of fractures and inhibiting fracture growth. For five perforations in Case V, the stress shadow and fracture shielding effects make multiple fractures difficult to propagate stably. When the *in-situ* stress ratio is small, the fractures shield each other and form a main fracture with short secondary minor fractures and little water injection; once the *in-situ* stress ratio becomes large, the multiple fractures are easy to interact with each other, making the minor fractures also experience a period of time propagation and inject a certain amount of fracturing fluid.

Table 6.10. Evolution of the shear stress for selected three stages in simultaneous fracturing using four perforation clusters in four-layer reservoir.

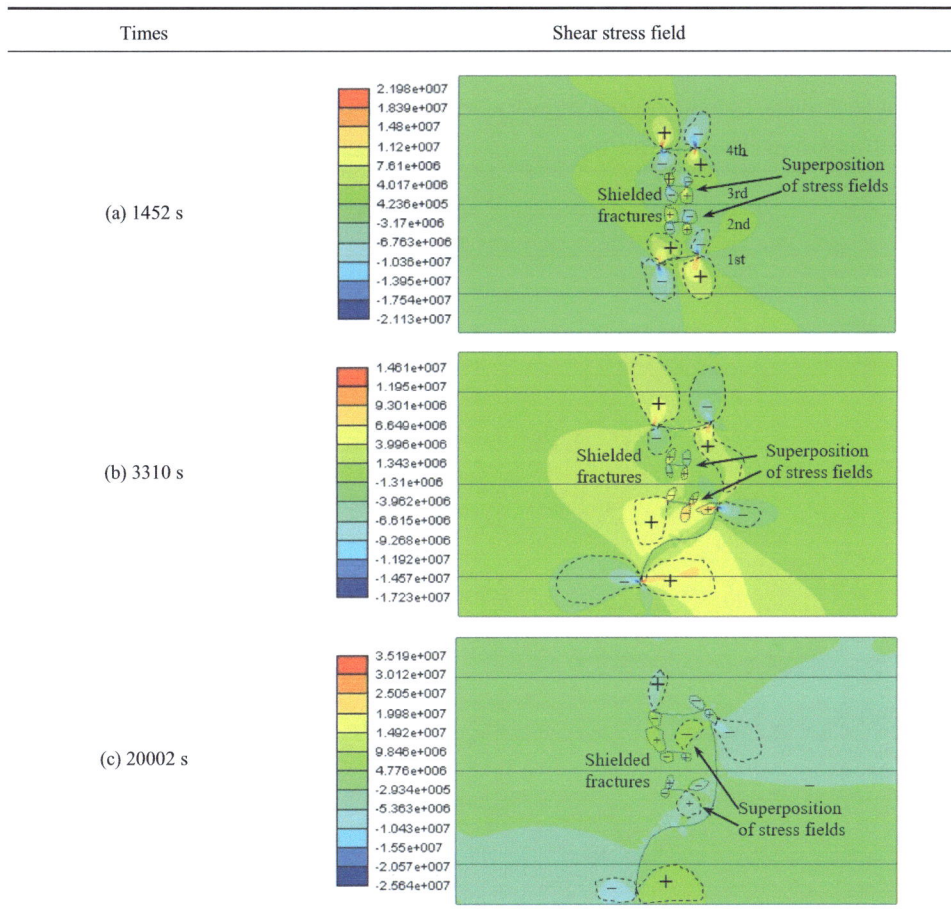

Times	Shear stress field
(a) 1452 s	
(b) 3310 s	
(c) 20002 s	

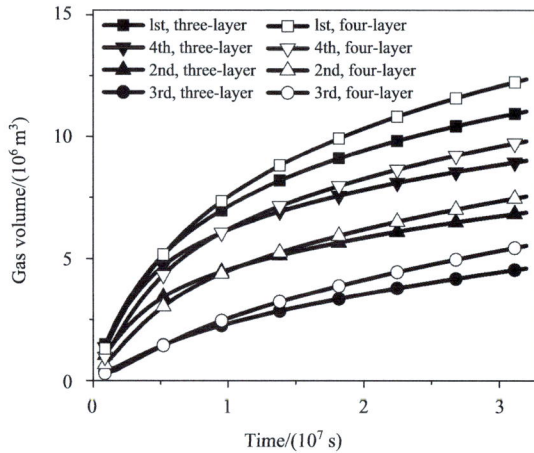

Figure 6.4. Gas production volume in three-layer and four-layer reservoirs.

Table 6.11. Evolution of fluid volume of hydraulic fracturing of brittle-enhanced reservoir stratum under different *in-situ* stress conditions.

Cases	*In-situ* stress conditions		
	(A) S_h:S_v =1.0:1.2	(B) S_h :S_v =1.0:1.3	(C) S_h:S_v =1.0:1.5
(I): Single perforation			
(III): Three perforations			
(V): Five perforations			

The evolution of fluid volume of hydraulic fracturing of *in-situ* stress $S_h : S_v = 1.0:1.2$ under different physical attributes for reservoir and interlayer strata is listed in Table 6.12. It can be seen that when a single perforation is applied, the fracture volume formed in plastic-enhanced reservoir is smaller than medium reservoir. When multiple perforations are used, brittle-enhanced and medium reservoir strata are more likely to form a main fracture and multiple secondary minor fractures, with obvious difference in fracture volume; plastic-enhanced reservoir stratum is more likely to form multiple fractures with small difference in fracture volume.

Table 6.12. Evolution of fluid volume of hydraulic fracturing of *in-situ* stress $S_h : S_v = 1.0 : 1.2$ under different physical attributes for reservoir and interlayer strata.

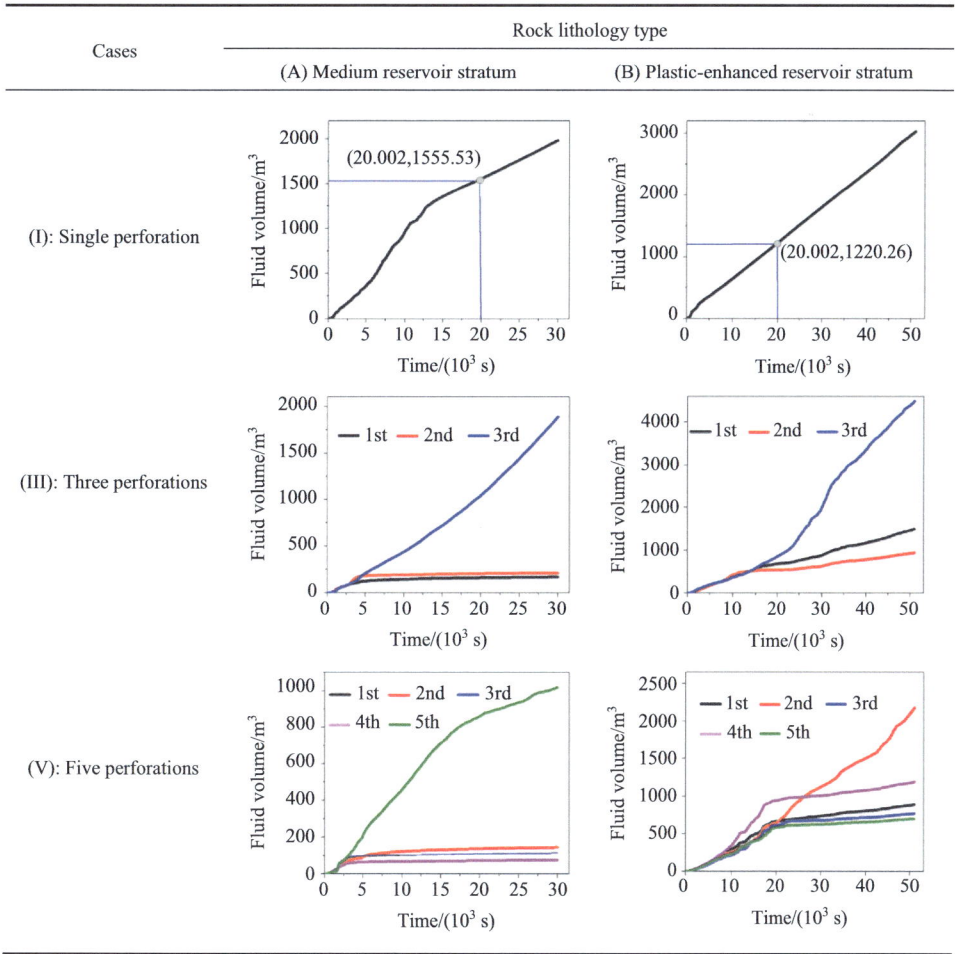

Cases	Rock lithology type	
	(A) Medium reservoir stratum	(B) Plastic-enhanced reservoir stratum
(I): Single perforation		
(III): Three perforations		
(V): Five perforations		

6.4.3   Influences of *in-situ* stress and bedded interfaces in multilayered reservoirs

For investigating of the influences of *in-situ stress* and physical attributes of reservoir and interlayer strata on fracturing results, we used fracture length to establish quantitative description of fracture network. Figure 6.5 shows the evolution of total fracture length of hydraulic fractures of brittle-enhanced reservoir stratum under different *in-situ* stress conditions. It can be seen that the fracture propagation in all cases can be divided into three stages: propagation in the reservoir, propagation through the bedded interface, and propagation in the interlayer. In the first stage, the fracture length increases linearly with time as the fracture propagates in the reservoir; when the fracture passes through the bedded interface, the fracture length passes through the bedded interface with a period of time, and the curve appears in the stage of platform; finally, the fracture propagates a short length in the interlayer. For single perforation, the total length of fracture is the shortest; for multiple perforations, the total length of fracture is the longest due to more fracturing fluid injected. With the increase of the *in-situ* stress ratio, the length of the fractures interface becomes shorter, in which the reason may be that the incidence angle θ between deflected fracture and bedded interface is gradually larger due to the larger *in-situ* stress ratio, which is more beneficial for the fracture to pass through the bedded interfaces. Once the incidence angle between fracture and interface becomes small, the fracture is not easy to pass through the reservoir, and even in some actual engineering projects, the fracture propagates along the interface (Cooke and Underwood, 2001; Fu *et al.*, 2016).

Figure 6.6 shows the final of total fracture length of hydraulic fracturing of brittle-enhanced reservoir stratum under different *in-situ* stress conditions. It can be seen that as the number of perforations increases, the fracture length increases, which is related to the increase of injected fracturing fluid caused by multiple perforations. When the *in-situ* stress ratio is $S_h : S_v = 1.0 : 1.3$, the total length of the fracture network is relatively the largest, $S_h : S_v = 1.0 : 1.2$ is the middle, and $S_h : S_v = 1.0 : 1.5$ is the smallest one, but there is little difference between these three conditions. It is worth noting that when the *in-situ* stress ratio is 1.5 for Case IV, the fracture length is greatly reduced due to the deflection of the fracture and the intersection and overlap of multiple fractures.

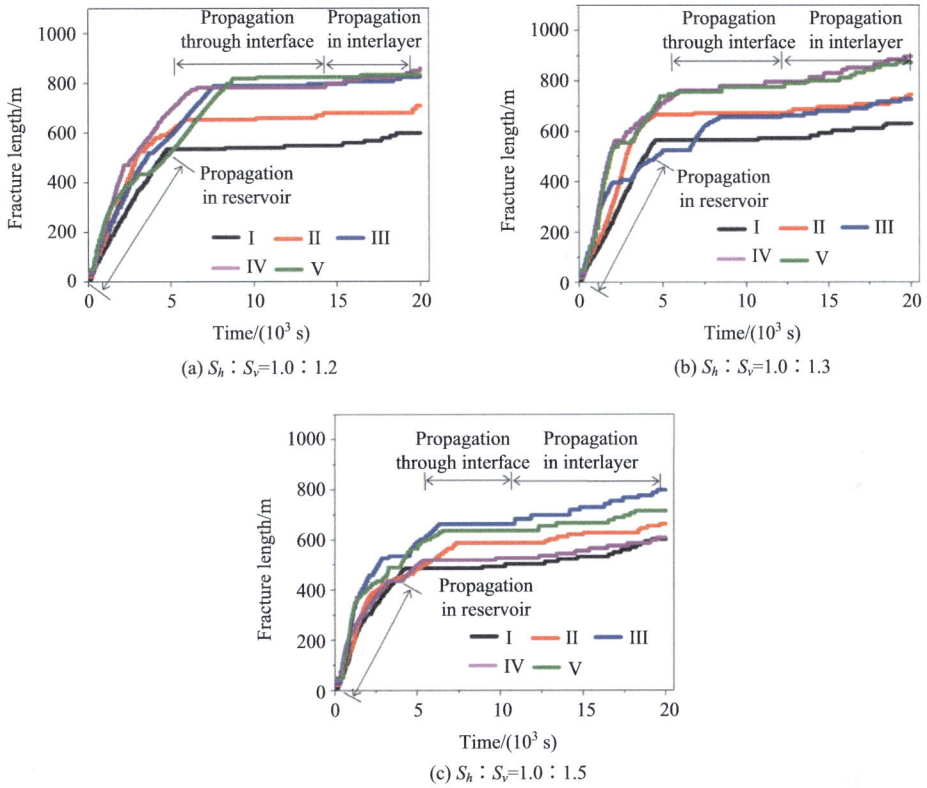

(a) $S_h : S_v = 1.0 : 1.2$

(b) $S_h : S_v = 1.0 : 1.3$

(c) $S_h : S_v = 1.0 : 1.5$

Figure 6.5. Evolution of total fracture length of hydraulic fracturing of brittle-enhanced reservoir stratum under different *in-situ* stress conditions.

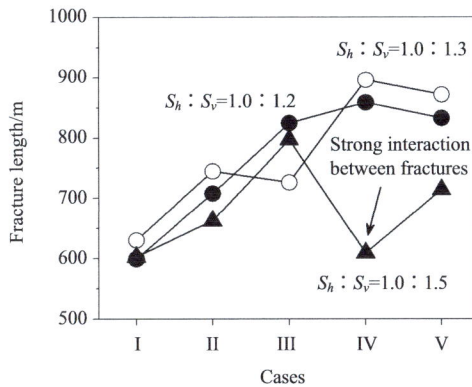

Figure 6.6. Final of total fracture length of hydraulic fracturing under different *in-situ* stress conditions.

Figure 6.7 shows the evolution of total fracture length of hydraulic fracturing of *in-situ* stress $S_h : S_v = 1.0 : 1.2$ under different physical attributes for reservoir and interlayer strata. Being different from the fracture propagation in brittle-enhanced reservoir as shown in Figure 6.5, there is no platform stage in the fracture length curve of medium reservoir, and all three stages of fracture propagation in reservoir, interface and interlayer are linear growth. In the case of plastic-enhanced reservoir, because of the influence of reservoir plasticity on fracture propagation, there is only a short period of fracture propagation, that is, the fracture only propagates in the first stage in the reservoir.

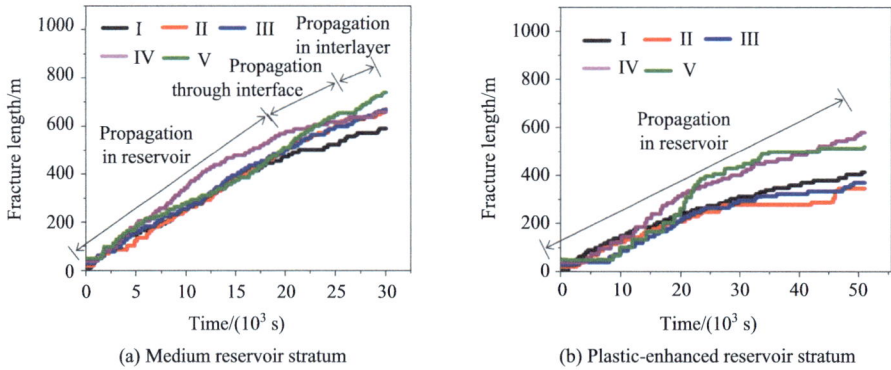

(a) Medium reservoir stratum   (b) Plastic-enhanced reservoir stratum

Figure 6.7. Evolution of total fracture length of hydraulic fracturing of *in-situ* stress $S_h : S_v = 1.0 : 1.2$ under different physical attributes for reservoir and interlayer strata.

The final total fracture length of hydraulic fracturing under different physical attributes for reservoir and interlayer strata is shown in Figure 6.8. As the number of perforations increases, the fracture length increases, which is related to the increase of injected fracturing fluid caused by multiple perforations. The fracture lengths in brittle-enhanced, medium and plastic-enhanced reservoir strata decreases gradually, and plastic-enhanced reservoir limit fracture growth and cannot form long fracture length. It should be noted that for plastic-enhanced reservoir, the total fracture length in Case IV is the largest because of the weak interaction between multiple fractures.

6.4.4   Microseismic event distributions and magnitudes induced by unstable fractures

The distributions and magnitudes of microseismic events will be derived by microseismic analysis below. Figure 6.9 shows the typical distribution of final damaged events under different *in-situ* stress conditions in brittle-enhanced reservoir stratum, in which the microseismic events represented by circular balls occur accompanying fracture propagation. For detecting the detail accumulation states of microseismic damaged and contact slip events, the selected representative local domains are selected as shown in Figure 6.10. In Figure 6.10 (a), there are a large number of microseismic events in the local domain A choose from the entire figure of Figure 6.9 (a), where the fracture passes through the bedded interface; there are a large number of microseismic events in the local domain B choose from the entire figure of Figure 6.9 (b), where multiple fractures intersect also generates cumulative microseismic events, as shown in Figure 6.10 (b). From the above results, it can be

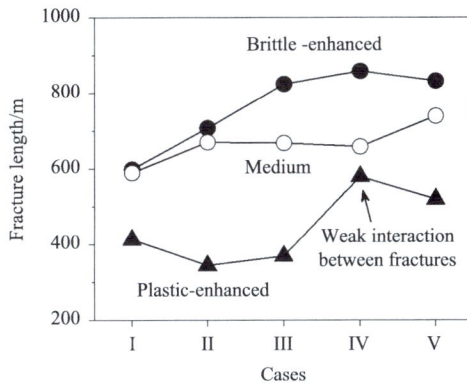

Figure 6.8. Final fracture length of hydraulic fracturing under different physical attributes for reservoir and interlayer strata.

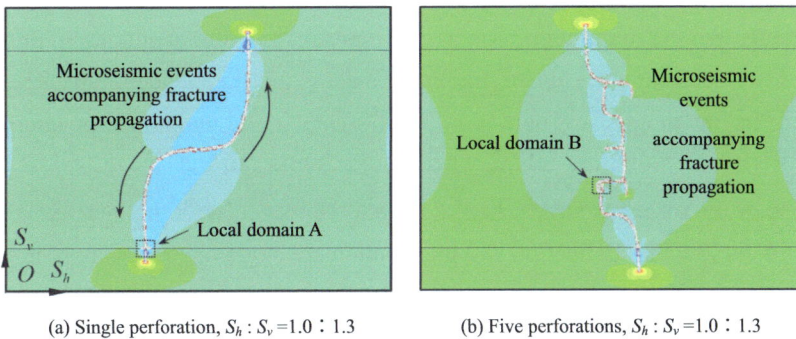

(a) Single perforation, $S_h : S_v =1.0 : 1.3$ (b) Five perforations, $S_h : S_v =1.0 : 1.3$

Figure 6.9. Typical distribution of final damaged events under different *in-situ* stress conditions in brittle-enhanced reservoir stratum by microseismic analysis.

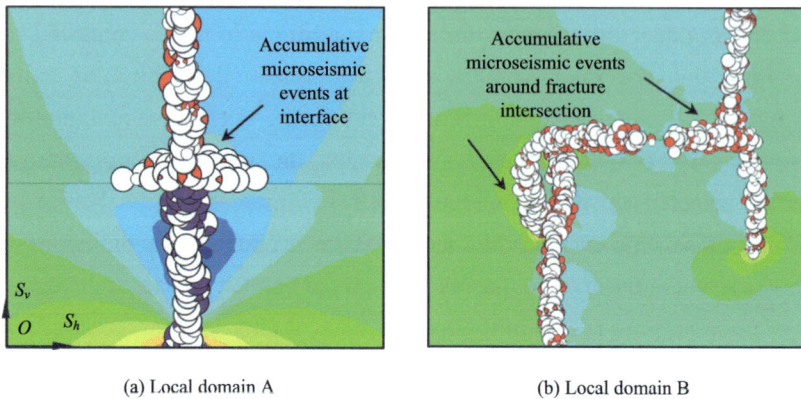

(a) Local domain A   (b) Local domain B

Figure 6.10. Distribution of final damaged events, contact slip events and the first principle stress (Pa) for selected local domains.

seen that hydrofracturing induced microseisms are more likely to occur at the bedded interfaces and the regions where multiple fractures meet, which are caused by the drastic change of stress fields in these local domains; furthermore, the microseisms in brittle-enhanced reservoir stratum have the coincident behaviours that the microseisms arise around domains of these bedded interfaces and crossed fractures, compared to above brittle-enhanced cases.

In order to analyse the influences of *in-situ* stress on microseisms, the evolution of maximum magnitude and accumulated magnitude of damaged events for single perforation of brittle-enhanced reservoir stratum under different *in-situ* stress conditions are selected as shown in Figure 6.11. For brittle reservoirs, the maximum values of microseismic events are higher than the cumulative values. In less than 5000 s, fractures increase gradually with time in the reservoir, and the magnitude in this period is very low; subsequently, when fractures meet with the bedded interface, larger magnitude begins to appear. There is little difference in the magnitude of microseisms under different *in-situ* stress ratios, when *in-situ* stress ratio is $S_h : S_v = 1.0 : 1.2$, the magnitude of microseisms almost keep constant. However, with the increase of *in-situ* stress ratio, the fracture deflection and intersection are aggravated, and more number of microseisms in different magnitudes arise.

To investigate the influences of physical attributes for reservoir and interlayer strata on microseisms, the evolution of maximum magnitude and accumulated magnitude of damaged events for single perforation of *in-situ* stress $S_h : S_v = 1.0 : 1.2$ under different physical attributes for reservoir and interlayer strata is selected as shown in Figure 6.12. Compared with the brittle-enhanced reservoir stratum (Figure 6.11 (a)), both of the medium reservoir (Figure 6.12 (a)) and plastic-enhanced reservoir (Figure 6.12 (b)) generate dramatically changing microseismic damaged events in reservoir stratum, and arise more number and larger magnitudes of dramatic changing microseisms in the later stages as the fracture propagation pass through the bedded interfaces and interlayers. As the plasticity is enhanced, the microseismic events become more and more accompanying the entire process of fracture propagation. Therefore, the plastic properties of reservoir are prone to induce more microseismic events with larger magnitude. Once the reservoir plasticity is ignored, it may cause microseism disasters induced by hydrofracturing.

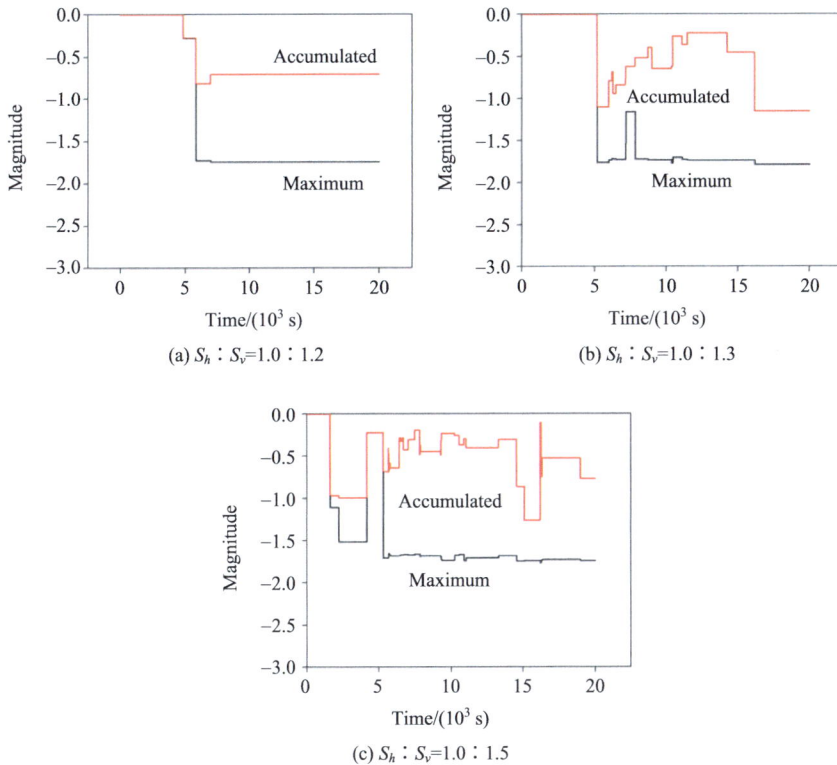

(a) $S_h : S_v=1.0 : 1.2$

(b) $S_h : S_v=1.0 : 1.3$

(c) $S_h : S_v=1.0 : 1.5$

Figure 6.11. Evolution of maximum magnitude and accumulated magnitude of damaged events for single perforation of brittle-enhanced reservoir stratum under different *in-situ* stress conditions.

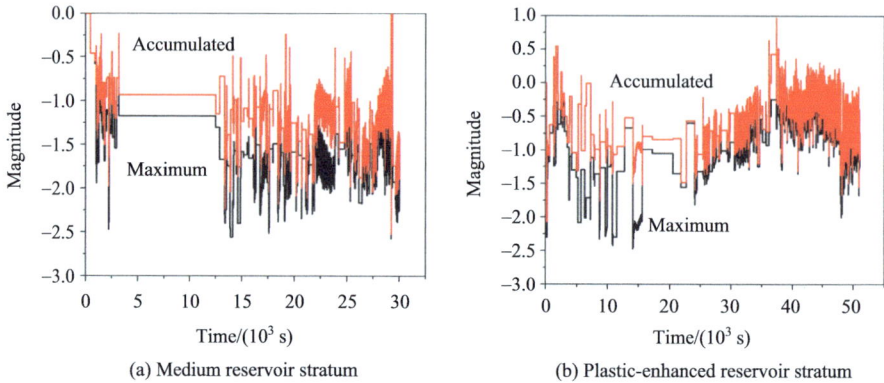

(a) Medium reservoir stratum

(b) Plastic-enhanced reservoir stratum

Figure 6.12. Evolution of maximum magnitude and accumulated magnitude of damaged events for single perforation of *in-situ* stress $S_h : S_v = 1.0 : 1.2$ under different physical attributes for reservoir and interlayer strata.

6.5   Conclusions

In this study, in order to evaluate the influences on unstable hydraulic fractures in plastic multilayered reservoirs containing the reservoir and interlayer strata, the unstable propagation and stress shadow effects of multiple hydraulic fractures are numerically investigated by engineering-scale multilayered FE-DE models. Some conclusions for the fracturing behaviours and mechanisms in simultaneous hydrofracturing of multiple perforation clusters in vertical wells can be summarized as follows:

(1) Domains of positive and negative values of shear stress fields between adjacent multiple hydraulic fractures are superposed to form stress shadows, which will shield propagation of these fractures. The initial *in-situ* stress field changes caused by fractures propagation and the superposition of stress shadows between them make the fractures deflect. Due to the influence of stress shadows, resulting in the shielding effects of fractures and inhibiting fractures growth to induce unstable propagation behaviours, a main single fracture and several minor fractures develop. As the *in-situ* stress ratio increases, fractures are more easily deflected towards the direction of maximum *in-situ* stress, and stress shadow and mutual interaction effects between them are intensified.

(2) Fracture propagation in brittle-enhanced reservoir can be divided into three stages, e.g., propagation in the reservoir stratum, propagation through the bedded

interface, and propagation in the interlayer. As the plasticity of reservoirs increases, there is no platform stage in the fracture length curve, and all three stages of fracture propagation in reservoir, interface and interlayer are linear growth. The fracture lengths in brittle-enhanced, medium reservoir and plastic-enhanced reservoir strata decrease gradually, and plastic-enhanced reservoir limit fracture growth and cannot form long fracture length.

(3) Accompanying the fracturing fracture propagation, microseismic behaviours induced in the process of energy exploitation are investigated. Simultaneous hydrofracturing in multilayered reservoirs induced microseisms are more likely to occur at the bedded interfaces and the regions where multiple fractures intersect, which are caused by the drastic change of stress fields in these local domains. As the *in-situ* stress ratio increases, the fracture deflection and intersection are aggravated, and more number of microseisms in different magnitudes arise. The plastic properties of reservoir are prone to induce more microseismic events with larger magnitude compared to brittle reservoir.

Taking above crucial controlling mechanisms as the research objects, this study focuses on the numerical analysis of stress shadow effects and unstable propagation mechanisms of multiple hydraulic fractures in plastic multilayered reservoirs, and successfully achieves the planer fracture networks, using two-dimensional numerical engineering-scale models. However, the three-dimensional spatial propagation and deflection behaviours of hydraulic fractures affected by stress shadow effects in heterogeneous multilayered reservoirs may occur and control the out-of-plane propagation of fractures in actual hydrofracturing processes, which need to be addressed to form optimized spatial fracturing fracture networks in the next study.

References

Bai, M. (2016), "Why are brittleness and fracability not equivalent in designing hydraulic fracturing in tight shale gas reservoirs", Petroleum, Vol. 2 No. 1, pp. 1–19.

Barati, R., Liang, J.T. (2014), "A review of fracturing fluid systems used for hydraulic fracturing of oil and gas wells", Journal of Applied Polymer Science, Vol. 131 No. 16, pp. 1–11.

Bažant, Z.P., Cedolin, L. (2010), "Stability of structures: elastic, inelastic, fracture and damage theories", 3rd Ed, World Scientific, Singapore.

Bažant, Z.P., Salviato, M., Chau, V.T., Viswanathan, H, Zubelewicz, A. (2014), "Why fracking works", Journal of Applied Mechanics, Vol. 81 No. 10, pp. 1–10.

Cai, M. (2008), "Influence of intermediate principal stress on rock fracturing and strength near excavation boundaries—insight from numerical modelling", International Journal of Rock

Mechanics & Mining Sciences, Vol. 45 No. 5, pp. 763–772.

Chen, Z. (2012), "Finite element modelling of viscosity-dominated hydraulic fractures", Journal of Petroleum Science and Engineering, Vol. 88, pp. 136–144.

Cheng, Y. (2012), "Mechanical interaction of multiple fractures--exploring impacts of the selection of the spacing/number of perforation clusters on horizontal shale-gas wells", SPE Journal, Vol. 17 No. 4, pp. 992–1001.

Cooke, M.L., Underwood, C.A. (2001), "Fracture termination and step-over at bedded interfaces due to frictional slip and interface opening", Journal of Structural Geology, Vol. 23, pp. 223–238.

Deng, S., Li, H., Ma, G., Huang, H., Li, X. (2014), "Simulation of shale–proppant interaction in hydraulic fracturing by the discrete element method", International Journal of Rock Mechanics & Mining Sciences, Vol. 70, pp. 219–228.

ELFEN TGR user and theory manual. (2016), Rockfield Software Ltd., United Kingdom, 2016.

Fischer, T., Hainzl, S., Eisner, L., Shapiro, S.A., Calvez, J.L. (2008), "Microseismic signatures of hydraulic fracture growth in sediment formations: observations and modeling", Journal of Geophysical Research: Solid Earth, Vol. 113 No. B2, pp. 1–12.

Fu, W., Ames, B.C., Bunger, A.P., Savitski, A.A. (2016), "Impact of partially cemented and non-persistent natural fractures on hydraulic fracture propagation", Rock Mechanics and Rock Engineering, Vol. 49 No. 11, pp. 4519–4526.

Gordeliy, E., Peirce, A. (2020), "Implicit level set schemes for modeling hydraulic fractures using the XFEM", Computer Methods in Applied Mechanics and Engineering, Vol.266, pp. 125–143.

Gudmundsson, A., Simmenes, T.H., Larsen, B., Philippc, S.L. (2010), "Effects of internal structure and local stresses on fracture propagation, deflection, and arrest in fault zones", Journal of Structural Geology, Vol. 32 No. 11, pp. 1643–1655.

Guo, J., Lu, Q., Zhu, H., Wang, Y., Ma, L. (2015a), "Perforating cluster space optimization method of horizontal well multi-stage fracturing in extremely thick unconventional gas reservoir", Journal of Natural Gas Science and Engineering, Vol. 26, pp. 1648–1662.

Guo, L., Xiang, J., Latham, J.P., Viré, A., Pavlidis, D., Pain, C.C. (2015b), "Numerical simulation of hydraulic fracturing using a three-dimensional fracture model coupled with an adaptive mesh fluid model", 49th US Rock Mechanics/Geomechanics Symposium. American Rock Mechanics Association, ARMA-15-397.

Gutenberg, B., Richter, C. (1956), "Magnitude and energy of earthquakes", Science, Vol. 112 No. 1–4, pp. 1–14.

He, J., Li, X., Yin, C., Zhang, Y., Lin, C. (2020), "Propagation and characterization of the micro cracks induced by hydraulic fracturing in shale", Energy, Vol. 191.

He, Q., Suorineni, F.T., Ma, T., Oh, J. (2017), "Effect of discontinuity stress shadows on hydraulic fracture re-orientation", International Journal of Rock Mechanics and Mining Sciences, Vol. 91, pp. 179–194.

Huang, C., Hou, H., Yu, G., Zhang, L., Hu, E. (2020), "Energy solutions for producing shale oil: Characteristics of energy demand and economic analysis of energy supply options", Energy, Vol. 192, 116603.

Ishida, T., Aoyagi, K., Niwa, T., Chen, Y., Murata, S., Chen, Q., Nakayama, Y. (2012), "Acoustic emission monitoring of hydraulic fracturing laboratory experiment with supercritical and liquid CO_2", Geophysical Research Letters, Vol. 39 No. 16, pp. 1–6.

Kresse, O., Weng, X., Gu, H., Wu, R. (2013), "Numerical modeling of hydraulic fractures interaction in complex naturally fractured formations", Rock Mechanics and Rock Engineering, Vol. 46 No. 3, pp. 555–568.

Latham, J.P., Xiang, J., Belayneh, M., Nick, H., Tsang, C., Blunt, M. (2013), "Modelling stress-dependent permeability in fractured rock including effects of propagating and bending fractures". International Journal of Rock Mechanics and Mining Sciences, Vol. 57, pp. 100–112.

Lecampion, B. (2009), "An extended finite element method for hydraulic fracture problems", Communications in Numerical Methods in Engineering, Vol. 25 No. 2, pp. 121–133.

Li, Y., Yang, S., Zhao, W., Zhang, J. (2018), "Experimental of hydraulic fracture propagation using fixed-point multistage fracturing in a vertical well in tight sandstone reservoir", Journal of Petroleum Science and Engineering, Vol. 171, pp. 704–713.

Masoudian, M.S., Hashemi, M.A., Tasalloti, A., Marshall, A. (2018), "Elastic-brittle-plastic behaviour of shale reservoirs and its implications on fracture permeability variation: an analytical approach", Rock Mechanics and Rock Engineering, Vol. 51 No. 5, pp. 1565–1582.

Miehe, C., Mauthe, S. (2016), "Phase field modeling of fracture in multi-physics problems. Part III. Crack driving forces in hydro-poro-elasticity and hydraulic fracturing of fluid-saturated porous media", Computer Methods in Applied Mechanics and Engineering, Vol. 304, pp. 619–655.

Munjiza, A., Owen, D.R.J., Bicanic, N. (1995), "A combined finite-discrete element method in transient dynamics of fracturing solids", Engineering Computations, Vol. 12 No. 2, pp. 145–174.

Olson, J.E. (2008), "Multi-fracture propagation modeling: applications to hydraulic fracturing in shales and tight gas sands", 42th US Rock Mechanics Symposium. American Rock Mechanics Association, ARMA-08-327.

Papanastasiou, P. (1997), "The influence of plasticity in hydraulic fracturing", International Journal of Fracture, Vol. 84 No. 1, pp. 61–79.

Peirce, A., Bunger, A. (2015), "Interference fracturing: nonuniform distributions of perforation clusters that promote simultaneous growth of multiple hydraulic fractures", SPE Journal, Vol. 20 No. 2, pp. 384–395.

Sone, H., Zoback, M.D. (2011), "Visco-plastic properties of shale gas reservoir rocks", 45th US Rock Mechanics/Geomechanics Symposium. American Rock Mechanics Association, ARMA-11-417.

Taghichian, A., Zaman, M., Devegowda, D. (2014), "Stress shadow size and aperture of hydraulic fractures in unconventional shales", Journal of Petroleum Science and Engineering, Vol. 124, pp. 209–221.

Van Dam, D.B., Papanastasiou, P., De Pater, C.J. (2000), "Impact of rock plasticity on hydraulic fracture propagation and closure", SPE Annual Technical Conference and Exhibition. Society of Petroleum Engineers, pp. 1–13.

Wang, H. (2016), "Poro-elasto-plastic modeling of complex hydraulic fracture propagation: simultaneous multi-fracturing and producing well interference", Acta Mechanica, Vol. 227 No.

2, pp. 507–525.

Wang, Q., Feng, Y.T., Wei, Z., Cheng, Y.G., Ma, G. (2020), "A phase-field model for mixed-mode fracture based on a unified tensile fracture criterion", Computer Methods in Applied Mechanics and Engineering, Vol. 370 113270.

Wang, Y. (2020), "Adaptive finite element-discrete element analysis for stratal movement and microseismic behaviours induced by multistage propagation of three-dimensional multiple hydraulic fractures", Engineering Computations, Vol. 38 No. 5, pp. 1350–1371.

Wang, Y. (2021), "Adaptive analysis of damage and fracture in rock with multiphysical fields coupling", Springer, Singapore.

Xiao, Y., Feng, X., Hudson, J.A., Chen, B., Feng, G., Liu, J. (2016), "ISRM suggested method for in-situ microseismic monitoring of the fracturing process in rock masses", Rock Mechanics and Rock Engineering, Vol. 49 No. 1, pp. 343–369.

Yan, X., Huang, Z., Yao, J., Song, W., Li, Y., Gong, L. (2016),"Theoretical analysis of fracture conductivity created by the channel fracturing technique", Journal of Natural Gas Science and Engineering, Vol. 31, pp. 320–330.

Yoon, J.S., Zimmermann, G., Zang, A. (2015), "Numerical investigation on stress shadowing in fluid injection-induced fracture propagation in naturally fractured geothermal reservoirs", Rock Mechanics and Rock Engineering, Vol. 48 No. 4, pp. 1439–1454.

Zhang, D., Ranjith, P.G., Perera, M.S.A. (2016), "The brittleness indices used in rock mechanics and their application in shale hydraulic fracturing: A review", Journal of Petroleum Science and Engineering, Vol. 143, pp. 158–170.

Zhang, X., Jeffrey, R.G. (2008), "Reinitiation or termination of fluid-driven fractures at frictional bedding interfaces", Journal of Geophysical Research: Solid Earth, Vol. 113 No. B8, pp. 1–16.

Zhang, X., Jeffrey, R.G., Thiercelin, M. (2007), "Deflection and propagation of fluid-driven fractures at frictional bedding interfaces: a numerical investigation", Journal of Structural Geology, Vol. 29 No. 3, pp. 396–410.

Zhou, J., Chen, M., Jin, Y., Zhang, G. (2008), "Analysis of fracture propagation behavior and fracture geometry using a tri-axial fracturing system in naturally fractured reservoirs", International Journal of Rock Mechanics and Mining Sciences, Vol. 45 No. 7, pp. 1143–1152.

Zhou, W., Liu, B., Wang, Q., Chang, X.L., Chen, X.D. (2020), "Formulations of displacement discontinuity method for crack problems based on boundary element method", Engineering Analysis with Boundary Elements, Vol. 115, pp. 86-95.

Appendix A. Fracture criteria

The fracture criteria based on damage evolution are expressed as follows:

$$H = \frac{\sigma_{\mathrm{t}}^2 C_1}{2 G_{\mathrm{f}}}, \qquad (A6.1)$$

$$\tilde{E} = E(1 - d), \qquad (A6.2)$$

where H is the slope of the stress-strain curve; $\tilde{E}$ is the Young's module considering the damage; d is the damage variable, for which a value of 0 indicates no damage and 1 indicates fully damaged (Wang, 2020); σ_t is the tensile strength; G_f is the fracture energy; and C_1 is the characteristic element length.

Appendix B. Microseismicity analysis by evaluating moment tensors

Once the stress solutions are obtained, the computed differential stress $\Delta\sigma^e$ from the pre-and post-events can be calculated over a single time step and the eigen decomposition of the decrease in the stress tensor is performed as follows (Wang, 2020):

$$\Delta\sigma^e v = \lambda v , \qquad (B6.1)$$

where λ and v are the eigenvalue and the eigenvector, respectively. The eigenvalues are used to scale the eigenvector, resulting in the definition of the moment tensor M as follows:

$$M = \lambda v . \qquad (B6.2)$$

The moment tensors can be computed if M is rotated into its principle direction using the unit eigenvectors, furthermore, the damaged events (Mode I, tensile failure) and contact slip events (Mode II and Mode III, shear failure) are distinguished.

For indicating the fault reactivation by determining the maximum magnitude of the expected seismic events, the Gutenberg Richter Law (Gutenberg and Richter, 1956) was used to provide the relationship between the number of events and magnitudes as follows:

$$\log_{10} N = a - b M_w , \qquad (B6.3)$$

where N is the cumulative number of seismic events, i.e. number of events with a magnitude greater than M_w, and M_w is the magnitude of the seismic event.

Chapter 7   Unstable propagation of multiple hydraulic fractures and shear stress disturbance in multi-well hydrofracturing

7.1   Introduction

Understanding the complexity of hydraulic fracture networks is crucial to increasing unconventional oil and gas production (Wang *et al.*, 2020). Multi-well hydrofracturing is an important technology for creating complex fracture networks and increasing reservoir permeability. Compared with single-well hydrofracturing, the interaction between fractures in multi-well hydrofracturing is more complex. In the process of multistage hydrofracturing of rock masses, the propagation of fractures is affected by various factors and deflected to different degrees, and the superposition of stress fields around the fractures produces a stress shadow effect during the hydrofracturing process (Olson and Taleghani, 2009; Bunger *et al.*, 2011). The stress shadow effect creates pressure on the surrounding rock and adjacent fractures, resulting in increased initiation pressure in the subsequent perforation, changes in the *in-situ* stress field, the direction of fracture propagation (Bunger *et al.*, 2011, 2012; He *et al.*, 2017; Manríquez, 2018), and formation of multiple fracture deflections and propagations. The stress shadow effect not only promotes fracture propagation and forms a more complex fracture network (Roussel and Sharma, 2011; Nagel *et al.*, 2013; Manchanda and Sharma, 2014; Wu and Olson, 2015), but inhibits the growth of local fractures (Manríquez, 2018; Bai *et al.*, 2000; Warpinski, 2000; Sobhaniaragh *et al.*, 2018). This effect inhibits reservoir rock mass hydrofracturing (Lu *et al.*, 2015; Tian *et al.*, 2019; Wang *et al.*, 2015). Compared with single-well hydrofracturing, the interaction between fractures in multi-well hydrofracturing is more complex. Figure 7.1 shows the unstable dynamic propagation and deflection of multiple hydraulic fractures in multi-well hydrofracturing in tight reservoirs. The dynamic propagation of multiple hydraulic fractures is simultaneously affected by adjacent perforation clusters (*a*: perforation cluster spacing) and adjacent wells (*b*: well spacing), resulting in a more complex fracture network. The hydrofracturing-induced stress field applies pressure on the surrounding rock, altering the initiation pressure in the subsequent perforation, the *in-situ* stress field, and the propagation direction of hydraulic fractures (Olson and Taleghani, 2009; Bunger *et al.*, 2011, 2012; He *et al.*, 2017; Manríquez, 2018). The hydrofracturing-induced stress field can promote fracture propagation and form a more complex fracture network (Roussel and Sharma, 2011; Nagel *et al.*, 2013; Manchanda and Sharma, 2014; Wu and Olson, 2015), and inhibit the growth of local fractures (Manríquez, 2018; Bai *et al.*, 2000; Warpinski, 2000; Sobhaniaragh *et al.*, 2018). This induced stress field is not helpful to controlling the fracture propagation process and increasing unconventional oil and gas production (Lu *et al.*, 2015; Tian *et al.*, 2019; Wang *et al.*, 2015). To evaluate and optimize the hydrofracturing process, it is vital to understand the relationship between the geometric distributions of multiple wells and the induced stress field and fracture propagation.

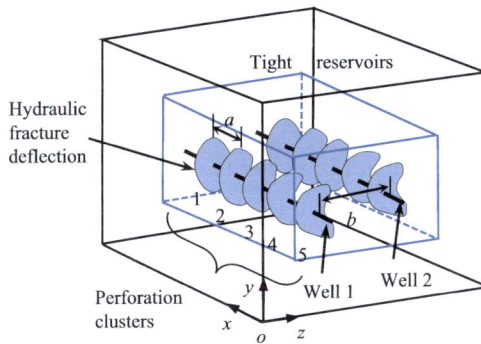

Figure 7.1. Schematic of unstable dynamic propagation and deflection of multiple hydraulic fractures in a tight reservoir.

The spacing of perforation clusters, fracturing sequence, and well spacing are the major factors causing the unstable propagation of parallel fractures to different degrees in the hydrofracturing process of multiple horizontal wells (Ghazal *et al.*, 2015; Kumar and Ghassemi, 2016; Ju *et al.*, 2020). When two adjacent horizontal wells were drilled, the hydraulic fractures in the subsequent well propagated away from the previous well (Wong *et al.*, 2013). Narrow adjacent well spacing may result in intersecting reservoir volumes, causing competition between wells and affecting fracture propagation and oil and gas production (Segatto and Colombo, 2011; Liu *et al.*, 2020). Well spacing should be carefully selected to avoid the connection of fracture tips in adjacent horizontal wells; some studies have carried out analyses from the perspective of hydrofracturing-induced stress field evolution. A small well spacing may create a drastic stress disturbance between wells, resulting in connections of hydraulic fractures from adjacent wells during hydrofracturing, which inevitably reduces the fracture complexity of the field. However, if the well spacing is too large, the natural fracture between the two wells cannot be reactivated (Li and Zhang, 2018; He *et al.*, 2020; Duan *et al.*, 2021). The spacing between adjacent wells can be optimized using the variation in the induced stress evolution and disturbance in the hydrofracturing process. Simulation results of the fracture path in the simultaneous fracturing of two horizontal wells show that when the well spacing exceeds 60 m, the stress disturbance can be effectively alleviated, and improve the fracturing effects (Duan *et al.*, 2021).

Multi-well hydrofracturing scenarios primarily include sequential fracturing, simultaneous fracturing, alternate fracturing, and zipper fracturing (cross-perforation) (Rafiee *et al.*, 2012; Yu and Sepehrnoori, 2013; Qiu *et al.* 2015; Li and Zhang, 2018; Liu *et al.*, 2020). The different fracture initiation sequences of multiple well hydrofracturing change the well spacing during the fracture initiation of horizontal wells, thereby affecting the hydraulic fracture propagation in multiple wells. Simultaneous fracturing refers to the simultaneous injection of fluids into all perforation clusters of two horizontal wells (Liu *et al.*, 2020). The geometry of each fractured wing was different on both sides of the well in simultaneous and sequential fracturing. The fractured wing between the two wells interacts with the other wing, limiting the fracture width and shortening the fracture length (Wu *et al.*, 2018). Simulation analysis of multiple well simultaneous fracturing shows that asymmetric fracture propagation exists in the process of multiple well simultaneous fracturing; the stress interference between wells is large, which inhibits the transverse propagation of internal fractures (Chen *et al.*, 2018). In some cases, the fracturing effect of a modified zipper fracturing is the best, followed by sequential fracturing, and the fracturing effect of simultaneous case is the worst (Li and Zhang, 2018). The total fracture length of alternate fracturing is longer than that in simultaneous and sequential fracturing; it is, however, the best method for volume fracturing (Liu *et al.*, 2020). By building a three-dimensional numerical model based on actual reservoir data, the continuous and simultaneous fracturing of multiple vertical wells in tight reservoirs can be analysed in three dimensions (Ju *et al.*, 2020).

The stress shadow effect between multiple wells affects the dynamic propagation of the fracture network; therefore, it is critical to determine the internal mechanisms of the stress shadow effect in multi-well hydrofracturing to optimize the fracturing scenarios (Ju *et al.*, 2020; Wu *et al.*, 2018). The stress shadow effect depends largely on the fracture height, mechanical properties (i.e., Young's modulus and Poisson's ratio), cluster spacing, well spacing, and fracturing scenario (Jo, 2012; Manchanda and Sharma, 2013; Qiu *et al.*, 2015). When multiple horizontal wells are adjacent, the stress shadows induced by perforation clusters affect the fracture propagation patterns in each well (Liu *et al.*, 2020). Increasing well spacing can effectively alleviate the stress shadow effect (Wu *et al.*, 2018; Duan *et al.*, 2021). In sequential and simultaneous fracturing, the stress interference between the wells is strong and fracture propagation is inhibited (Chen *et al.*, 2018). Rafiee *et al.* (2012) reported that zipper fracturing (cross-perforation) increases the stress shadow effect between multiple well fractures. However, the effect of spacing between adjacent wells and fracturing scenarios on fracture propagation, deflection, connection, and the quantitative relationship between the induced stress field and fracture propagation is not well understood.

To further investigate the effect of the perforation fracture initiation sequence and adjacent well spacing on the dynamic propagation and shear stress disturbance in multi-well hydrofracturing. The main contents of this study are as follows: Section 7.2 introduces the combined finite element-discrete element method considering thermal-hydro-mechanical coupling. Section 7.3 introduces numerical models and cases of multiple horizontal wells, including geometric models, finite element models, conditions, and parameter settings. Section 7.4 details the results and analysis of the unstable propagation of hydraulic fractures with variable well spacings. Section 7.5 details the results and analyses of the unstable propagation of hydraulic fractures with variable initiation sequences. Section 7.6 summarizes the conclusions and directions for future research.

7.2   Combined finite element-discrete element method considering thermal-hydro-mechanical coupling

In the hydrofracturing of reservoir rock, the thermal, hydraulic, and mechanical fields form a mutual coupling process (Tsang *et al.*, 2000; Rutqvist *et al.*, 2005). The main mechanism of the coupling process is shown in Figure 7.2 (Wang *et al.*, 2019). Through the transfer between the solid stress, fluid pressure, heat, and fracture network, the thermal-hydro-mechanical coupling and fracture network mechanisms are introduced (Wang *et al.*, 2018).

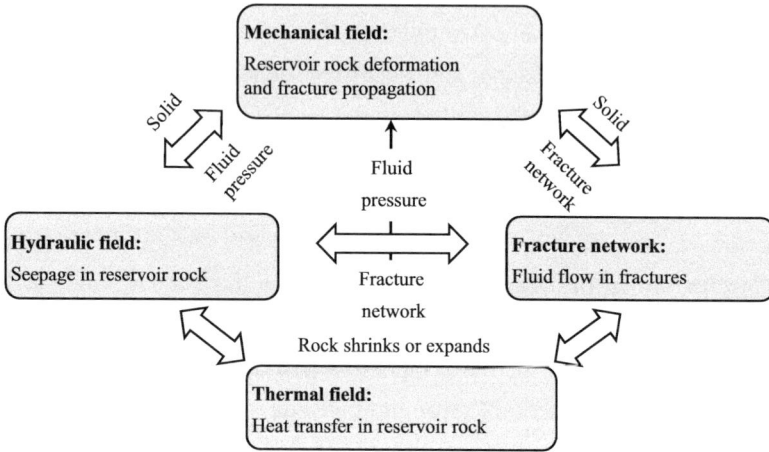

Figure 7.2. Thermal-hydro-mechanical coupling and fracture propagation mechanisms.

The governing equations for solid deformation, fluid flow, and heat transfer are as follows:

(1) Solid deformation

The governing equation of rock mass matrix deformation is:

$$L^{\mathrm{T}}\left(\boldsymbol{\sigma}' - \alpha \boldsymbol{m} p_1\right) + \rho_{\mathrm{B}} \boldsymbol{g} = \boldsymbol{0} \tag{7.1}$$

where L is the differential operator, $\boldsymbol{\sigma}'$ is the effective stress tensor, α is Biot's coefficient, $\boldsymbol{m}$ denotes the identity tensor, p_1 is the rock mass pore fluid pressure, ρ_{B} is the saturated bulk density of the rock mass, and $\boldsymbol{g}$ is the gravity vector.

(2) Fluid flow

The governing equation of seepage in rock matrix is:

$$\mathrm{div}\left[\frac{k}{\mu_1}\left(\nabla p_1 - \rho_1 \boldsymbol{g}\right)\right] = \left(\frac{\phi}{K_1} + \frac{\alpha - \phi}{K_{\mathrm{s}}}\right)\frac{\mathrm{d}p_1}{\mathrm{d}t} + \alpha\frac{\mathrm{d}\varepsilon_v}{\mathrm{d}t} \tag{7.2}$$

where k is the intrinsic permeability of the rock mass pore structure, μ_1 is the viscosity of the pore liquid, ρ_1 is the pore liquid density, ϕ is the porosity of the porous medium, K_1 is the pore fluid stiffness, K_{s} is the solid skeleton stiffness, ε_v is the volumetric strain of the rock mass pore structure, and t is the current moment.

The governing equation of fluid flow in the fracture is:

$$\frac{\partial}{\partial x}\left[\frac{k^{\mathrm{fr}}}{\mu_n}\left(\nabla p_n - \rho_{\mathrm{fn}} \boldsymbol{g}\right)\right] = S^{\mathrm{fr}}\frac{\mathrm{d}p_n}{\mathrm{d}t} + \alpha\left(\Delta \dot{e}_\varepsilon\right) \tag{7.3}$$

where k^{fr} is the fracture intrinsic permeability, μ_n is the viscosity of the hydrofracturing fluid, p_n is the fluid pressure in the fracture, ρ_{fn} is the fluid density in the fracture, S^{fr} is the parameter describing rock mass compressibility under fluid action, and $\dfrac{\mathrm{d}\Delta\dot{e}_\varepsilon}{\mathrm{d}t}$ is the aperture strain rate. According to the plate theory of fluid, the permeability of the fluid in the fracture is

$$k^{fr} = \frac{e^2}{12} \tag{7.4}$$

where e is the fracture width. The parameters for compressibility are expressed as:

$$S^{fr} = \left(\frac{1}{e}\right)\left[\left(\frac{1}{K_n^{fr}}\right) + \left(\frac{e}{K_f^{fr}}\right)\right] \tag{7.5}$$

where K_n^{fr} is normal stiffness of fracture and K_f^{fr} is bulk modulus of the hydrofracturing fluid.

(3) Heat transfer

The governing equation of heat transfer between rock matrixes, hydrofracturing fluids is:

$$\mathrm{div}\left[k_b \nabla T_f\right] = \rho_b c_b \frac{\partial T_f}{\partial t} + \rho_f c_f q_f \nabla T_f \tag{7.6}$$

where k_b is the thermal conductivity coefficient, T_f is the fluid temperature, ρ_b is the volume density, c_b is the specific heat coefficient, ρ_f is the fluid density, c_f is the specific heat coefficient of the fluid, and q_f is the Darcy fluid flux.

The heat transfers between the finite element nodes in the formation and the network is shown in Figure 7.3.

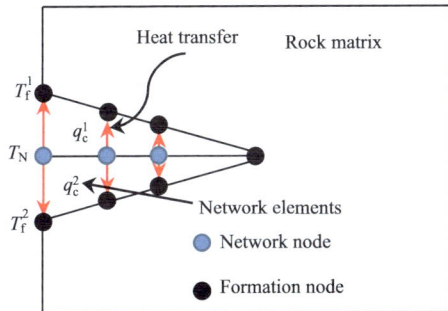

Figure 7.3. Heat transfer between network and formation nodes.

The heat transfer from the fracture network element from the fluid in the fracture zone to the rock matrix is expressed as follows (Inui *et al.*, 2014):

$$q_c^1 = \alpha_c \left(T_N \right) \left(T_N - T_f^1 \right) \tag{7.7}$$

$$q_c^2 = \alpha_c \left(T_N \right) \left(T_N - T_f^2 \right) \tag{7.8}$$

where q_c^1 and q_c^2 are the contact heat flows between the network and formation nodes, respectively; T_N is the temperature value of the node within the fracture; α_c is the contact thermal conductivity; T_f^1 and T_f^2 are the corresponding formation node temperatures.

Heat transfer through the reservoir rock causes the rock to shrink or expand, causing stress changes. The volume change depends on the linear thermal expansion coefficient:

$$\frac{\Delta V}{V} = \alpha_T \Delta T \tag{7.9}$$

where V is the initial volume, ΔV is the incremental volume, ΔT is the incremental temperature, and α_T is the linear thermal-expansion coefficient of the rock matrix.

The numerical discretization of the coupled governing equations and the fracture criterion based on the form of fracture energy can be used from the literatures of previous relevant researches (Wang *et al.*, 2018, 2019).

7.3   Numerical models and cases of multiple horizontal wells

In this study, multi-well multistage hydrofracturing in a deep tight reservoir was modelled, as shown in Figure 7.4. Three horizontal wells (denoted as wells 1, 2, and 3) were set in this model, and five perforation clusters (numbered 1–5 in sequence) were set for each well. There are two geometric variables in the model: *a* is the perforation cluster spacing and *b* is the well spacing.

In the study of unstable propagation of fractures at different well spacings, the perforation cluster spacing *a* was 75 m, and the well spacings *b* were 100, 75, and 50 m, respectively. The numerical cases of different well spacings are listed in Table 7.1, where the perforation clusters in each well use sequential fracturing (1→2→3→4→5). Three different perforation initiation sequences were used to study the unstable propagation of fractures in different initiation sequences of horizontal wells: sequential, simultaneous, and alternate fracturing. There are three alternate fracturing scenarios: alternate scenario 1 implements hydrofracturing process in the order of well

2→well 1→well 3, and alternate case scenario 2 in the order of well 2→well 1→well 3. To study the effect of crossed staggered perforation clusters between multiple wells on the unstable propagation of fractures, the arrangement scenario of multi-well crossed perforation clusters shown in Figure 7.5 is set here, and alternate scenario 3 implements hydrofracturing process in the order of well 1→well 3→well 2. The numerical cases of different initiation sequences are listed in Table 7.2, where the perforation cluster spacing a and well spacing b are both set to 75 m. The fracturing fluid temperature was set to 20 °C, and the rock matrix temperature was set to 60 °C for all the numerical cases. The basic physical parameters of the model were set as listed in Table 7.3 and were tested from the tight rock samples in the Shengli Oilfield in Shandong Province, China. Figure 7.6 shows the initial mesh refinement of the finite element model, where an initial dense mesh is used around the local perforation domains to guarantee a reliable fracture propagation path in the initial stage.

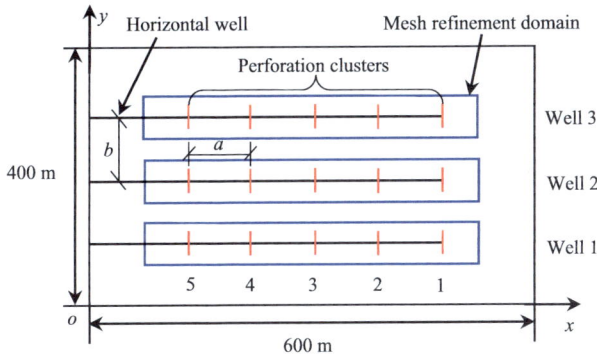

Figure 7.4. Geometric model of multi-well multistage hydrofracturing.

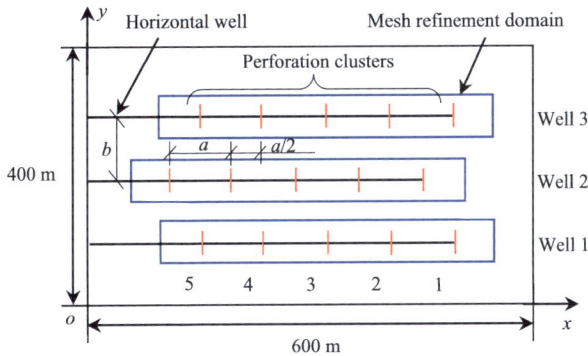

Figure 7.5. Geometric model of multi-well multistage hydrofracturing using cross perforation clusters.

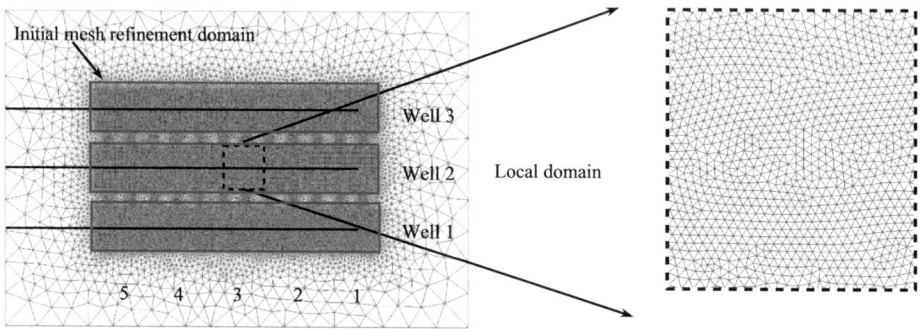

Figure 7.6. Initial mesh refinement of finite element model.

Table 7.1. Numerical cases of different well spacings.

Case	Perforation cluster spacing a /m	Well spacing b /m	Fracturing fluid temperature /℃	Rock matrix temperature /℃
1-I	75	100	20	60
1-II	75	75	20	60
1-III	75	50	20	60

Table 7.2. Numerical cases of different fracturing scenarios.

Case	Fracturing scenarios	Initiation sequences
2-I	Sequential	Well 1→Well 2→Well 3
2-II	Simultaneous	Well 2→Well 1→Well 3
2-III	Alternate	Well 2→Well 1→Well 3
2-IV	Alternate	Well 2→Well 1→Well 3 (crossed perforation clusters)
2-V	Alternate	Well 1→Well 3→Well 2

Table 7.3. Basic physical parameters of the numerical model.

Parameters	Value
Horizontal minimum ground stress (x-direction) S_h /MPa	40
Horizontal max ground stress (y-direction) S_H /MPa	44
Fluid injection rate Q/(m³/s)	0.5
Pore pressure p_s /MPa	10
Biot coefficient α	0.75
Elastic modulus E /GPa	31
Poisson ratio v	0.22

Continued

Parameters	Value
Penetration k /nD	50
Porosity ϕ	0.05
Kinematic viscosity coefficient μ_n /(Pa·s)	1.67×10^{-3}
Fracture fluid bulk modulus K_f^{fr} /MPa	2000
Tensile Strength σ_t /MPa	5.26
Fracture energy G_f /(N·m)	165

7.4   Results and analysis of unstable propagation of hydraulic fractures with variable well spacings

7.4.1   Mesh refinement and thermal diffusion in fracture propagation process

In this study, the finite element-discrete element model was used to simulate fracture propagation. The mesh of the fracture tip was optimized to ensure that the fracture tip attained a high-precision solution and a reliable fracture propagation path. Figure 7.7 shows the mesh division around the fracture after fracturing of well 1 at a well spacing of b=100 m. The local mesh at the fracture tip was subdivided and refined to obtain a high-precision solution for the stress-concentration domain. Mesh refinement was no longer required in the domain fractures that have propagated through, and the mesh in this domain was coarsened to improve the computational efficiency. This technology was used for the computation of each case.

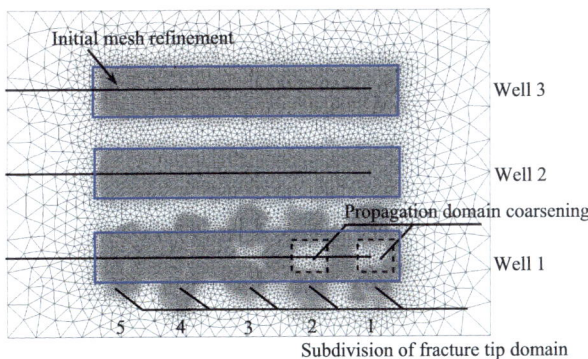

Figure 7.7. Mesh refinement and coarsening around hydraulic fractures.

The thermal diffusion behaviours between the rock matrix (initial temperature of 60 °C) and hydrofracturing fluid (initial temperature of 20 °C) in the propagation process of hydrofracturing are examined below. Figure 7.8 shows the temperature evolution around the hydraulic fractures in multi-well hydrofracturing. Figures 7.8 (a) – (c) show that the propagation of fractures in perforation 1 of well 1, thermal diffusion occurs between the perforation cluster and the surrounding rock matrix owing to the existence of a temperature gradient. The temperature of the hydrofracturing fluid was gradually increased to reach the same temperature as that achieved during fracture propagation. Figure 7.8 (d) is the temperature field distribution of the three wells in the final stage, and thermal diffusion occurs near each perforation; with the termination of hydrofracturing fluid injection in the perforation cluster, the thermal diffusion between the perforation cluster and the surrounding rock matrix weakens.

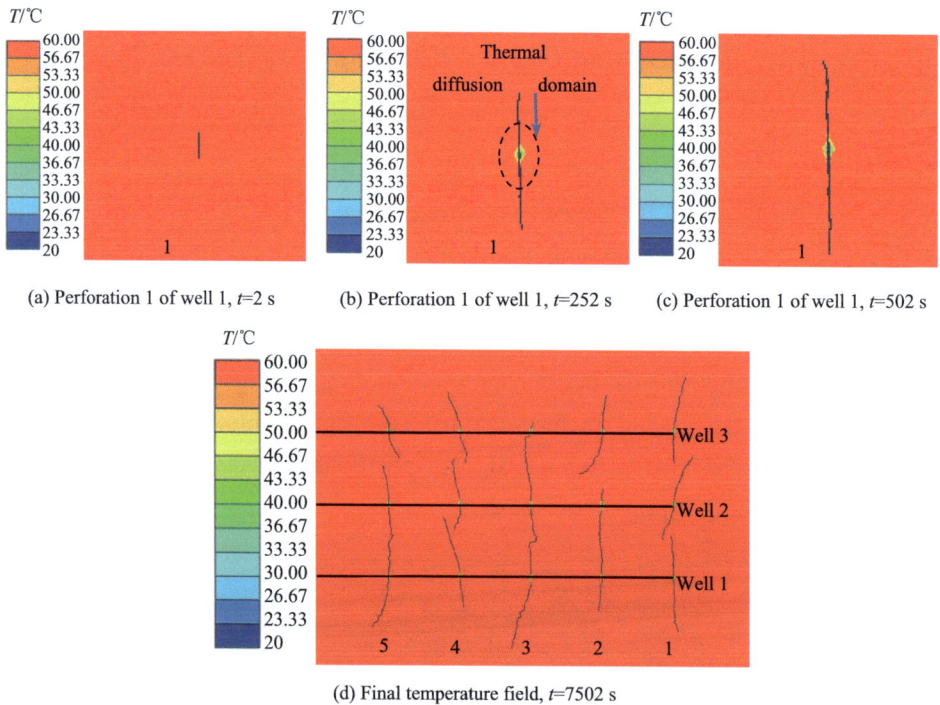

(a) Perforation 1 of well 1, t=2 s   (b) Perforation 1 of well 1, t=252 s   (c) Perforation 1 of well 1, t=502 s

(d) Final temperature field, t=7502 s

Figure 7.8. Temperature evolution around hydraulic fractures in multi-well hydrofracturing.

7.4.2   Fracture network propagation and shear stress shadows

7.4.2.1   Case 1-I: well spacing b=100 m

The fracture propagation of multiple well multistage hydrofracturing is analysed below. Figure 7.9 shows the distribution of hydraulic fractures when the well spacing is 100 m; the displacement (m) in x-direction is shown to detect the deformation in the reservoir. Fractures 3, 4, and 5 of well 1 were affected by previous fractures 1 and 2 and deflected away from the previous fractures. This is identical to the multistage hydrofracturing in a single horizontal well (Wang et $al.$, 2021). Owing to the influence of well 1 on the in-$situ$ environment of the reservoir, fractures 1 and 2 of well 2 penetrated well 1, and fractures 3, 4, and 5 of well 2 were deflected to different extents from well 1. Owing to the influence of wells 1 and 2 on the in-$situ$ environment of the reservoir, fractures 1 and 2 in well 3 were connected with well 1, and fractures 3, 4, and 5 were deflected dramatically away from well 2. Meanwhile, both sides of the stratum were deformed to the left and right, respectively, owing to the hydrofracturing fluid injected into the reservoir.

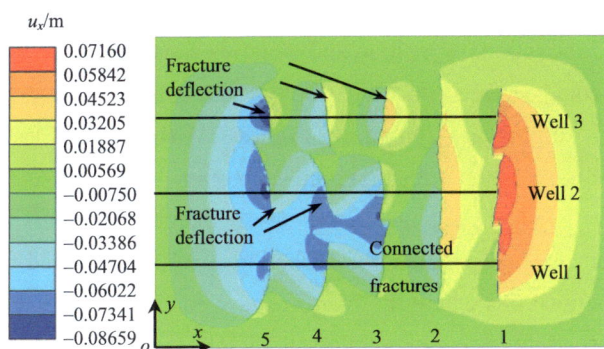

Figure 7.9. Final fracture network morphology and deformation in multi-well hydrofracturing, $b = 100$ m.

The deformation field of the rock matrix derived below was used to analyse the reservoir compression during fracture propagation. Figure 7.10 shows the displacement vector evolution of the reservoir rock mass in the local domain of well 1 fracture 1 during multi-well sequential fracturing. In the first stage, the displacement vector is mainly concentrated in the middle of the perforation cluster; in the second stage, the displacement vectors are distributed on both sides of the fracture tip; and in the third stage, the displacement vectors are distributed on both sides of the fracture. During fracture propagation, the compressional deformation of the rock matrix on both sides of the fracture increased gradually and became the main factor in the expansion of the next-stage perforation cluster.

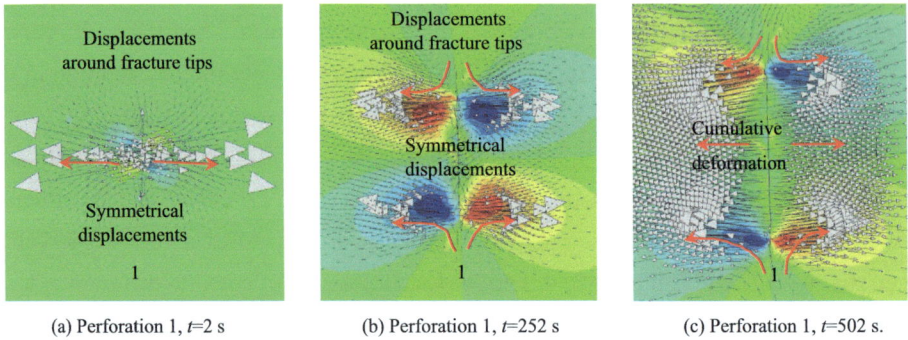

(a) Perforation 1, t=2 s   (b) Perforation 1, t=252 s   (c) Perforation 1, t=502 s.

Figure 7.10. Displacement vector evolution of single perforation in hydrofracturing.

Figure 7.11 shows the evolution of shear stress τ_{xy} in sequential fracturing at 100 m well spacing, at t=2502 s, t=5002 s, t=7502 s, respectively. Figure 7.11 (a) shows the shear stress results of well 1 after the sequential fracturing is completed. The superposition and reduction of positive and negative shear stress fields around fracture tips result in unequal stresses on both sides of these fracture tips, and the fracture deflects toward the high stress domain. In Figure 7.11 (b), in the local domains of the two adjacent wells, the superposition and reduction of shear stresses occur (e.g. fractures 5 in wells 1 and 2); and these stress disturbances deflect the fractures. Meanwhile, the stress area (indicated by dashed line) of the disturbed formation in well 2 has covered the positions where the fractures in well 3 propagate, which is the stress shadow area between multiple wells. In Figure 7.11 (c), the fractures 1 and 2 in well 3 and well 2 are penetrated and connected, and the deflection of fractures 3, 4, and 5 occurs due to the influence of stress shadows.

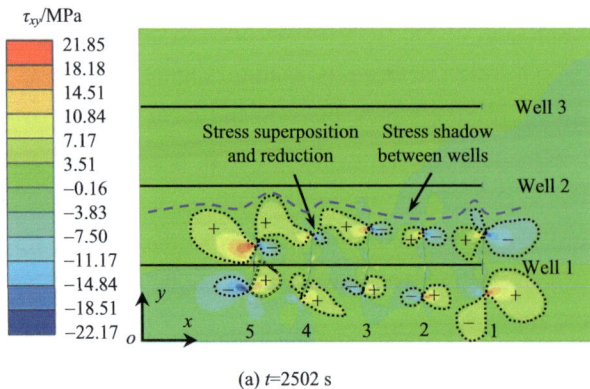

(a) t=2502 s

τ_{xy}/MPa

(b) t=5002 s

τ_{xy}/MPa

(c) t=7502 s

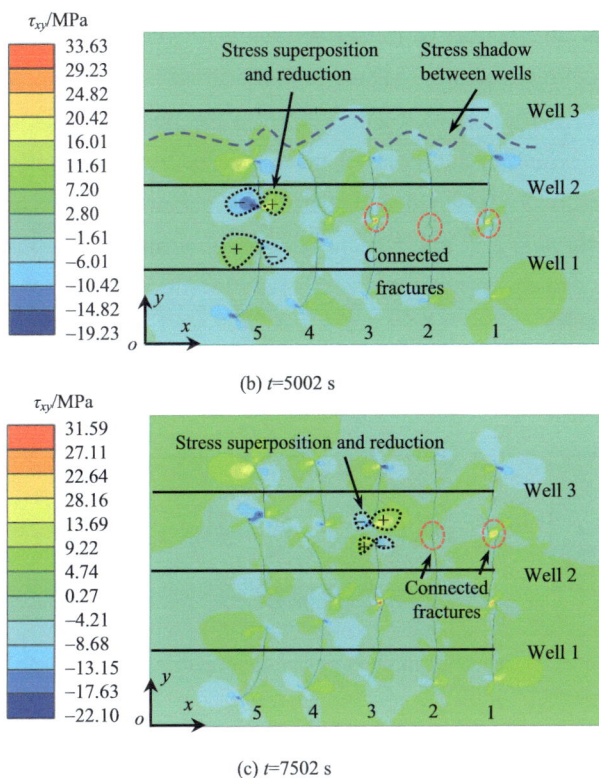

Figure 7.11. Evolution of shear stress τ_{xy} (MPa) in sequential fracturing at 100 m well spacing.

7.4.2.2   Case 1-II: well spacing b=75 m

Figure 7.12 shows the final fracture network morphology and deformation in multi-well hydrofracturing when the well spacing is 75 m. After the propagation of five fractures in well 1, owing to the influence of well 1 on the *in-situ* environment of the reservoir, each fracture in well 2 was connected to well 1, and the fractures were deflected to varying extents in the direction away from well 1. The fractures in well 3 deflected significantly away from well 2.

Figure 7.13 shows the evolution of shear stress τ_{xy} in sequential fracturing at 75 m well spacing, at t=2502 s, t=5002 s, t=7502 s, respectively. Figure 7.13 (a) shows the shear stress results of well 1 after the sequential fracturing is completed. The superposition and reduction of positive and negative shear stress fields around fracture tips result in unequal stresses on both sides of these fracture tips, and the fracture deflects toward the high stress domain, which is similar to Figure 7.13 (a). In Figure 7.13 (b), in the local domains of the two adjacent wells, the superposition and reduction of shear stresses occur (e.g. fractures 5 in wells 1 and 2); and these stress disturbances deflect the fractures. Meanwhile, the stress area (indicated by dashed line) of the disturbed formation in well 2 has covered the positions where the fractures in well 3 propagate, which is the stress shadow area between multiple wells. In Figure 7.13 (c), the fracture 3 in well 3 and well 2 are penetrated and connected, and the deflection of fractures 1, 2, 4, and 5 occurs due to the influence of stress shadows.

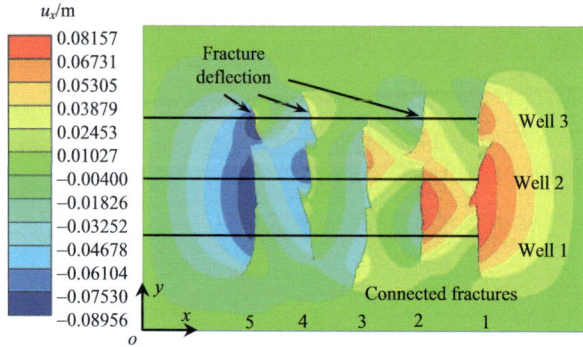

Figure 7.12. Final fracture network morphology and deformation in multi-well hydrofracturing, $b=$ 75 m.

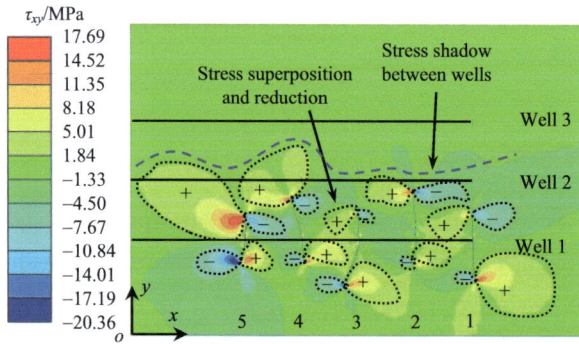

(a) $t=2502$ s

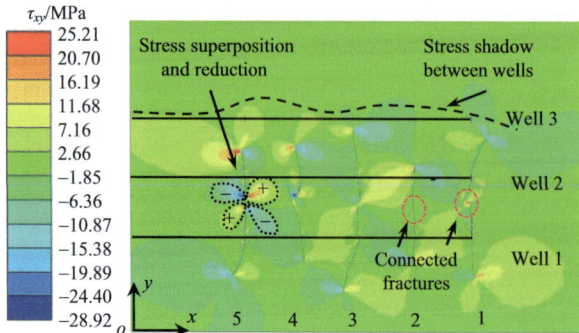

(b) $t=5002$ s

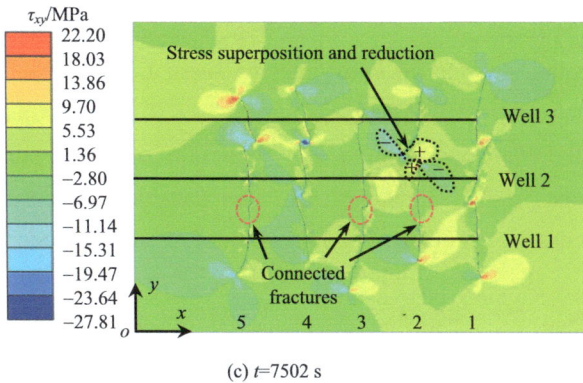

(c) *t*=7502 s

Figure 7.13. Evolution of shear stress τ_{xy} (MPa) in sequential fracturing at 75 m well spacing.

7.4.2.3   Case 1-III: well spacing *b*=50 m

Figure 7.14 shows the final fracture network morphology and deformation in multi-well hydrofracturing when the well spacing is 50 m. Subsequent to the propagation of the five fractures in well 1, the fractures in wells 1 and 2 were connected, causing a narrow well spacing. In well 3, except for fracture 3, which was connected with the adjacent fracture, the other fractures were deflected to varying extents by the fractures in well 2. Compared with the results for well spacings of 100 m and 75 m, continuing to reduce the well spacing will lead to increased fracture connections and penetration.

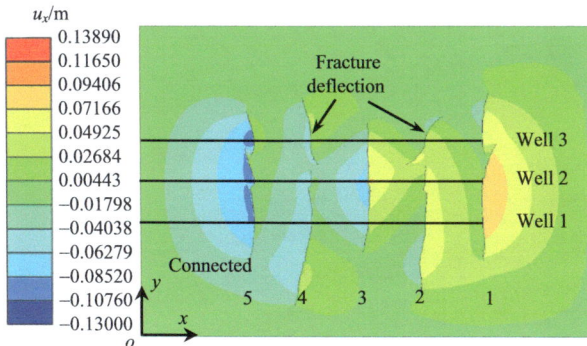

Figure 7.14. Final fracture network morphology and deformation in multi-well hydrofracturing, *b*= 50 m.

Figure 7.15 shows the evolution of shear stress τ_{xy} in sequential fracturing at 50 m well spacing, at *t*=2502 s, *t*=5002 s, *t*=7502 s, respectively. Figure 7.15 (a) shows the shear stress results of well 1 after the sequential fracturing is completed. The superposition and reduction of positive and negative shear stress fields around fracture tips result in unequal stresses on both sides of these fracture tips, and the fracture deflects toward the high stress domain. In Figure 7.15 (b), in the local domains of the two adjacent wells, the superposition and reduction of shear stresses occur (e.g. fractures 3 in wells 1 and 2); and these stress disturbances deflect the fractures. Meanwhile, the stress area (indicated by dashed line) of the disturbed formation in well

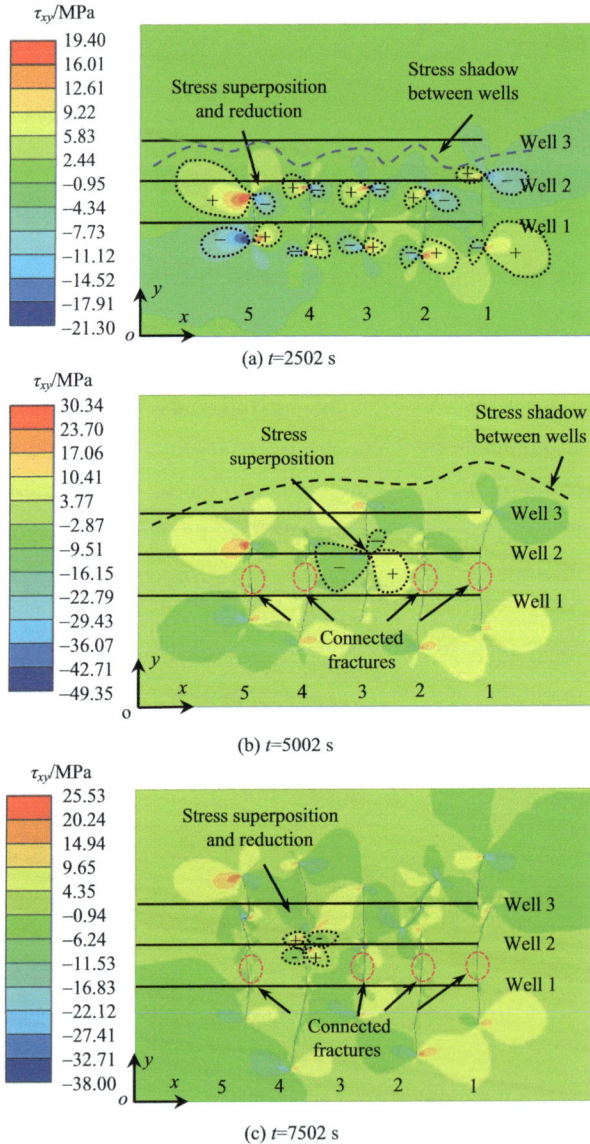

(a) t=2502 s

(b) t=5002 s

(c) t=7502 s

Figure 7.15. Evolution of shear stress τ_{xy} (MPa) in sequential fracturing at 50 m well spacing.

2 has covered the positions where the fractures in well 3 propagate, which is the stress shadow area between multiple wells. In Figure 7.15 (c), the fractures 1, 2, 3, and 5 in well 3 and well 2 are penetrated and connected, and the deflection of fracture 4 occurs due to the influence of stress shadows.

Comparing the fracture network propagation and shear stress evolution of the above different well spacings, with the decrease in well spacing, the stress shadow area between adjacent wells increases; the stress superposition and reduction increase, the fracture propagation is inhibited; the propagation length of the connected fractures decreases, and the deflection extent of the deflection fracture increases. The stress field disturbance generated by the fracture propagation of the previous horizontal well gradually increases the inhibitory effect on the fracture propagation of the subsequent horizontal well.

Comparing the fracture network propagation and shear stress evolution of the above different well spacings, with the decrease in well spacing, the stress shadow area between adjacent wells increases, the stress superposition and reduction increase, the fracture propagation is inhibited, the propagation length of connected fractures decreases, and the deflection degree of the deflection fracture increases. The stress field disturbance generated by the fracture propagation of the previous horizontal well gradually increased the inhibitory effect on the hydraulic fracture propagation of the subsequent horizontal well.

7.4.3   Quantitative analysis of propagation length and volume

To quantitatively analyse the propagation behaviour of the fracture network in the multi-well hydrofracturing model with varying adjacent spacings, the fracture length and volume of the fracture network for different stages and well spacings were derived, as listed in Table 7.4. With the continuous injection of the hydrofracturing fluid, the total length and volume of the fractures increase. For convenience of comparison and analysis Figures 7.16 (a), and (b) display the curves of the length and volume evolution of the fracture network under different well spacings with time. The length of the hydraulic fracture from the initial stage shows a decreasing trend with decreasing well spacing b (b =100 m → 75 m → 50 m) because the reduction in well spacing intensifies the stress shadow between fractures. In contrast, the fracture volume from the initial stage shows an increasing trend with decreasing well spacing b because fractures that are not easy to propagate forward may hold more hydrofracturing fluid. The final total fracture length was the largest when the well spacing was 100 m; the final total fracture length was the smallest when the well spacing was 50 m. Figures 7.17 (a) and (b) show the comparison curves of the final total length and volume of the fracture network with different well spacings. The final total fracture length was the smallest when the well spacing was 50 m and the total fracture length was the largest when the well spacing was 100 m. The fracturing fluid was filtered off in longer fractures, resulting in a larger fracture volume supported by the fracturing fluid at the well spacing of 100 m.

Table 7.4. Fracture length and volume of fracture network for different stages and well spacings.

b /m	Time t /s	Fracture length L /m	Fracture volume V/m^3
100		480.76	193.58
75	Stage 1 (t=2502 s)	529.62	195.12
50		498.00	193.73
100		992.45	371.39
75	Stage 2 (t=5002 s)	903.94	395.10
50		763.11	407.22
100		1438.69	556.38
75	Stage 3 (t=7502 s)	1271.25	595.70
50		1174.75	608.84

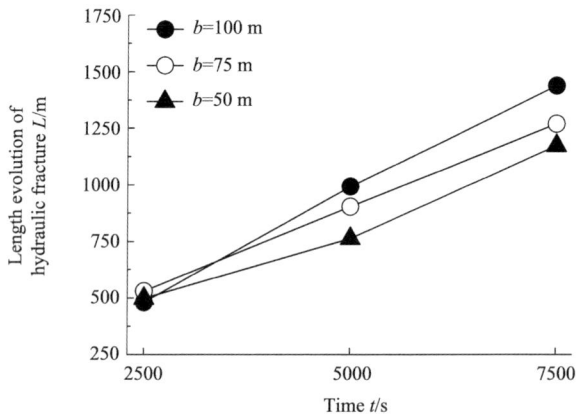

(a) Length evolution of the hydraulic fracture

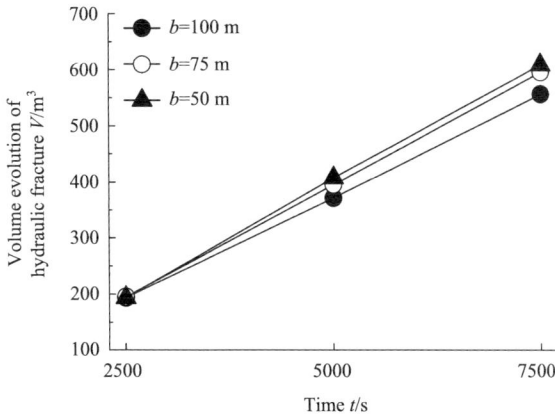

(b) Volume evolution of the hydraulic fracture

Figure 7.16. Length and volume evolution of hydraulic fracture for different stages and well spacings.

The linear influence of well spacing on the propagation length and volume of the fracture network is reflected above, which can provide a reference for the evaluation of fracture networks in multi-well hydrofracturing. However, this evaluation alone is not sufficient; in the future, it is necessary to further investigate the influence of the smaller spacing of multiple wells (the linear relationship may no longer be valid). In addition, the impact of fracture distribution on oil and gas production should be evaluated because under the same fracture length and volume, different distribution forms of the fracture network will produce different oil and gas, as it connects with different reservoir areas. The above is some in-depth research work planned for the future.

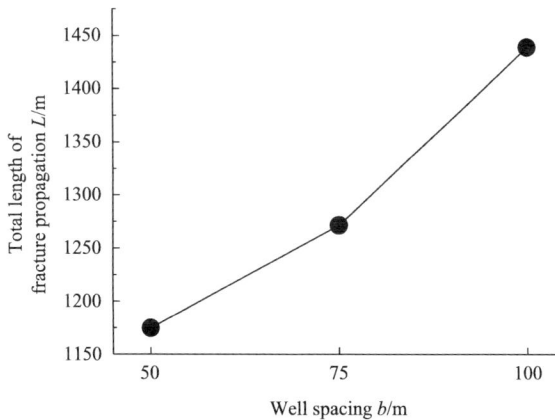

(a) Total length of the hydraulic fracture

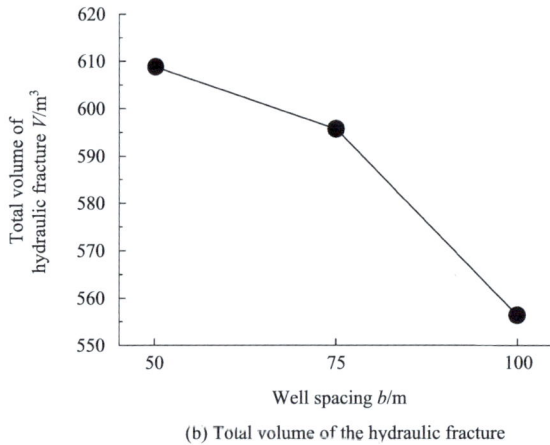

(b) Total volume of the hydraulic fracture

Figure 7.17. Final length and volume of hydraulic fracture for different well spacings.

7.5　Results and analysis of unstable propagation of hydraulic fractures with variable initiation sequences

7.5.1　Fracture network propagation and shear stress shadows

7.5.1.1　Case 2-II: simultaneous fracturing Well 1 → Well 2 → Well 3

Figure 7.18 shows the distribution of hydraulic fractures in simultaneous fracturing, and the displacement (m) in x-direction is shown to detect the deformation in the reservoir. In the simultaneous fracturing process of multiple well perforations from right to left, the first fracture in each well is straight and connected, the subsequent fractures are disturbed and deflected by the first initiation fracture on the right, and the fractures of wells 1 and 3 are deflected to a large extent away from well 2.

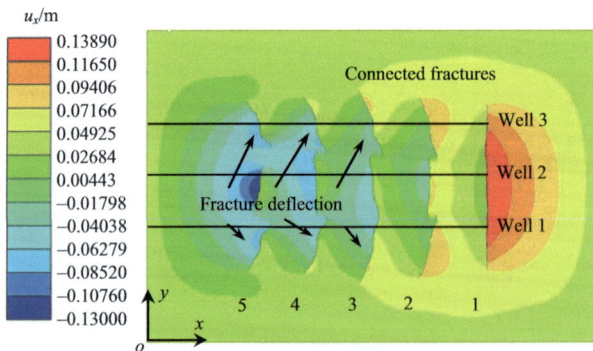

Figure 7.18. Final fracture network morphology and deformation in multi-well simultaneous fracturing.

The deformation field of the rock mass matrix is derived to analyse the reservoir compression in the fracture propagation. Figure 7.19 shows the displacement vector of the reservoir rock mass in the local domain of fracture 1 in wells 1, 2, and 3 during typical multi-well simultaneous fracturing. Compared with the evolution of the displacement vectors of a single fracture (as shown in Figure 7.10), the displacement vectors were primarily distributed in the fracture intersection domain after the three fractures were connected. In the local domain, where the multi-well fractures are connected, a large compressional deformation of the reservoir matrix occurs, which has a severe effect on the deflection disturbance of fracture 2 in the next stage.

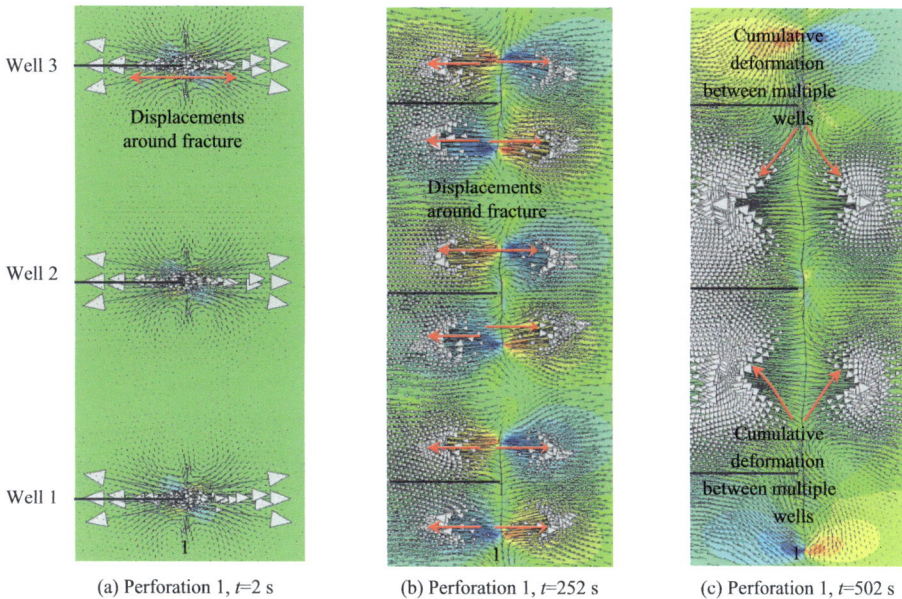

(a) Perforation 1, t=2 s　　　　(b) Perforation 1, t=252 s　　　　(c) Perforation 1, t=502 s

Figure 7.19. Displacement vector evolution of perforations 1 of multiple wells in hydrofracturing

Figure 7.20 shows the result of the shear stress τ_{xy} in simultaneous fracturing, at t=502 s, t=1502 s, t=2502 s, respectively. As shown in Figure 7.20 (a), after the simultaneous fracturing of the first fractures of the three horizontal wells, the three fractures are connected and the shear stress at the fracture tip is almost symmetrical, and the fractures propagate straightly. With subsequent fractures fractured, the superposition and reduction of positive and negative shear stress fields around fracture tips result in unequal stresses on both sides of these fracture tips, and the fracture deflects toward the high stress domain. In Figure 7.20 (b), in the local domains of the two adjacent wells, the superposition and reduction of shear stresses occur (e.g.

fractures 3 in wells 1 and 2); and these stress disturbances deflect the fractures. Meanwhile, the stress area (indicated by dashed line) of the disturbed formation in well 2 has covered the positions where the fractures in wells 1 and 3 propagate, which is the stress shadow area between multiple wells. In Figure 7.20 (c), the shear stress around well 2 is superimposed and reduced with the adjacent domains of wells 1 and 3

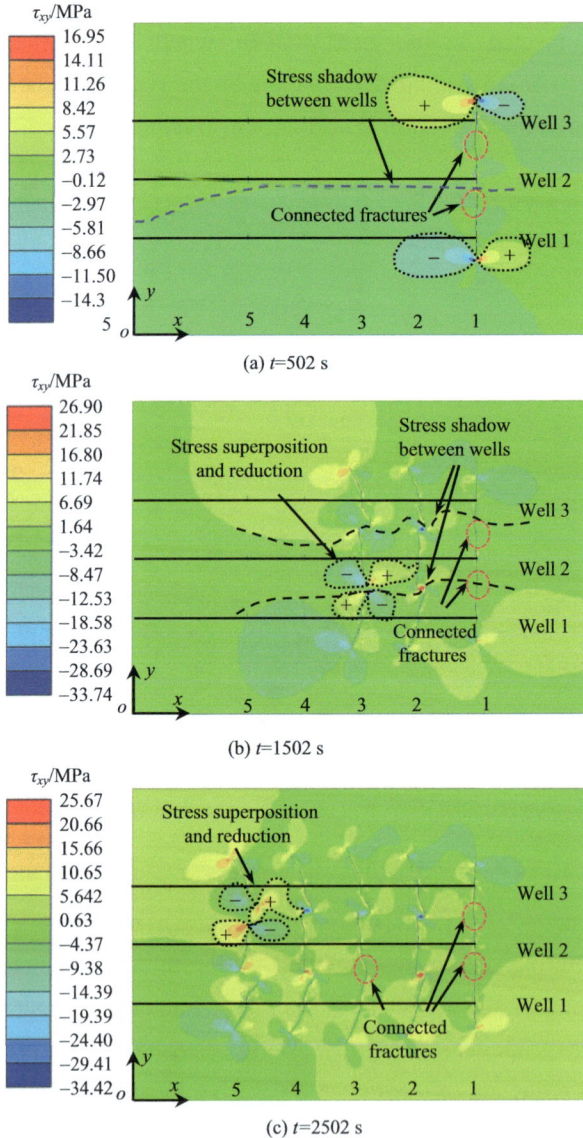

Figure 7.20. Evolution of shear stress τ_{xy} (MPa) in simultaneous fracturing of multiple wells.

simultaneously. The shear stress on both sides of the fractures tip of well 2 is almost symmetrical. The fracture 3 in wells 1 and 2 are penetrated and connected, and the deflection of fractures in wells 1 and 3 occur due to the influence of stress shadows.

7.5.1.2   Case 2-III: alternate fracturing Well 2→ Well 1 →Well 3

Figure 7.21 shows the final fracture network morphology and deformation in multi-well alternate fracturing case 2-III. The impact of well 2 fracturing on the *in-situ* environment of the reservoir is that the fractures of wells 1 and 3 are deflected to a large extent away from well 2, and the first two fractures of each well are connected.

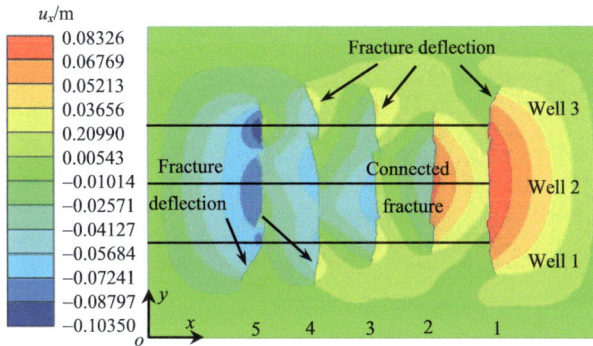

Figure 7.21. Final fracture network morphology and deformation in multi-well alternate fracturing case 2-III.

Figure 7.22 shows the result of the shear stress τ_{xy} in alternate fracturing case 2-III, at t=2502 s, t=5002 s, t=7502 s, respectively. Figure 7.22 (a) shows the shear stress results of well 2 after the sequential fracturing is completed. The superposition and reduction of positive and negative shear stress fields around fracture tips result in unequal stresses on both sides of these fracture tips, and the fracture deflects toward the high stress domain. In Figure 7.22 (b), in the local domains of the two adjacent wells, the superposition and reduction of shear stresses occur (e.g. fractures 4 in wells 1 and 2); and these stress disturbances deflect the fractures. Meanwhile, the stress area (indicated by dashed line) of the disturbed formation in well 2 has covered the positions (the stress shadow area between multiple wells) where the fractures in wells 1 and 3 propagate. In Figure 7.22 (c), the fracture 2 in wells 1, 2, and 3, and fractures 1 and 3 in wells 1 and 2 are penetrated and connected; the deflection of fractures 4 and 5 occurs due to the influence of stress shadows.

(a) t=2502 s

(b) t=5002 s

(c) t=7502 s

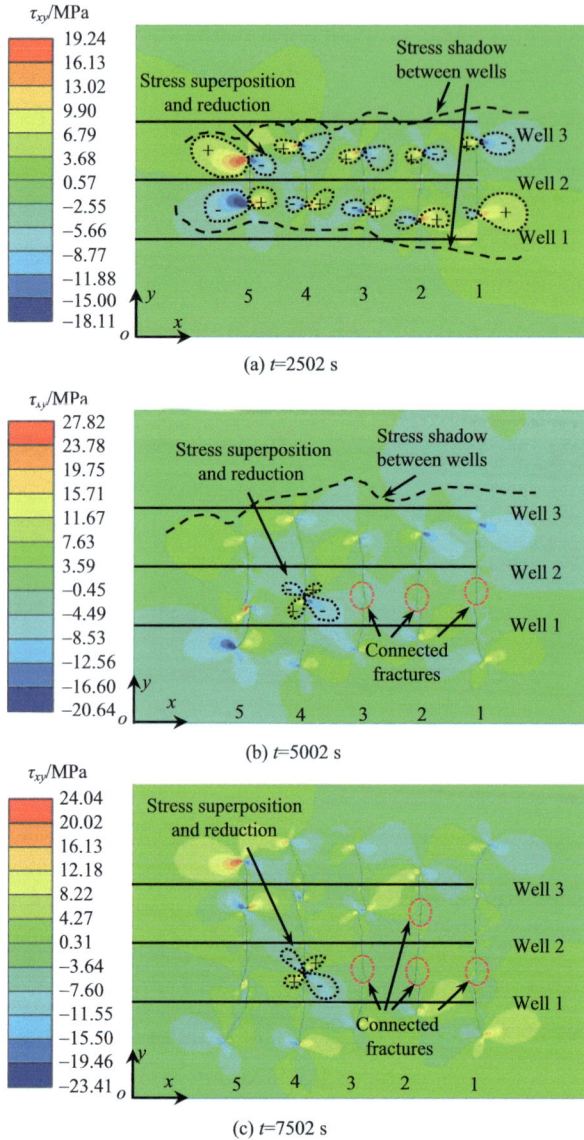

Figure 7.22. Evolution of shear stress τ_{xy} (MPa) in alternate fracturing case 2-III of multiple wells.

7.5.1.3　Case 2-IV: alternate fracturing Well 2 → Well 1 → Well 3

Figure 7.23 shows the final fracture network morphology and deformation in multi-well alternate fracturing case 2-IV. It is observed that the fractures of wells 1 and 3 are only deflected to a small extent away from well 2 owing to the influence of well

2 fracturing on the *in-situ* environment of the reservoir and the cross-staggered distribution of perforation clusters. None of the fractures in each well was connected. Compared with the parallel distribution of perforation clusters, this method can alleviate the unstable propagation behaviour of fractures in multi-well hydrofracturing.

Figure 7.24 shows the result of the shear stress τ_{xy} in alternate fracturing case 2-IV at $t=2502$ s, $t=5002$ s, $t=7502$ s, respectively. Figure 7.24 (a) shows the shear stress results of well 2 after sequential fracturing is completed. The superposition and reduction of the positive and negative shear stress fields around the fracture tips result in unequal stresses on both sides of these fracture tips, and the fracture deflects towards the high-stress domain. In Figure 7.24 (b), in the local domains of the two adjacent wells, superposition of shear stresses occurs (e.g., fractures 2 and 3 in wells 1 and 2 in well 2), and these stress disturbances deflect the fractures. Meanwhile, the stress area (indicated by the dashed line) of the disturbed formation in well 2 covers the positions where the fractures in wells 1 and 3 propagate, which is the stress shadow area between multiple wells. In Figure 7.24 (c), in the local domains of wells 2 and 3, the superposition and reduction of shear stresses occur (e.g., fractures 2 and 3 in wells 3 and 2 in well 2), and the deflection of fractures 1, 3, 4, and 5 in wells 1 and 3 occurs owing to the influence of stress shadows.

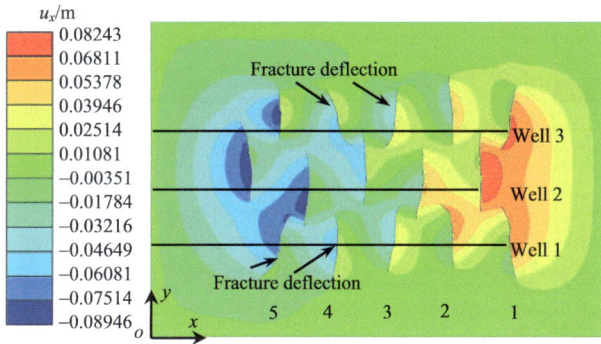

Figure 7.23. Final fracture network morphology and deformation in multi-well alternate fracturing case 2-IV.

7.5.1.4   Case 2-V: alternate fracturing Well 1 → Well 3 → Well 2

Figure 7.25 shows the final fracture network morphology and deformation in multi-well alternate fracturing case 2-V. The influence of well 1 fracturing on the reservoir *in-situ* environment can be observed. During the continuous fracturing of well 3, the distance from well 1 was 150 m (75 m × 2), such that the fractures in well 3 were less disturbed and close to stable propagation. Under the influence of well 3 on the *in-situ* environment of the reservoir, a connection behaviour was observed in many fractures.

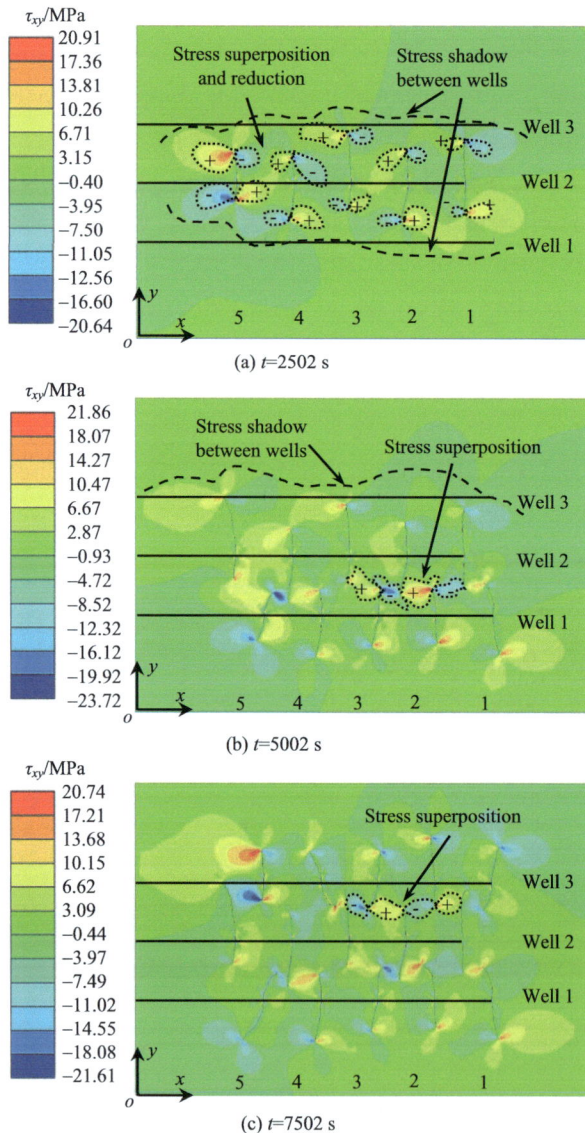

(a) *t*=2502 s

(b) *t*=5002 s

(c) *t*=7502 s

Figure 7.24. Evolution of shear stress τ_{xy} (MPa) in alternate fracturing case 2-IV of multiple wells.

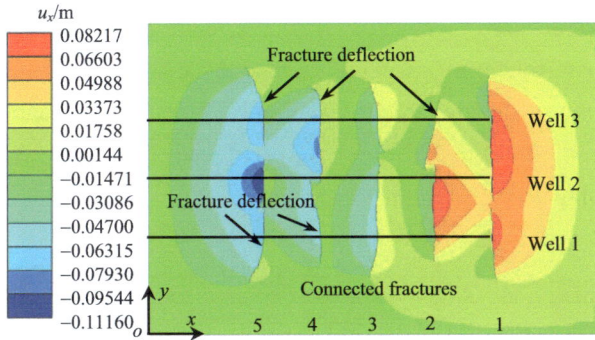

Figure 7.25. Final fracture network morphology and deformation in multi-well alternate fracturing case 2-V.

Figure 7.26 shows the evolution of shear stress τ_{xy} in alternate fracturing case 2-V, at t=2502 s, t=5002 s, t=7502 s, respectively. Figure 7.26 (a) shows the shear stress results of well 1 after the sequential fracturing is completed. The superposition and reduction of positive and negative shear stress fields around fracture tips result in unequal stresses on both sides of these fracture tips, and the fracture deflects toward the high stress domain. In Figure 7.26 (b), in the local domains of the two adjacent wells, the superposition and reduction of shear stresses occur (e.g. fractures 4 in wells 1 and 3); and these stress disturbances deflect the fractures. Meanwhile, the stress area (indicated by dashed line) of the disturbed formation in wells 1 and 3 has covered the positions where the fractures in well 2 propagate, which is the stress shadow area between multiple wells. In Figure 7.26 (c), the fractures 2 and 3 in wells 1 and 2, fractures 1 and 4 in well 2 and well 3 are penetrated and connected, respectively, and the deflection of fractures 2 and 3 occurs due to the influence of stress shadows. The spacing of two wells in the early stage of the multiple wells alternate fracturing case 2-V is 150 m (75 m × 2), which is equivalent to double the original spacing. The stress shadow area around the fracture of the first two wells is small, and the stress disturbance around the fractures of the two wells is weak.

The fracture network propagation and shear stress evolution under different fracture initiation sequences for multiple wells were compared. In multi-well sequential fracturing, the stress field disturbance caused by the previous horizontal well gradually accumulates, and the fracture deflection increases. Stress superposition and reduction occur simultaneously between extended, adjacent, and previous fractures. The stress shadow effect is enhanced, resulting in an increase in fracture deflection and a decrease in fracture propagation length. In the alternate fracturing process, there were more connected fractures, and the fracture interaction was weakened. Alternate fracturing case 2-V increased the spacing between the two wells and reduced the stress shadow area. In alternative fracturing scenario 2, the fracture tips were staggered, stress disturbances were reduced, and single fractures could be fractured and propagated.

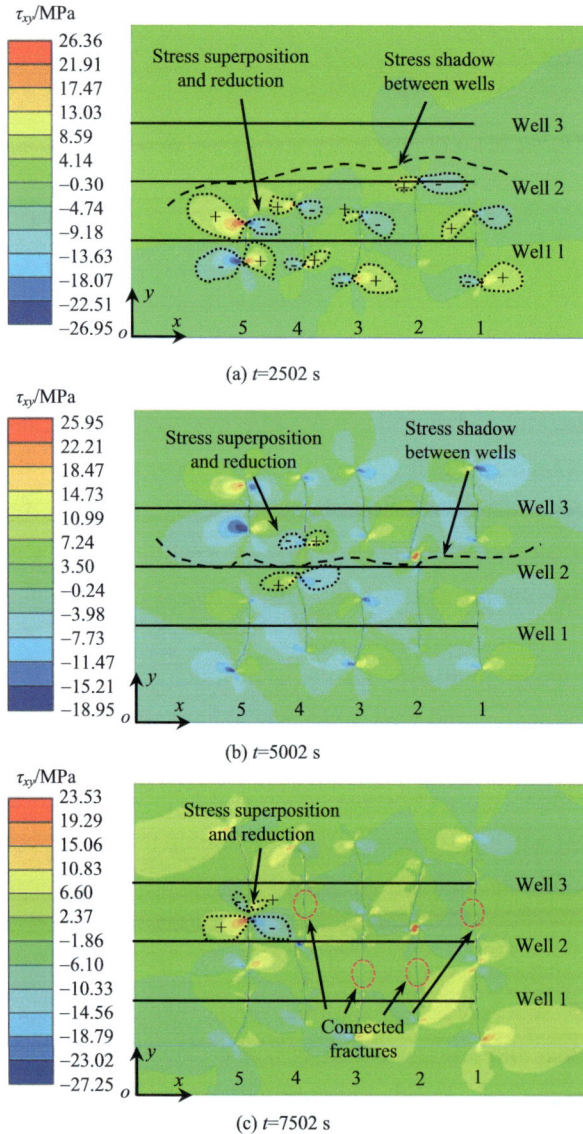

Figure 7.26. Evolution of shear stress τ_{xy} (MPa) in alternate fracturing case 2-V of multiple wells.

7.5.2   Quantitative analysis of propagation length and volume

To quantitatively analyse the fracture network propagation behaviour under variable initiation sequences (sequential, simultaneous, and alternate fracturing), the fracture length and volume of the fracture network for different stages and horizontal well initiation sequences were derived, as listed in Table 7.5. With The continuous injection of fluid increases the total length and volume of the fractures.

Table 7.5. Fracture length and volume of fracture network for different stages and different initiation sequences.

Fracturing scenarios	Time t	Fracture length L /m	Fracture volume V /m^3
2-I, sequential	Stage 1 (t=2502 s)	529.62	195.12
	Stage 2 (t=5002 s)	903.94	395.10
	Stage 3 (t=7502 s)	1271.25	595.70
2-II, simultaneous	Stage 1 (t=502 s)	305.35	129.08
	Stage 2 (t=1502 s)	765.88	374.10
	Stage 3 (t=2502 s)	1204.93	609.72
2-III, alternate	Stage 1 (t=2502 s)	508.63	194.66
	Stage 2 (t=5002 s)	872.73	397.61
	Stage 3 (t=7502 s)	1235.46	598.85
2-IV, alternate	Stage 1 (t=2502 s)	520.63	191.48
	Stage 2 (t=5002 s)	954.02	386.07
	Stage 3 (t=7502 s)	1401.90	576.98
2-V, alternate	Stage 1 (t=2502 s)	510.79	196.52
	Stage 2 (t=5002 s)	998.90	386.61
	Stage 3 (t=7502 s)	1277.50	601.84

For convenience of visual comparison and analysis, Figure 7.27 (a) and (b) show the curves of the length and volume of the fracture network under different initiation sequences with time. In the first stage, with the continuous injection of fracturing fluid, the length and volume of the fracture remained the same under different initiation sequences. Subsequently, the length of the hydraulic fracture generally increased with further injection of fracturing fluid. Figure 7.28 (a) and (b) show the comparison curves of the final total length and total volume of the fracture network under different fracturing scenarios. By comparing different initiation sequences, we can observe that the hydraulic fracture length of simultaneous fracturing was the smallest, which was due to the strong stress shadow effect of simultaneous fracture propagation. The fracture lengths of alternate fracturing cases 2-IV and 2-V were longer because they reduced both the stress shadow area and stress shadow effect between wells. In contrast, the fracture volume of simultaneous fracturing is the largest, because the fractures that are not easy to propagate forward may hold more fracturing fluid. The final total fracture length in alternate fracturing case 2-IV was the largest, and the final total fracture length was the smallest in simultaneous fracturing.

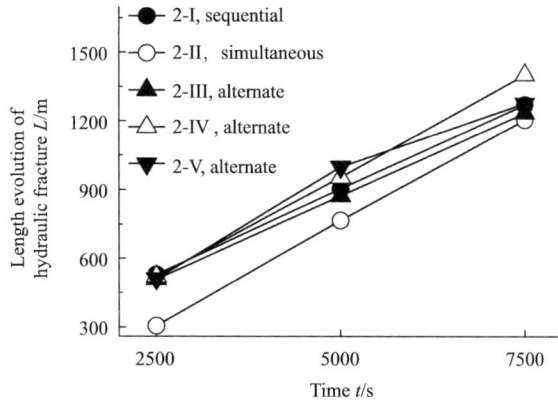

(a) Length evolution of the hydraulic fracture

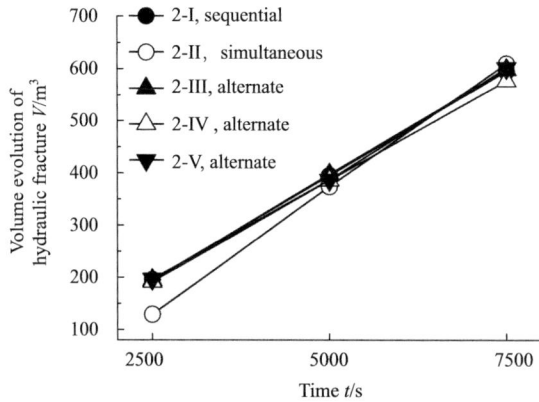

(b) Volume evolution of the hydraulic fracture

Figure 7.27. Length and volume evolution of hydraulic fracture for different stages and different initiation sequences.

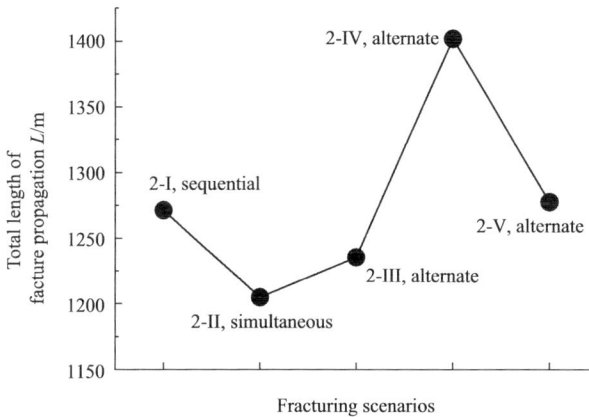

(a) Total length of the hydraulic fracture

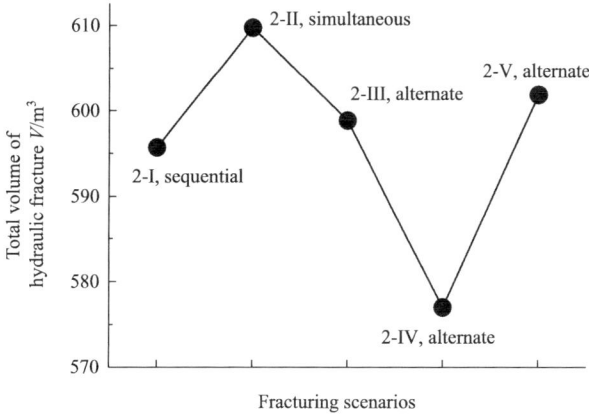

(b) Total volume of the hydraulic fracture

Figure 7.28. Final length and volume of hydraulic fracture for different fracturing scenarios.

The influence of different initiation sequences on the propagation length and volume of the fracture network is reflected above, which can provide a reference for the fracture network evaluation of multi-well hydrofracturing. However, this evaluation is not sufficient; in the future, we plan to investigate the influence of different initiation sequences at other well spacings. In addition, we plan to evaluate the impact of fracture distribution on oil and gas production because different fracture network distribution patterns can produce different oils and gases at the same fracture length and volume because they connect different reservoir zones.

7.6   Conclusions

In this study, a multi-well hydrofracturing model was used to simulate and analyse the hydrofracturing process of multiple horizontal wells under different well spacings and initiation sequences. The unstable propagation and stress evolution of the fracture network under thermal-hydro-mechanical coupling were studied through the disturbed reservoir shear stress field, reservoir displacement vector distribution, fracture propagation length, and volume. The following conclusions were drawn:

(1) In multi-well hydrofracturing, the stress around the fracture interferes with the adjacent fractures in adjacent wells. The shear stress fields around the fractures of the horizontal wells are superimposed, and the fractures are deflected to the side with a larger shear stress; multi-well hydrofracturing leads to fracture connectivity between wells.

(2) Varying well spacing affects the unstable propagation of hydraulic fractures. With a decrease in the well spacing, the disturbance of the stress field and the stress shadow area between wells gradually increase, the number of connected fractures increases, the propagation length of the connected fractures gradually decreases, and the unconnected fractures deflect. The degree of deflection increased, and the well spacing became an important factor affecting fracture propagation in multi-well hydrofracturing.

(3) The analysis of the unstable propagation of multiple well variable initiation sequence fractures shows that the multiple well sequential fracturing gradually accumulates as the horizontal wells are fractured in sequence, resulting in strong stress interference for subsequent fracture initiation, and the deflection gradually increases. Simultaneous fracturing is interfered by the stress caused by the adjacent well and the first fractured fractures, and the subsequent fractures are interfered by stronger stress. The degree of deflection increases in turn, and the length of fracture propagation decreases accordingly. There are more connected fractures in alternate fracturing, and the mutual interference of fractures is weakened. The fracture tips of alternate fracturing case 2-IV (crossed perforation clusters) are staggered with each other, the stress interference is clearly weakened, and a single fracture can be fractured and propagated.

(4) In the quantitative analysis of the length and volume of the fracture network, the total length of the hydraulic fractures decreases with a decrease in well spacing.

The total volume of hydraulic fractures increases with a decrease in well spacing. In this study, when the well spacing was set to 75 m under field conditions, a larger total length and volume of fracture propagation was observed. The total fracture length was larger under alternate fracturing, among them, the total length of hydraulic fractures of alternate fracturing case 2-IV was the largest, and simultaneous fracturing was the smallest; the total volume of fracture propagation of synchronous fracturing was the largest, and alternate fracturing case 2-IV was the smallest. Among different initiation sequences studied in this study, when the initiation sequence was set to alternate fracturing case 2-V, the larger total length and volume of fracture propagation was observed.

The simulation of multi-well hydrofracturing is based on a two-dimensional plane model. It is necessary to establish a three-dimensional space model to further study the propagation behaviour of the fracture network in multi-well hydrofracturing. By comparing the fracture network propagation and stress field disturbance of multiple wells under different working conditions, some hydrofracturing process settings that can improve the complexity of multiple well fracture networks were found, which can provide a reference for optimizing multiple wells, multiple perforations, multiple sequences, and the perforation layout of horizontal wells.

References

Bai, T., Pollard, D.D., Gao, H. (2000), "Explanation for fracture spacing in layered materials", Nature, Vol. 403, pp. 753–756.

Bunger, A.P., Jeffrey, R.G., Kear, J., Zhang, X., Morgan, M. (2011), "Experimental investigation of the interaction among closely spaced hydraulic fractures", 45th U.S. Rock Mechanics / Geomechanics Symposium, San Francisco, California. USA, ARMA-11-318.

Bunger, A.P., Zhang, X., Jeffrey, R.G. (2012), "Parameters affecting the interaction among closely spaced hydraulic fractures", SPE Journal, Vol. 17 No. 1, pp. 292–306.

Chen, X., Li, Y., Zhao, J., Xu, W., Fu, D. (2018), "Numerical investigation for simultaneous growth of hydraulic fractures in multiple horizontal wells", Journal of Natural Gas Science and Engineering, Vol. 51, pp. 44–52.

Duan, K., Li, Y., Yang, W. (2021), "Discrete element method simulation of the growth and efficiency of multiple hydraulic fractures simultaneously-induced from two horizontal wells", Geomechanics and Geophysics for Geo-Energy and Geo-Resources, Vol. 7 No. 1, pp. 1–20.

Ghazal, I., Michael, G., Leonardo, C., Christine, B., Daniel, M., Fu, P.C. (2015), "Fully 3D hydraulic fracturing model: optimizing sequence fracture stimulation in horizontal wells", 49th U.S. Rock Mechanics/Geomechanics Symposium, San Francisco, California, USA, ARMA-2015-119.

He, Q., Suorineni, F. T., Ma, T., Oh, J. (2017), "Effect of discontinuity stress shadows on hydraulic fracture re-orientation", International Journal of Rock Mechanics and Mining Sciences, Vol. 91, pp. 179–194.

He, Y., Yang, Z., Li, X., Song, R. (2020), "Numerical simulation study on three-dimensional fracture propagation of synchronous fracturing", Energy Science & Engineering, Vol. 8 No. 4, pp. 944–958.

Inui, S., Ishida, T., Nagaya, Y., Nara, Y., Chen, Y., Chen, Q. (2014), "AE monitoring of hydraulic fracturing experiments in granite blocks using supercritical CO_2, water and viscous oil", 48th U.S. Rock Mechanics/Geomechanics Symposium, Minneapolis, Minnesota, USA, ARMA-2014-7163.

Jo, H. (2012), "Optimizing fracture spacing to induce complex fractures in a hydraulically fractured horizontal wellbore", SPE Americas Unconventional Resources Conference, Pittsburgh, Pennsylvania, USA. SPE-154930-MS.

Ju, Y., Li, Y., Wang, Y., Yang, Y. (2020), "Stress shadow effects and microseismic events during hydrofracturing of multiple vertical wells in tight reservoirs: a three-dimensional numerical model", Journal of Natural Gas Science and Engineering, Vol. 84, pp. 103–684.

Kumar, D., Ghassemi, A. (2016), "A three-dimensional analysis of simultaneous and sequential fracturing of horizontal wells", Journal of Petroleum Science and Engineering, Vol. 146, pp. 1006–1025.

Li, S., Zhang, D. (2018), "A fully coupled model for hydraulic-fracture growth during multiwell-fracturing treatments: enhancing fracture complexity", SPE Production & Operation, Vol. 33 No. 2, pp. 235–250.

Lu, C., Guo, J. C., Liu, Y. X., Yin, J., Deng, Y., Lu, Q. L., Zhao, X. (2015), "Perforation spacing optimization for multi-stage hydraulic fracturing in Xujiahe formation: a tight sandstone formation in Sichuan Basin of China", Environmental Earth Sciences, Vol. 73 No. 10, pp. 5843–5854.

Liu, X., Rasouli, V., Guo, T., Qu, Z., Sun, Y., Damjanac, B. (2020), "Numerical simulation of stress shadow in multiple cluster hydraulic fracturing in horizontal wells based on lattice modelling", Engineering Fracture Mechanics, Vol. 238, 107278.

Manchanda, R., Sharma, M.M. (2013), "Time-delayed fracturing: a new strategy in multi-stage, multi-well pad fracturing", SPE Annual Technical Conference and Exhibition, New Orleans, Louisiana, USA, SPE-166489-MS.

Manchanda, R., Sharma, M.M. (2014), "Impact of completion design on fracture complexity in horizontal shale wells", SPE Drilling & Completion, Vol. 29 No 1, pp. 78–87.

Manríquez, A. L. (2018), "Stress behavior in the near fracture region between adjacent horizontal wells during multistage fracturing using a coupled stress-displacement to hydraulic diffusivity model", Journal of Petroleum Science and Engineering, Vol. 162, pp. 822–834.

Nagel, N., Zhang, F., Sanchez-Nagel, M., Lee, B., Agharazi, A. (2013), "Stress shadow evaluations for completion design in unconventional plays", SPE Unconventional Resources Conference Canada, Calgary, Alberta, Canada, SPE-167128-MS.

Olson, J.E., Taleghani, A.D. (2009), "Modeling simultaneous growth of multiple hydraulic fractures and their interaction with natural fractures", SPE Hydraulic Fracturing Technology Conference, The Woodlands, Texas, USA, SPE-119739-MS.

Qiu, F., Porcu, M.M., Xu, J., Malpani, R., Pankaj, P., Pope, T.L. (2015), "Simulation study of zipper fracturing using an unconventional fracture model", SPE/CSUR Unconventional Resources Conference, Calgary, Alberta, Canada, SPE-175980-MS.

Roussel, N.P., Sharma, M.M. (2011), "Optimizing fracture spacing and sequencing in horizontal-well fracturing", SPE Production & Operations, Vol. 26 No. 2, pp. 173–184.

Rutqvist, J., Barr, D., Datta, R., Gens, A., Millard, A., Olivella, S., Tsang, C.F., Tsang, Y. (2005), "Coupled thermal–hydrological–mechanical analyses of the Yucca Mountain Drift Scale Test—Comparison of field measurements to predictions of four different numerical models", International Journal Rock Mechanics and Mining Sciences, Vol. 42 No. 5–6, pp. 680–697.

Rafiee, M., Soliman, M.Y., Pirayesh, E. (2012), "Hydraulic fracturing design and optimization: a modification to zipper frac", SPE Annual Technical Conference and Exhibition, San Antonio, Texas, USA, SPE-159786-MS.

Sobhaniaragh, B., Mansur, W.J., Peters, F.C. (2018), "The role of stress interference in hydraulic fracturing of horizontal wells", International Journal of Rock Mechanics and Mining Sciences, Vol. 106, pp. 153–164.

Segatto, M., Colombo, I. (2011), "Use of reservoir simulation to help gas shale reservoir estimation", International Petroleum Technology Conference, Bangkok, Thailand, IPTC-14798-MS.

Tian, W., Li, P., Dong, Y., Lu, Z., Lu, D. (2019), "Numerical simulation of sequential, alternate and modified zipper hydraulic fracturing in horizontal wells using XFEM", Journal of Petroleum Science and Engineering, Vol. 183, pp. 1–12.

Tsang, C.F., Stephansson, O., Hudson, J.A. (2000), "A discussion of thermos-hydro-mechanical (THM) processes associated with nuclear waste repositories", International Journal of Rock Mechanics and Mining Sciences, Vol. 37 No. 1–2, pp. 397–402.

Wang, Y., Li, X., Wang, J.B., Zheng, B., Zhang, B., Zhao, Z.H. (2015), "Numerical modeling of stress shadow effect on hydraulic fracturing", Natural Gas Geoscience, Vol. 26 No. 10, pp. 1941–1950.

Wang, Y., Ju, Y., Yang, Y., (2018), "Adaptive finite element-discrete element analysis for microseismic modelling of hydraulic fracture propagation of perforation in horizontal well considering pre-existing fractures", Shock and Vibration, Vol. 2018, 2748408

Wang, Y., Ju, Y., Chen, J., Song, J. (2019), "Adaptive finite element–discrete element analysis for the multistage supercritical CO_2 fracturing and microseismic modelling of horizontal wells in tight reservoirs considering pre-existing fractures and thermal-hydro-mechanical coupling", Journal of Natural Gas Science and Engineering, Vol. 61, pp. 251–269.

Wang, Y., Ju, Y., Zhang, H., Gong, S., Song, J., Li, Y., Chen, J. (2021), "Adaptive finite element-discrete element analysis for the stress shadow effects and fracture interaction behaviours in three-dimensional multistage hydrofracturing considering varying perforation cluster spaces and fracturing scenarios of horizontal wells", Rock Mechanics and Rock Engineering, Vol. 54 No. 4, pp. 1815–1839.

Wang, T.Y., Tian, S.C., Zhang, W.H., Ren, W.X., Li, G.S. (2020), "Production model of a fractured horizontal well in shale gas reservoirs", Energy & Fuels, Vol. 35 No. 1, pp. 493–500.

Warpinski, N.R. (2000), "Analytic crack solutions for tilt fields around hydraulic fractures", Journal of Geophysical, Vol. 105 No. B10, pp. 23463–23478.

Wong, S.W., Geilikman, M., Xu, G. (2013), "Interaction of multiple hydraulic fractures in horizontal wells", SPE Unconventional Gas Conference and Exhibition, Muscat, Oman, SPE-163982-MS.

Wu, K., Olson, J.E. (2015), "Simultaneous multifracture treatments: fully coupled fluid flow and fracture mechanics for horizontal wells", SPE Journal, Vol. 20 No. 2, pp. 337–346.

Wu, K., Olson, J.E. (2017), "Numerical investigation of complex hydraulic-fracture development in naturally fractured reservoirs", SPE Production & Operations, Vol. 31 No. 4, pp. 300–309.

Wu, K., Wu, B., Yu, W. (2018), "Mechanism analysis of well interference in unconventional reservoirs: insights from fracture-geometry simulation between two horizontal wells", SPE Production & Operations, Vol. 33 No. 1, pp. 12–20.

Yu, W., Sepehrnoori, K. (2013), "Optimization of multiple hydraulically fractured horizontal wells in unconventional gas reservoirs", Journal of Petroleum Engineering, Vol. 2013, 151898.

Chapter 8   Unstable propagation of multiple three-dimensional hydraulic fractures and shear stress disturbance in heterogeneous reservoirs

8.1   Introduction

Shale gas reservoirs are widely distributed and contain large reserves, and their exploration and development potential exceed that of conventional and unconventional natural gas. Shale gas has a great commercial value and has become a global hotspot of exploration. However, shale gas reservoirs have strong compactness and poor porosity and permeability. Thus, it is difficult to realise economic and effective development using conventional hydrofracturing. Years of field practice have proven that multistage fracturing of oil and gas reservoirs is an effective means of developing such reservoirs and increasing the production of oil and gas fields (Bazant *et al.*, 2014; Wu *et al.*, 2016). The unstable dynamic propagation diagram of three-dimensional (3D) multiple hydraulic fractures in a heterogeneous reservoir is shown in Figure 8.1. The figure shows that during multistage fracturing, horizontal wells are drilled in heterogeneous reservoirs and perforation clusters are set, where the heterogeneity includes hard rock particles, joints, weak planes, and natural micro-fractures. After hydrofracturing, multiple hydraulic fractures begin to propagate and form a spatial fracture network. These fractures contribute more reservoir volume and improve reservoir permeability and oil and gas production. Therefore, the propagation form of the spatial fracture network has an important impact on the final oil and gas production. Therefore, it is necessary to study the propagation mechanism and form of the spatial fracture network. In the actual fracturing process of a deep tight reservoir rock mass, the fracturing fractures often form unstable and spatial deflection propagation of multiple fractures owing to the influence of various factors such as perforation cluster space, fracturing sequence, and reservoir heterogeneity. These factors restrict the manual transformation and optimisation of fracturing fracture networks.

The fracture propagation direction of multistage fracturing does not always follow the perforation direction but exhibits a spatial deflection behaviour. The deflection of the fracture direction may cause premature water breakthrough of production wells along fracture direction, preventing the actual production performance to achieve the design effect and introducing risks to oilfield development. However, deflection of the fracture direction can also increase reservoir volumes and improve production. The direction of the hydraulic fracture mainly depends on the *in-situ* stress state of the reservoir, while its geometric size is affected by the rock mechanical properties and construction parameters. The mechanism of fracture steering propagation has been widely studied, and the influence mechanism of perforation on hydrofracturing has been analysed using a numerical simulation method (Shan *et al.*, 2017). Based on continuous-discontinuous numerical models and cubic law, a fracture propagation model was established, and the flow pressure distribution and fracture initiation and propagation process of porous materials under hydrofracturing were studied (Yan and Jiao, 2018). Through triaxial hydrofracturing experiments, the intersection behaviour of hydraulic and natural fractures was studied, and the effects of natural fracture shear strength and *in-situ* stress on fracture intersection behaviour were analysed (Zhou *et al.*, 2008, 2010; Cheng *et al.*, 2015a, b). Triaxial hydrofracturing experiments were also conducted on large-scale shale rocks to study the effects of fracturing fluid injection rate, fracturing fluid viscosity, horizontal *in-situ* stress difference, and perforation parameters on fracture initiation and deflection; moreover, the studies have shown that reservoirs can experience the effect of complex fracture networks through hydrofracturing (Guo *et al.*, 2014).

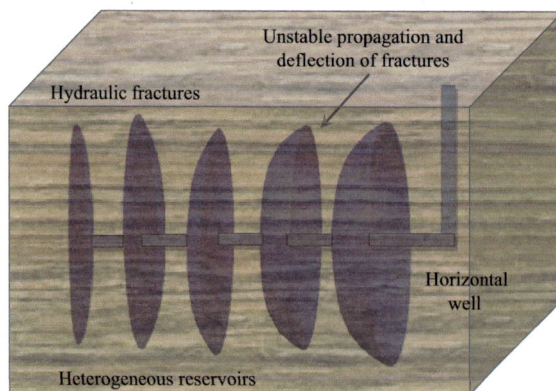

Figure 8.1. Schematic diagram of unstable dynamic propagation of 3D multiple hydraulic fractures in heterogeneous reservoirs.

An important factor contributing to fracture deflection is stress shadow effects. It is related to the stress redistribution area caused by water pressure inside the fracture that produces an induced stress field in the adjacent area, which superimposes with the original stress field, resulting in stress redistribution around the fracture. The adjacent fracture can deflect when it propagates to this stress redistribution area. The stress shadow effect is an important factor in controlling the spatial deflection behaviour of fractures and plays an important role in the development of hydraulic fracture networks. The stress shadow includes the local compressive stress area perpendicular to the fracture surface near the fracture centre. This causes the direction of the maximum stress to be reoriented in the stress shadow area. By determining the location of the next treatment in this area, the fracture growth even become parallel to the wellbore axis. Therefore, the fracture space must be optimised to obtain the maximum number of fractures perpendicular to the wellbore (Roussel and Sharma, 2011a). Conversely, other studies showed that the inhomogeneity of stimulation measures may also be related to the stress shadow produced by the development of hydraulic fractures (Germanovich et al., 1997; Abass et al., 2009). Stress shadows may also include effects from previously placed hydraulic fractures, occasionally resulting in the deflection of subsequent fracture paths (Roussel and Sharma, 2011b; Bunger et al., 2012). Many studies have demonstrated that accurate geomechanical information about rocks and their changes is also important, because stress is the main factor that controls the initiation and development of faults (Abousleiman et al., 2007). The stress shadow effect between multistage hydrofracturing fractures affects the fracture network and fracturing (Bai et al., 2007; Steacy et al., 2005; Taghichian et al., 2014; Yoon et al., 2015; He et al., 2017; Gutierrez et al., 2019).

The perforation cluster space and initiation sequence between multiple fractures are factors that influence stress shadows. For low-permeability unconventional oil and gas reservoirs, in order to pursue a better fracture spatial propagation form, the perforation cluster space is becoming increasingly denser. A smaller perforation cluster space will aggravate the stress shadow effect between fractures and help to form a complex fracture network to improve the reservoir conductivity. However, the stress shadow effect may also cause difficulty in fracture initiation and propagation to the formation of sand plugging. Therefore, the perforation cluster space of horizontal wells in tight rock reservoirs is the key factor affecting the fracture interaction behaviour in oil production (Bazant et al., 2014; Zhang and Jefrey, 2012). Theoretical studies have shown that when the conductivity of the main fracture reaches a certain value, increasing the number of perforation clusters or decreasing the cluster space has little effect on the final stimulation effect (Cipolla *et al.*, 2010). Therefore, the number of perforating clusters should be properly controlled. The displacement discontinuity method was used to establish the mathematical model of multistage fracturing in horizontal wells, analyse the stress field of multiple cluster fractures, and study the inter-fracture interference to optimise the cluster space. Other studies have discussed the coupling of pore pressure and stress in the design of dense hydrofracturing with adjacent fractures to optimise multistage treatment and hydraulic fracture space (Olson and Taleghani, 2009; Manríquez *et al.*, 2017; Roussel *et al.*, 2012; Sobhaniaragh *et al.*, 2018b) During hydrofracturing, the fracturing sequence generally includes sequential, alternate, and simultaneous fracturing. Different fracturing sequences produce different stress shadow effects. In simultaneous fracturing, each perforation cluster starts fracturing at the same time, resulting in the strongest stress shadow effect. Therefore, the fracturing sequence will also affect the final shape of the fracture spatial propagation by affecting the stress shadow effect. It is difficult to control and optimise the formation of the fracture network when the influencing factors and mechanism of perforation cluster space and fracturing sequence cannot be well-controlled in the model. To overcome this, it is very important to carry out research on reservoir heterogeneity. Owing to the complexity of the geological environment, reservoirs are generally heterogeneous. Understanding reservoir heterogeneity mainly focuses on porosity, physical properties, characteristics of organic matter, and mechanical properties. This heterogeneity affects the hydraulic fracture propagation and stress shadow effect, and, as a result, the final shape of the fracture spatial propagation. Understanding the influencing factors and mechanism of reservoir heterogeneity is of vital importance for optimisation of the formation of the fracture network and requires further studies. At present, some progress has been made in research on reservoir heterogeneity. Studies on lacustrine delta reservoir have shown that the pore structure has an important impact on the heterogeneity of the reservoir, referred to as micro-heterogeneity where the bound water is mainly distributed in the primary pores (Marco and Moraes, 1991). After studying the micro-heterogeneity of reservoirs, it was found that carbonate cement has an important impact, mainly reflected in sandstone reservoirs, which is closely related to the failure porosity (Sylvia and Luiz, 2000). This hinders the migration of oil and gas in the reservoir, resulting in more prominent reservoir heterogeneity, especially in oil and gas accumulations. Long-term studies have shown that the formation of calcite cement is usually accompanied by diagenesis, which has a certain impact on the permeability of the reservoir and hinders oil and gas migration in the reservoir (Sylvia and Luiz, 2003).

Several field observations, laboratory experiments, and theoretical models have been developed to analyse the spatial deflection and stress shadow effects among multiple hydraulic fractures (Sarmadivaleh and Rasouli, 2014; Lecampion and Desroches, 2015; Ding *et al.*, 2017). However, these methods have great limitations: it is difficult to implement complete monitoring in the field, indoor experiments cannot carry out large-scale physical models, while theoretical models cannot analyse the coupling problem between multistage fracture propagation and multiple physical fields. Therefore, additional numerical methods and models have been proposed to study these complex effects (Sobhaniaragh *et al.*, 2018a; Yu *et al.*, 2015; Hossain and Rahman, 2008). The influence of the stress shadow effect and other fracturing parameters on fracture interaction behaviour was studied using the extended finite element method (XFEM) (Gutierrez *et al.*, 2019; Saberhosseini *et al.*, 2019). Other studies have combined the XFEM and discrete element method (DEM) to study the propagation law of hydraulic fractures in porous media containing natural fracture blocks (Ghaderi *et al.*, 2018). The finite element method (FEM) (Manríquez, 2018) or combined finite element-discrete element method (FE-DE method or FDEM) were used to simulate fracture propagation (Paluszny *et al.*, 2013; Lisjak *et al.*, 2018; Munjiza *et al.*, 1995; Profit *et al.*, 2015; Wang *et al.*, 2018a, 2019), based on the adaptive mesh refinement techniques (Wang *et al.*, 2018b; Wang, 2021b). Several of these methods have developed analysis models with high-precision stress solutions in hydraulic fracture propagation. To simulate and analyse the fracture propagation behaviour and stress field evolution in hydrofracturing, numerical models should be flexibly built according to the specific conditions of the research object (Wong *et al.*, 2013; Yu and Sepehrnoori, 2013). The stress shadow size and pore size of unconventional shale in hydrofracturing were analysed by combining the FEM of a simple fracture geometry with an analytical solution of 2D hydraulic fractures (Taghichian *et al.*, 2014). In addition, a 2D displacement discontinuity model (Kresse *et al.*, 2013) and linear elastic fracture mechanics method (Kumar and Ghassemi, 2016) were developed to investigate the stress shadows and fracture propagation. We have previously established 3D numerical models combined with the FE-DE method to study the stratal movement and stress shadow effects during the propagation process of multiple hydraulic fractures in homogeneous reservoirs (Wang, 2021b; Wang *et al.*, 2021). In this study, these strategies and procedures are extended to an improved fracturing sequence, perforation cluster space, reservoir heterogeneity, and stress shadow effects on the unstable propagation of hydraulic fractures.

The remainder of this chapter proceeds as follows. In Section 2, a numerical method of fracturing considering hydro-mechanical coupling is introduced, and the corresponding governing equations and models are presented (i.e. the mechanical equation of porous rock, the liquid seepage equation of porous rock seepage, the liquid network equation of fluid flow in the fracture area, fracturing fluid leak-off, fracture criterion, numerical discretization scheme of governing equation, local remeshing, and coarsening strategy of fracture propagation). Section 3 introduces the 3D numerical model and modelling process, the fracturing sequence is set as sequential, alternate, and simultaneous fracturing, two types of material models of homogeneous and heterogeneous reservoirs are introduced, two improved alternate fracturing methods are added at 25 m intervals, and a global programme of 3D multistage fracturing. Section 4 presents the results and discussion of the 3D fracture network, shear stress field evolution, fracture area, and volume. Finally, Section 5 summarises the main conclusions of this study.

8.2　Combined finite element-discrete element method for hydrofracturing

8.2.1　Geomechanical equations in hydrofracturing considering hydro-mechanical coupling

Bellow we introduce the governing equations for the related solid deformation and fluid flow in porous rocks and fracture networks.

The mechanical governing equation for the solid stress field is:

$$\boldsymbol{L}^{\mathrm{T}}(\boldsymbol{\sigma}^{\mathrm{e}} - \alpha \boldsymbol{m} p_{\mathrm{s}}) + \rho_{\mathrm{b}} \boldsymbol{g} = \boldsymbol{0}, \tag{8.1}$$

where $\boldsymbol{L}$ is the spatial differential operator, $\boldsymbol{\sigma}^{\mathrm{e}}$ is the effective stress tensor, α is Biot's coefficient, $\boldsymbol{m}$ is the identity tensor, p_{s} is the pore fluid pressure in the rock formation, ρ_{b} is the wet bulk density, and $\boldsymbol{g}$ is the gravity vector. The meanings and values of the physical parameters in the governing equations are listed in Table 8.1. The fracture propagation criteria for solids are used in this study, and the details can be found in previous study (Wang, 2021a; Wang *et al.*, 2021). Once the fracture criterion is met, the hydraulic fracture begins to initiate or propagate by the separation of FE node. Once three adjacent FE nodes simultaneously satisfy the fracture criteria, the nodes will separate and one surface of new 3D fracture will form. Actually, the node separation and detection techniques of the classical DEM were used in this study. When fracture is predicted at failure by the fracture criteria, the solid is ruptured upon separation at the node.

The governing equations for liquid seepage and fracture fluid flow are given as follows:

$$\mathrm{div}\left[\frac{k}{\mu_1}(\nabla p_1 - \rho_1 g)\right] = \left(\frac{\phi}{K_1} + \frac{\alpha - \phi}{K_s}\right)\frac{\partial p_1}{\partial t} + \alpha\frac{\partial \varepsilon_v}{\partial t}, \tag{8.2}$$

$$\frac{\partial}{\partial x}\left[\frac{k^{\mathrm{fr}}}{\mu_n}(\nabla p_n - \rho_{\mathrm{fn}}g)\right] = S^{\mathrm{fr}}\frac{\mathrm{d}\,p_n}{\mathrm{d}\,t} + \alpha\left(\Delta \dot{e}_\varepsilon\right), \tag{8.3}$$

where k is the intrinsic permeability of the porous media; μ_1 is the viscosity of the pore liquid; p_1 is the pore liquid pressure; ρ_1 is the density of the pore liquid; ϕ is the porosity of the porous media; K_1 is the bulk stiffness of the pore liquid; K_s is the bulk stiffness of the solid grains; ε_v is the volumetric strain of the porous media; k^{fr} is the intrinsic permeability of the fractured region (according to the parallel plate theory, the intrinsic permeability k^{fr} of a fractured region is related to the fracture aperture e by $k^{\mathrm{fr}} = \frac{e^2}{12}$); μ_n is the viscosity of the fracturing fluid; p_n is the fracturing fluid pressure; ρ_{fn} is the density of the fracture fluid; S^{fr} is the storage coefficient (which is an effective measure of the compressibility of the fractured region when a fluid is present); and $\Delta \dot{e}_\varepsilon$ is the aperture strain rate. There is a term $\alpha\left(\Delta \dot{e}_\varepsilon\right)$ at the right end of the above fluid flow equation for describing the fluid in the fracture network, which considers the influence of solid deformation with time.

Table 8.1. Basic physical parameters for solid and fluid in reservoirs for reservoir properties and hydrofracturing settings.

Parameter	Value
Vertical *in-situ* stress (z direction) S_v /MPa	60
Horizontal minimum *in-situ* stress (y direction) S_h /MPa	40
Horizontal maximum *in-situ* stress (x direction) S_H /MPa	46
Fluid injection rate Q/ (m^3/s)	0.5
Leak-off coefficient C_1 / (m^3/s$^{1/2}$)	1×10^{-16}

	Continued
Parameter	Value
Leak-off coefficient C_{II} /($m^3/s^{1/2}$)	1×10^{-16}
Pore pressure p_s /MPa	15
Biot's coefficient α	0.75
Elastic modulus E /GPa	38
Poisson's ratio v	0.15
Porosity ϕ	0.05
Permeability k /nD	50
Tensile strength σ_t /MPa	1.5
Fracture energy G_f /(N·m)	25
Gravity g/(m/s^2)	9.81
Friction angle ψ /(°)	45
Cohesion c /MPa	25
Density ρ_b / (kg/m^3)	2.71×10^3
Dynamic viscosity coefficient of the pore fluid μ_g /(Pa·s)	1.00×10^{-3}
Dynamic viscosity coefficient of the fracturing fluid μ_n /(Pa·s)	1.67×10^{-3}
Liquid density of the pore fluid ρ_g /(kg/m^3)	1×10^3
Liquid density of the fracturing fluid ρ_{fn} /(kg/m^3)	1×10^3
Bulk modulus of the pore fluid K_g /MPa	2050
Bulk modulus of the fracturing fluid K_f^{fr} /MPa	2000

8.2.2   Leak-off of fracturing fluid

The Carter leak-off model and bulk flow rate (Carter, 1957) were utilised for considering the proppant. According to the experimental results of fluid loss (Williams, 1970), the model assumes an initial volume loss V_{sp} per unit area over a spurt time t_{sp} followed by a two-term leak-off coefficient C. The formulas are as follows:

$$t - t_{exp} < t_{sp}, \quad q_1 = \frac{V_{sp}}{t_{sp}}, \tag{8.4}$$

$$t - t_{exp} \geqslant t_{sp}, \quad q_1 = \frac{C}{\sqrt{t - t_{sp}}}, \tag{8.5}$$

with the two-term parameters given by:

$$C = \frac{2C_{\mathrm{I}} C_{\mathrm{II}}}{C_{\mathrm{I}} + \sqrt{C_{\mathrm{I}}^2 - 4C_{\mathrm{II}}^2}}, \quad C_{\mathrm{I}} = \left[\frac{k_{\mathrm{f}} \phi_{\mathrm{f}} \Delta p}{2\mu_{\mathrm{f}}} \right]^{0.5}, \quad C_{\mathrm{II}} = \left[\frac{k_{\mathrm{r}} \phi_{\mathrm{r}} c_{\mathrm{T}}}{\mu_{\mathrm{r}}} \right]^{0.5} \Delta p, \quad (8.6)$$

where t is the current time, t_{exp} is the time at which a fracture surface is exposed for leak-off, and q_{l} is the one-dimensional normal leak-off velocity, which can be found in the reference (Profit *et al.*, 2015).

8.2.3   Poroelastic effective medium model

This study investigates the influence of reservoir heterogeneity on the propagation behaviour of hydraulic fractures. Heterogeneous reservoir modelling is based on a poroelastic effective medium model (Sriram *et al.*, 2014; Hornby *et al.*, 1994; Ebigbo *et al.*, 2016; Terry and Knapp, 2018), and the heterogeneity includes hard rock particles, joints, weak planes, and natural micro-fractures. This poroelastic effective medium model contains heterogeneity which is equivalent to the homogenised reservoir. A schematic diagram of transforming a heterogeneous reservoir into an equivalent medium reservoir is shown in Figure 8.2. These heterogeneous planes were then treated into the reservoir. These heterogeneous planes have a small spacing in the reservoir and disperse into the reservoir rock. These planes were assigned the specified material properties.

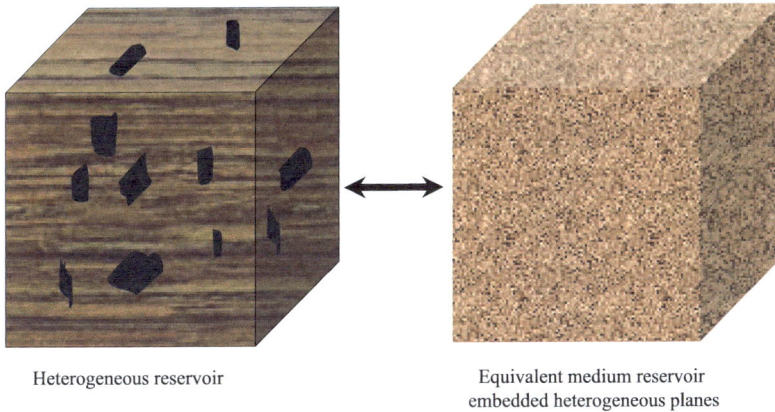

Heterogeneous reservoir   Equivalent medium reservoir
                                                embedded heterogeneous planes

Figure 8.2. Schematic diagram of transforming a heterogeneous reservoir into an equivalent medium reservoir.

The heterogeneous planes do not construct discontinuous planes in reservoir rocks, but some weak or strong planes with different properties, so as to characterise the rocks containing different materials. In this way, the discontinuous plane is not set, but the rock is considered as a continuous material, which is convenient for modelling and computation analysis in continuum media. In this study, the material properties set on the heterogeneous plane were given different parameter values. The ratio relationship between these values and the parameter values in the entire domain is given below:

$$r_\alpha = \frac{\alpha^e}{\alpha}, \quad r_E = \frac{E^e}{E}, \quad r_v = \frac{v^e}{v}, \quad r_\sigma = \frac{\sigma_t^e}{\sigma_t}, \quad r_G = \frac{G_f^e}{G_f}, \quad r_\psi = \frac{\psi^e}{\psi}, \quad r_c = \frac{c_f^e}{c_f}, \quad (8.7)$$

Deformation at the heterogeneous plane changes its permeability. Here, the permeability after the deformation of the heterogeneous plane is computed using the following formula:

$$k^e = (45\varepsilon_{max} + 1) \times k, \quad \varepsilon_{max} \geqslant 0, \text{ tensile deformation,} \quad (8.8)$$

$$k^e = (4.95\varepsilon_{max} + 1) \times k, \quad \varepsilon_{max} < 0, \text{ compressive deformation,} \quad (8.9)$$

where ε_{max} is the maximum strain at the heterogeneous plane, k is the original permeability of the reservoir rock, and k^e is the permeability after considering the deformation at the heterogeneous plane.

8.2.4　Numerical discretization

The governing equations were discretized using the fem. The shape functions were assumed to be independent of the structure N_u, seepage N_s, and network N_n fields.

$$B_u = L_u N_u, \quad B_s = L_s N_s, \quad B_n = L_n N_n, \quad (8.10)$$

where L_u, L_s, and L_n are the gradient operators, and B_u, B_s, and B_n are the shape function spatial gradient matrices for the structure, seepage, and network fields, respectively. The coupled governing equations for the structure and network fields can be written in matrix form as:

$$\begin{bmatrix} M & 0 & 0 \\ 0 & 0 & 0 \\ 0 & 0 & 0 \end{bmatrix} \begin{Bmatrix} \ddot{u} \\ \ddot{p}_s \\ \ddot{p}_n \end{Bmatrix} + \begin{bmatrix} Q_n^T & 0 & 0 \\ 0 & S_s & 0 \\ 0 & 0 & S_n \end{bmatrix} \begin{Bmatrix} \dot{u} \\ \dot{p}_s \\ \dot{p}_n \end{Bmatrix} + \begin{bmatrix} K & 0 & 0 \\ 0 & H_s & 0 \\ 0 & 0 & H_n \end{bmatrix} \begin{Bmatrix} u \\ p_s \\ p_n \end{Bmatrix} = \begin{Bmatrix} f_u \\ f_s \\ f_n \end{Bmatrix} \quad (8.11)$$

using the matrices and vectors defined by convertional FEM (Profit *et al.*, 2015).

8.3   Three-dimensional numerical models of multistage hydrofracturing

8.3.1   Three-dimensional geometrical and finite element models

To study the influence of reservoir heterogeneity on the stress shadow effect and fracture propagation patterns between fractures, two homogeneous and heterogeneous reservoir rock material models were selected. The basic physical parameters for reservoir properties and hydrofracturing settings are listed in Table 8.1. Figure 8.3 shows the geometric 3D model of multiple perforations of the horizontal wells in this study. The side lengths of the model along the x, y, and z coordinate axes were 400 m, 400 m, and 600 m, respectively. The spacing between the initial five perforations was equal, and the numbers from right to left were 1, 2, 3, 4, and 5, respectively. The circular plane radius representing the initial perforation was 2 m, and the perforation cluster space was a modifiable variable in this study, which has a significant impact on the hydraulic fracture propagation and stress shadow effect between adjacent fractures. In practical engineering, multiple perforations are distributed along the horizontal well in a spiral staggered angle to activate fracture propagation. After the initiation and propagation of perforations cluster in one stage, the fracturing fracture forms an approximate circle plane as the main fracture. In the engineering-scale analyses, the initial spiral perforation cluster is almost close to one point, and the initial perforation is close to circular plane distribution. In the numerical simulation, it is challenging to establish each perforation and simulate the fracture propagation. In this study, the spiral perforation cluster is simplified as an initial circular fracture; the circular fracture formed after fracturing is close to the engineering practice. The perforation cluster space was set to 100 m, 75 m, 50 m, and 25 m, the corresponding remeshing domain of different perforation cluster space needed to be modified, and the local dense domains of detailed meshes for perforation clusters are shown in Table 8.2. These local dense domains of detailed meshes for perforation clusters ensured high solution accuracy in this area after perforation initiation, as well as the reliability of perforation initiation and fracture path.

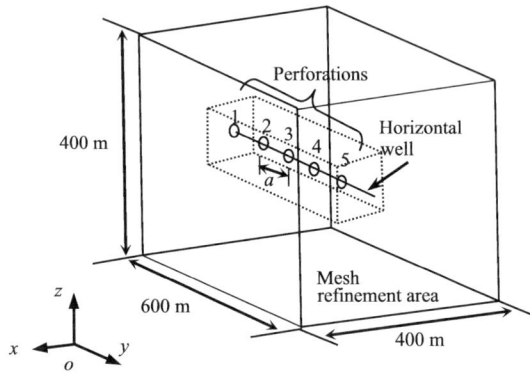

Figure 8.3. 3D geometrical model of multiple perforations of horizontal well in reservoir.

Table 8.2. Local dense domains of detailed meshes for perforation clusters.

Perforation cluster spaces /m	Local dense domains /m
100	$70 \leqslant y \leqslant 530$, $150 \leqslant y \leqslant 250$, $150 \leqslant y \leqslant 250$
75	$70 \leqslant y \leqslant 530$, $150 \leqslant y \leqslant 250$, $150 \leqslant y \leqslant 250$
50	$70 \leqslant y \leqslant 530$, $150 \leqslant y \leqslant 250$, $150 \leqslant y \leqslant 250$
25	$170 \leqslant y \leqslant 430$, $150 \leqslant y \leqslant 250$, $150 \leqslant y \leqslant 250$

8.3.2　Cases study for varying fracturing scenarios, perforation cluster spaces, and heterogeneous properties

8.3.2.1　Improved alternate fracturing sequences

Different fracturing sequences also have a significant impact on the stress shadow effect between hydraulic fractures. In this study, three types of 3D numerical models of different fracturing scenarios were used: sequential, alternate, and simultaneous fracturing. The sequence of the fracturing fluid injection was 1→2→3→4→5, while the perforating sequence of conventional alternate fracturing fluid injection was 1→3→2→5→4 in previous study (Wang, 2021a; Wang *et al.*, 2021). This conventional alternate fracturing scenario was named alternate fracturing I so that the results can be compared and distinguish from other improved alternate fracturing sequence methods. According to the existing results, the shadow effect of stress between perforations was very significant when the perforation cluster space was 25 m. As shown in Table 8.3, two improved alternate fracturing conditions are supplemented for 25 m perforation cluster space. The fracturing sequence of the alternate fracturing II was 2→4→1→

3→5, whereas that of alternate fracturing III was 4→2→1→3→5. Simultaneous fracturing injects fracturing fluid into five perforations at the same time, that is, 1–2–3–4–5. The perforation cluster spaces and fracturing scenarios of the numerical cases are listed in Table 8.3. For each fracturing method, four types of perforation cluster spaces were set: 100 m, 75 m, 50 m, and 25 m. Because the numerical simulation of simultaneous fracturing in this study mainly focused on the stress shadow effect and fracture interaction behaviour in the process of 3D multistage hydrofracturing, the influence of fracturing fluid diversion between perforation clusters on the complex fracture network shape was not considered.

Table 8.3. Perforation cluster spaces and fracturing scenarios of numerical cases.

Fracturing scenarios	Sequential fracturing 1→2→3→4→5	Simultaneous fracturing 1→2→3→4→5	Alternate fracturing I 1→3→2→5→4	Improved alternate fracturing II 2→4→1→3→5	Improved alternate fracturing III 4→2→1→3→5
	100	100	100	–	–
Cluster spaces	75	75	75	–	–
	50	50	50	–	–
	25	25	25	25	25

8.3.2.2   Heterogeneous reservoirs based on poroelastic effective medium model

In this study, heterogeneous reservoirs were characterised using the above-mentioned poroelastic effective medium model. The size of the heterogeneous model was the same as that of the homogeneous model. The heterogeneous model was formed by setting dense heterogeneous planes with a space of 1 m. The basic physical parameters for heterogeneous planes in reservoirs based on the poroelastic effective medium model are listed in Table 8.4. The physical parameters not given in the table are identical to those of the homogeneous model. The case condition settings of the heterogeneous model were the same as the fracturing sequence and space of the homogeneous model. The purpose of setting a heterogeneous model was to analyse the influence of reservoir heterogeneity on fracture propagation and stress field, and to compare the results with those of the homogeneous model.

Table 8.4. Basic physical parameters for heterogeneous planes in reservoirs based on poroelastic effective medium model.

Parameter	Ratio	Value
Biot coefficient α^e	$r_\alpha = \dfrac{\alpha^e}{\alpha} = 1.2$	0.9
Elastic modulus E^e /GPa	$r_E = \dfrac{E^e}{E} = 0.84$	32
Poisson's ratio v^e	$r_v = \dfrac{v^e}{v} = 1.33$	0.2
Tensile strength σ_t^e /MPa	$r_\sigma = \dfrac{\sigma_t^e}{\sigma_t} = 0.67$	1.0
Fracture energy G_f^e /(N·m)	$r_G = \dfrac{G_f^e}{G_f} = 2.0$	50
Friction angle ψ^e / (°)	$r_\psi = \dfrac{\psi^e}{\psi} = 1.0$	45
Cohesion c_f^e /MPa	$r_c = \dfrac{c_f^e}{c_f} = 1.0$	25

8.3.2.3   Numerical cases settings

Sequential fracturing and three types of alternate fracturing adopt multistage fracturing with multiple fracturing stages. According to the set fracturing sequence, only a single perforation was injected with the fracturing fluid each time. The duration and total time of multiple fracturing stages are listed in Table 8.5. The initial equilibrium stage presented was used to ensure the simulation of the initial *in-situ* stress and realise the uniform equilibrium of the entire reservoir area in the numerical model. The injection speed of the fracturing fluid for each perforation was set to be the same to ensure that the volume of fracturing fluid injected in each working condition is equal when comparing the final fracturing results and avoiding the interference of different volumes of fracturing fluid injected into the results. In the process of hydrofracturing, there is water pressure inside the hydraulic fracture, which produces an induced stress field in the adjacent area. The superposition of the original *in-situ* stress and induced stress fields in this area causes stress redistribution around the hydraulic fracture, which then affects the final fracture shape. Table 8.1 lists the *in-situ* stress conditions of tight reservoirs with *in-situ* stress. Because this study did not focus on *in-situ* stress, the initial *in-situ* stress of all numerical models adopted these values. The mesh refinement technique (Wang, 2021a; Wang *et al.*, 2021a) for high-precision stress solutions and reliable hydraulic fracture propagation was introduced in this study. The computational parameters for mesh refinement and coarsening in the simulation are listed in Table 8.6. The schematic process of local remeshing and coarsening for fracture propagation using a tetrahedral element is shown in Figure 8.4. Figure 8.4 (a) shows the initial dense mesh around the perforation cluster and in the coarse mesh for the external structure. Figure 8.4 (b) shows the remeshing implemented around the external fracture tips to refine the elements. Figure 8.4 (c) shows the mesh coarsening after fracture propagation, where the remeshing remains implemented around the external fracture tips, along with the fracture propagation. In the internal structure, mesh coarsening is implemented once the dense elements are no longer required. The numerical cases were computed using the proposed models with the program package ELFEN (Rockfield Software Ltd., 2016). The above program package has been verified and tested by some examples, and has shown good results in the previous research, therefore, it is introduced in this study into the application.

Table 8.5. Duration and total time of multiple fracturing stages.

Fracturing scenarios	Stage	Duration /s	Total time /s
Sequential and alternate fracturing	Initial balance	10	10
	1st stage fracturing	400	410
	2nd stage fracturing	400	810
	3rd stage fracturing	400	1,210
	4th stage fracturing	400	1,610
	5th stage fracturing	400	2,010
Simultaneous fracturing	Initial balance	10	10
	Single stage fracturing	400	410

Table 8.6. Computational parameters for mesh refinement and coarsening.

Parameter	Value	Parameter	Value
Small detail size	5/3	Coarsening density factor	2
Fracture mesh size factor	0.2	Coarsening threshold	4.5
Mesh density	1	Coarsening threshold factor	0.9
Mesh density factor	1	Non-coarsening zone	25
Bubble size	3	Non-coarsening zone factor	5
Coarsening frequency	10	Max coarsening zone	10
Coarsening density	5	Max coarsening zone factor	2

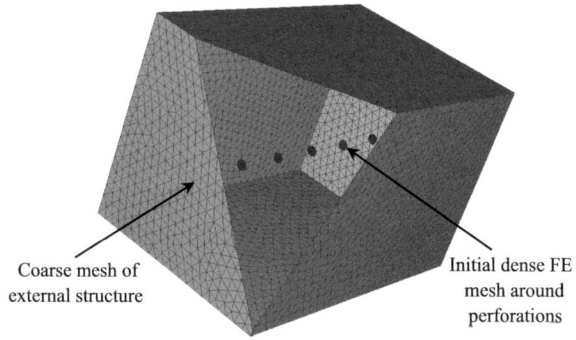

(a) Initial dense finite element mesh.

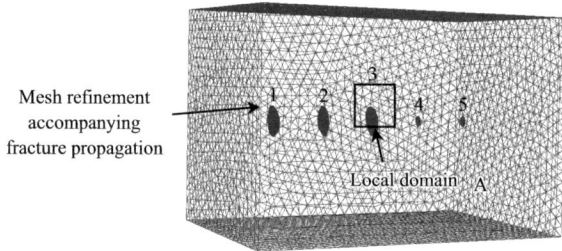

(b) Mesh refinement during fracture propagation (sequential fracturing, 3rd stage).

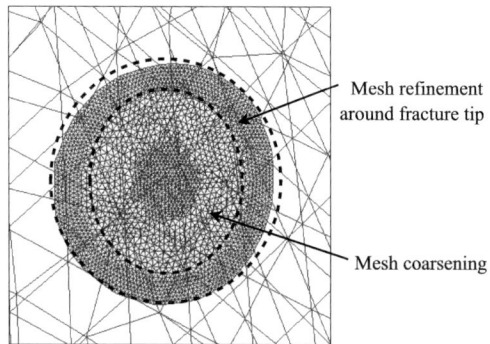

(c) Mesh refinement and coarsening during fracture propagation.

Figure 8.4. Initial mesh and mesh refinement of 3D finite element model.

8.4   Results and discussion

8.4.1   Spatial propagation morphology of fracturing fracture network

Figure 8.5 shows the results of the final propagation mode and the first principal stress field of hydraulic fractures in the heterogeneous and homogeneous reservoirs models during sequential fracturing. We set the perforation cluster space at 100 m, 75 m, 50 m, and 25 m. It can be seen that the fracture stress changes continuously during fracture propagation, but all fractures in the figure have a stress concentration near the fracture tip. Figures 8.5 (a) and (b) show the fracture results of the two reservoirs with a perforation space of 100 m. All fractures propagated along the initial perforation plane, while the final and fractured forms were flat without obvious deflection. Figures 8.5 (c) and (d) show the fracture results of the two reservoirs with a perforation space of 75 m, where the first three fractures deflected slightly, but the deflection of fractures 4 and 5 was more obvious because of the deformation accumulation of the first three fractures. Figures 8.5 (e) and (f) show the fracture results of the two reservoirs with a perforation space of 50 m. In the heterogeneous reservoir shown in Figure 8.5 (e), all fractures began to deflect in the initial stage; in the homogeneous reservoir shown in Figure 8.5 (f), fracture 3 began to deflect, whereas fractures 2 and 1 deflected substantially. The influence of reservoir heterogeneity on the final fracture propagation form appeared. Figures 8.5 (g) and (h) show the fracture results of the two reservoirs with a perforation space of 25 m. Figure 8.5 (g) shows the results of the heterogeneous reservoir, while Figure 8.5 (h) shows the results of the homogeneous reservoir. As shown in Figure 8.5 (g), the deflection starts from fracture 2. Owing to the cumulative effect of fracture propagation, the subsequent fractures are increasingly more oriented along the x-axis, and the deflection degree gradually increases. As shown in Figure 8.5 (h), all fractures have obvious deflection, the upper part of the fracture is under the positive deflection of the x-axis, and the half part is negative to the x-axis. In conclusion, in sequential fracturing, the fracture is more prone to deflection, particularly with a decrease in the perforation cluster space. The effect of fracture propagation on the fracture can be superimposed. The existence of reservoir heterogeneity has an important impact on the fracture deflection direction and final fracture propagation. The first fracture 1 will not deflect in the case of a homogeneous reservoir, whereas fracture 1 will deflect in the case of a heterogeneous reservoir.

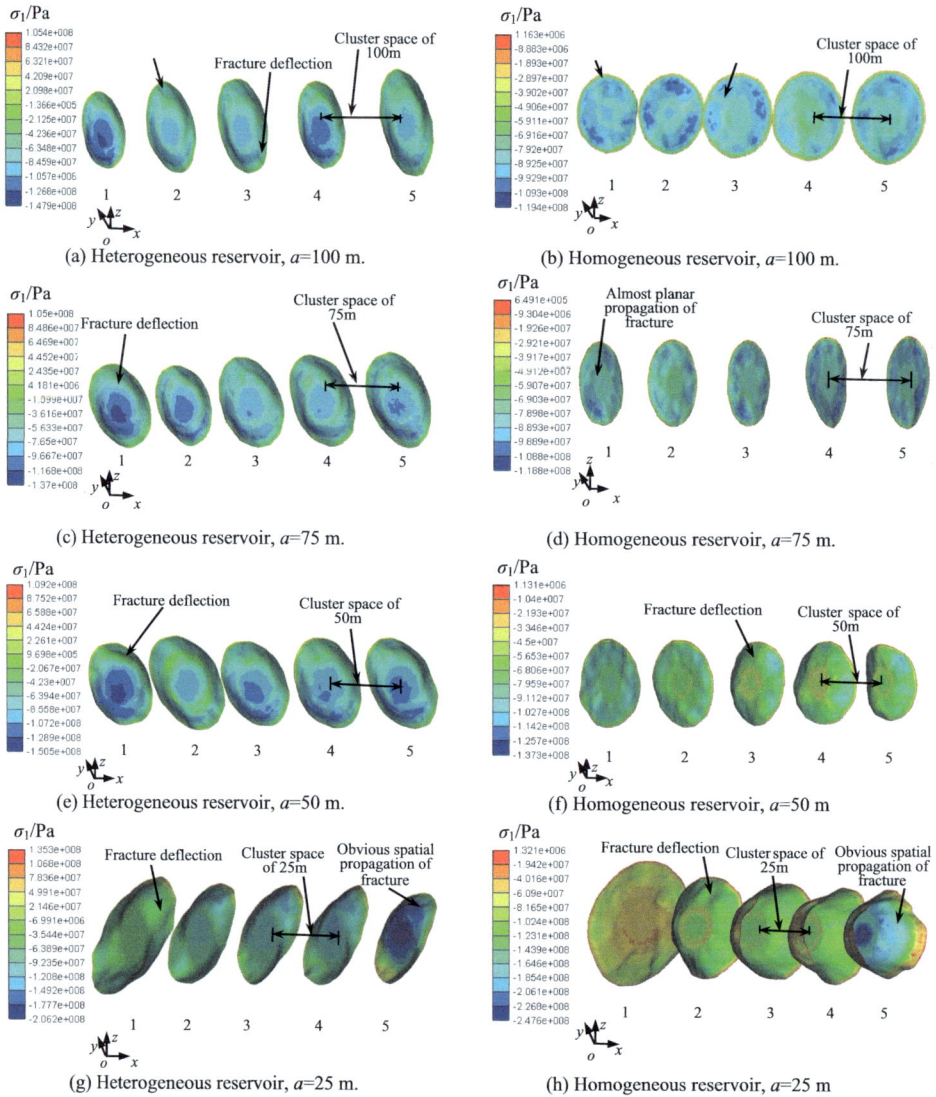

Figure 8.5. Final morphologies of fracture propagation and first principal stress σ_1 (Pa) for sequential fracturing with varying perforation cluster space.

Figure 8.6 shows the final propagation form and the first principal stress field of the hydraulic fracture when alternate fracturing I was selected as the fracturing sequence. Figures 8.6 (b), (d), (f), and (h) show the results for homogeneous reservoirs, while Figures 8.6 (a), (c), (e), and (g) show the results for heterogeneous reservoirs. As shown in Figures 8.6 (a), (b), (c), and (d), for the perforation cluster spaces of 100 m and 75 m, almost all fractures propagate along the initial perforation plane without deflection. Figures 8.6 (e) and (f) show the fracture results of the two reservoirs with a perforation space of 50 m. In the heterogeneous reservoir (Figure 8.6 (f)), all fractures began to deflect in the initial stage, whereas in the homogeneous reservoir shown in Figure 8.6 (e), the final fracture shape was nearly completely planar. Figures 8.6 (g) and (h) show the fracture results of the two reservoirs with a perforation space of 25 m. In the heterogeneous reservoir shown in Figure 8.6 (g), the deflection of fracture 5 is more obvious, and fracture 4 hardly deflected owing to the compression of fractures 3 and 5. In the homogeneous reservoir (Figure 8.6 (h)), fractures 1, 3, and 5 propagated

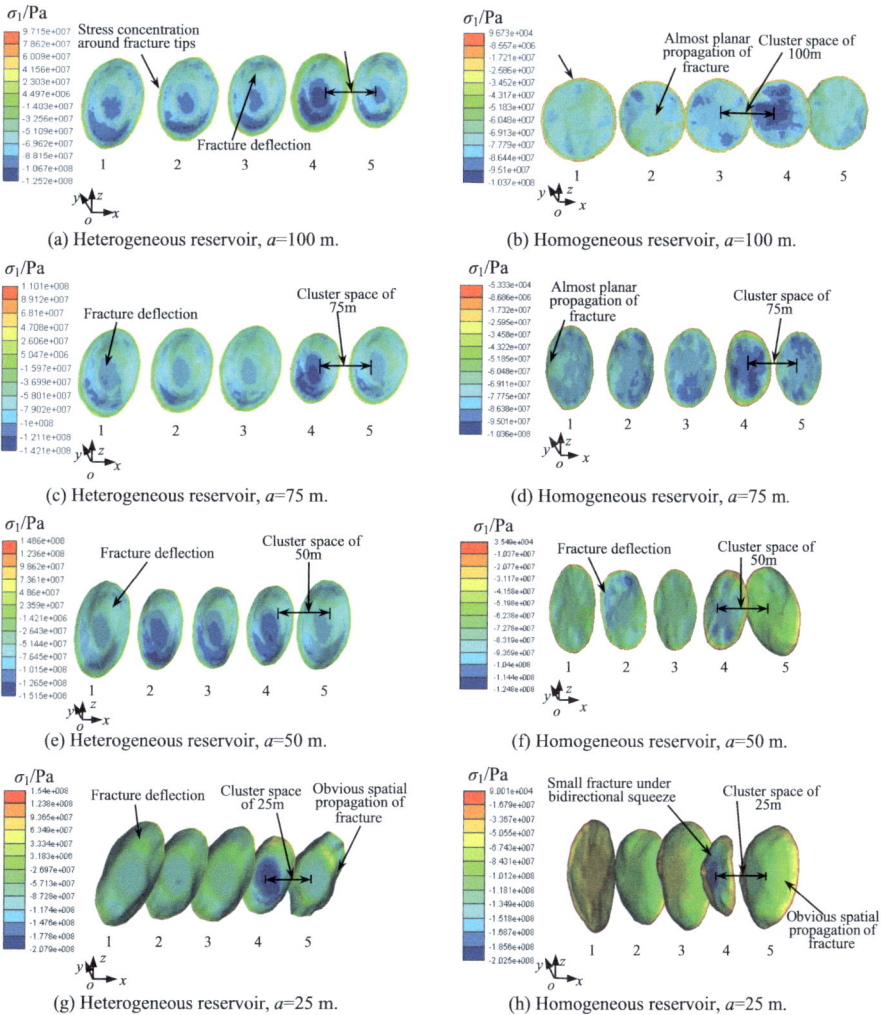

Figure 8.6. Final morphologies of fracture propagation and first principal stress σ_1 (Pa) for alternate fracturing I with varying perforation cluster space.

completely, fractures 3 and 5 were prominently deflected, and fractures 2 and 4 were affected by the action of the adjacent fractures. The influence of fracture 2 is not obvious. Owing to the comprehensive action of the first four fractures, the propagation of fracture 4 is restrained. The results show that if the reservoir is heterogeneous, alternate fracturing can significantly improve the deformation of fractures and the interaction between fractures.

Figure 8.7 shows the final propagation form and the first principal stress field of the hydraulic fracture when simultaneous fracturing was selected as the fracturing sequence. Figures 8.7 (b), (d), (f), and (h) show the results for homogeneous reservoirs, while Figures 8.7 (a), (c), (e), and (g) show the results for heterogeneous reservoirs. For the perforation cluster spaces of 100 m, 75 m, and 50 m, almost all fractures propagated along the initial perforation plane without deflection. In the heterogeneous reservoir with a perforation cluster space of 25 m (Figure 8.7 (g)), each fracture had an obvious deflection and no evident symmetry. In a homogeneous reservoir with a perforation cluster space of 25 m (Figure 8.7 (h)), the compression of fractures 1, 2, 4, and 5 caused fracture 3 to completely propagate along the initial perforation plane. Fractures 1 and 5 as well as 2 and 4 were symmetrical to the plane of fracture 3. Based on these results, in the case of a homogeneous reservoir, simultaneous fracturing had a symmetrical reservoir structure and symmetrical fracturing mode, resulting in spatial symmetrical deflection. In contrast, in the case of a heterogeneous reservoir, there was no spatial migration due to the asymmetry of the reservoir structure.

The final propagation pattern and the first principal stress field of the hydraulic fracture under two types of improved and conventional alternate fracturing supplemented at 25 m perforation cluster space are shown in Figure 8.8 (a) and (b) show the fracturing results when alternate I is selected for the fracturing sequence. In the heterogeneous reservoir (Figure 8.8 (a)), the deflection of fracture 5 is more obvious, while fracture 4 hardly deflected owing to the compression of fractures 3 and 5. In the homogeneous reservoir (Figure 8.8 (b)), fractures 1, 3, and 5 propagated completely, fractures 3 and 5 were obviously deflected, and fractures 2 and 4 were affected by the action of the adjacent fractures. The influence of fracture 2 was not obvious. Owing to the comprehensive action of the first four fractures, the propagation of fracture 4 was restrained. Figures 8.8 (c) and (d) show the fracturing results when alternate II was selected for the fracturing sequence. In the heterogeneous reservoir (Figure 8.8 (c)), the first four fractures propagated completely, with the identical deflection degree and direction. The upper end of the fracture deflected to the right, and the lower end of the fracture deflected to the left. Owing to the cumulative influence of the first four fractures on the last fracture 5, the propagation of fracture 5 was significantly restrained. In the homogeneous reservoir (Figure 8.8 (d)), fracture 2 propagated along the plane, while other fractures deflected significantly. The deflection directions on both sides of fracture 2 are opposite. Figures 8.8 (e) and (f) show the

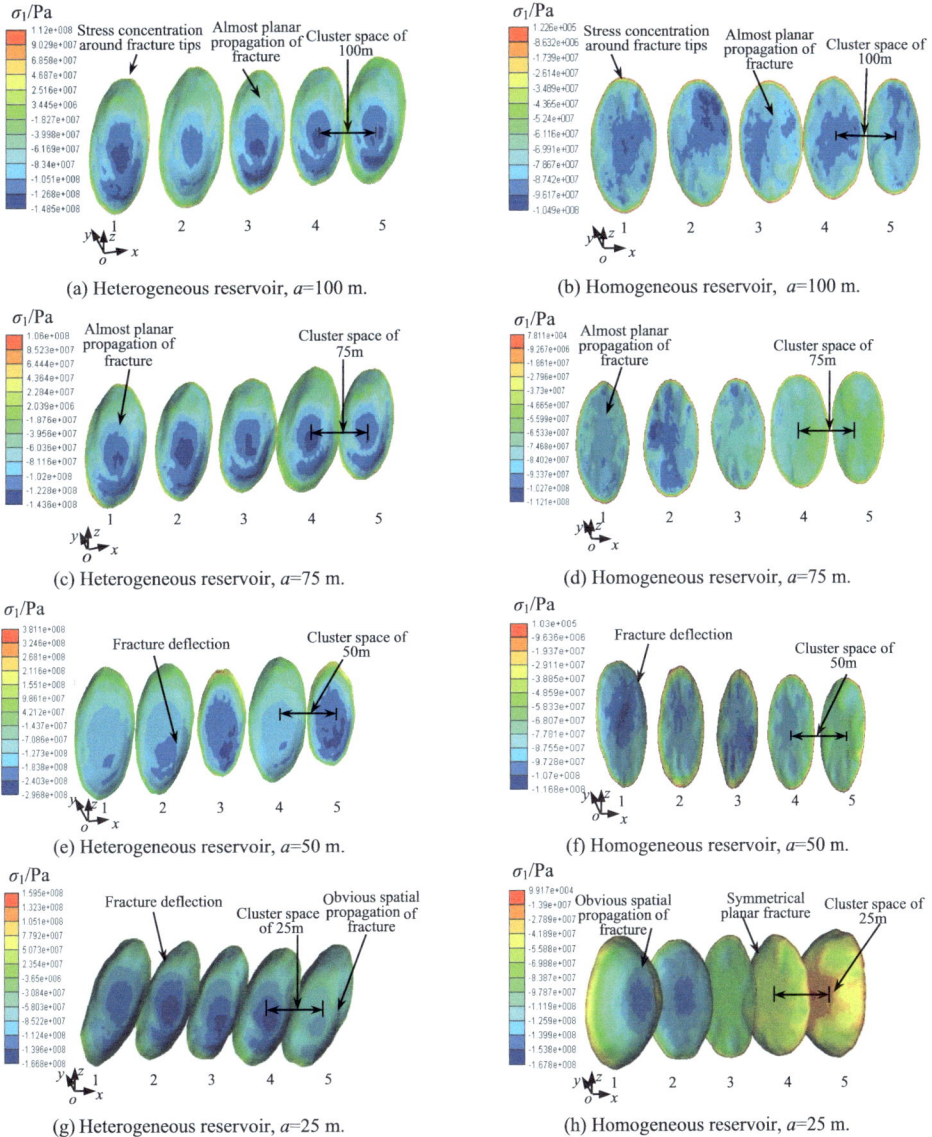

(a) Heterogeneous reservoir, a=100 m.

(b) Homogeneous reservoir, a=100 m.

(c) Heterogeneous reservoir, a=75 m.

(d) Homogeneous reservoir, a=75 m.

(e) Heterogeneous reservoir, a=50 m.

(f) Homogeneous reservoir, a=50 m.

(g) Heterogeneous reservoir, a=25 m.

(h) Homogeneous reservoir, a=25 m.

Figure 8.7. Final morphologies of fracture propagation and first principal stress σ_1 (Pa) for simultaneous fracturing with varying perforation cluster space.

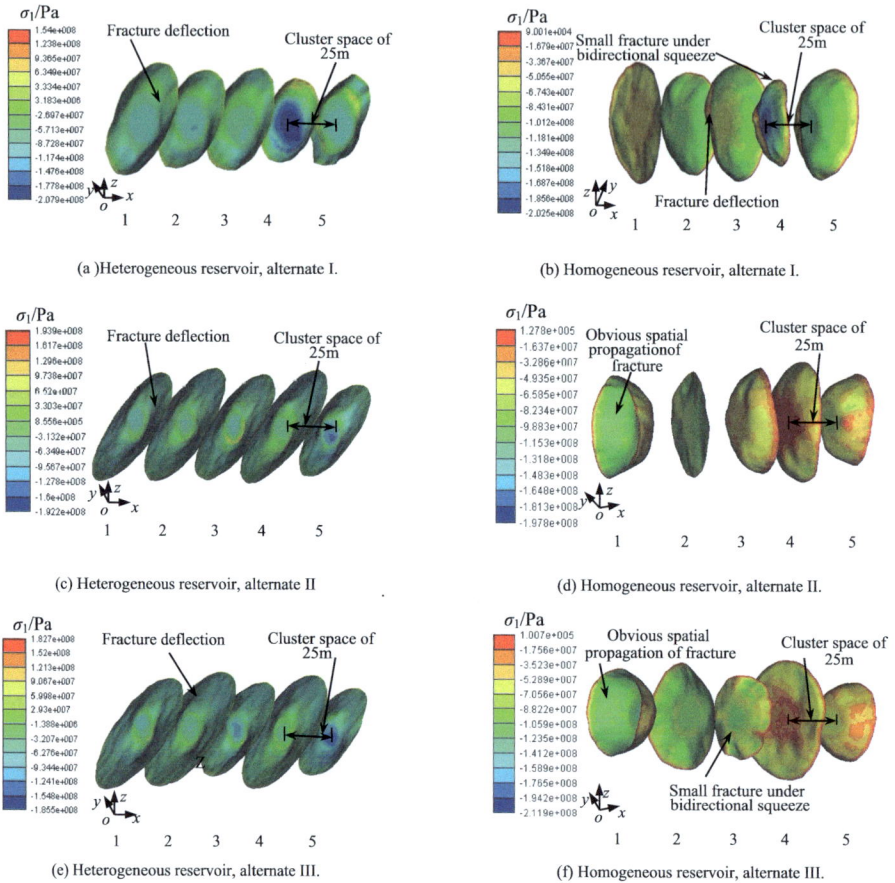

(a)Heterogeneous reservoir, alternate I.

(b) Homogeneous reservoir, alternate I.

(c) Heterogeneous reservoir, alternate II

(d) Homogeneous reservoir, alternate II.

(e) Heterogeneous reservoir, alternate III.

(f) Homogeneous reservoir, alternate III.

Figure 8.8. Final morphologies of fracture propagation and first principal stress σ_1 (Pa) for improved scenarios of alternate fracturing II and III, for a perforation cluster space of 25 m.

fracturing results when alternate III was selected for the fracturing sequence. In the heterogeneous reservoir (Figure 8.8 (e)), the first four fractures propagated completely, with identical degree and direction of deflection. The upper end of the fracture deflected to the right, while the lower end deflected to the left. Owing to the cumulative influence of the first four fractures, the propagation of fracture 5 was restrained. In the homogeneous reservoir shown in Figure 8.8 (f), the propagation and deflection of fracture 3 were restrained owing to the compression of fractures 2 and 4 on fracture 3. In conclusion, compared with conventional alternate fracturing, the improved alternate fracturing can significantly improve the fracture propagation pattern. However, in the case of a homogeneous reservoir, the gain effect of this improved method was not obvious.

8.4.1.1   Sequential, simultaneous, and alternate fracturing

In this section, the results of a heterogeneous reservoir with a 25 m perforation cluster space were taken as an example to discuss the spatial propagation pattern of the pressure fracture network with sequential, simultaneous, and alternate fracturing. Figure 8.5 (h) shows the final propagation pattern and first principal stress field of the hydraulic fracture with a perforation cluster space of 25 m under the condition of sequential fracturing. The stress on the fracture surface changed continuously during the fracture propagation process, while all the fractures exhibited stress concentration near the fracture tip (Figure 8.5 (h)). The fracture began to deflect from fracture 2 (Figure 8.6 (h)). Owing to the cumulative effect of the first fracture, the subsequent fracture deflected more toward the x-axis, and the degree of deflection increased gradually.

Figure 8.7 (h) shows the final propagation pattern and the first principal stress field of the hydraulic fracture at a 25 m perforation cluster space during simultaneous fracturing. The compression of fractures 5, 4, 2, and 1 caused fracture 3 to propagate nearly completely along the initial perforation plane. Fractures 5, 1, 4, and 2 were symmetrical to the plane of fracture 3. Because the fracturing fluid was injected into five perforations simultaneously, the five fractures began to propagate at the same time, and the stress shadow effect was significant. Because the tight reservoir was set as a homogeneous reservoir, the fracturing behaviour of simultaneous fracturing was symmetrical, so the final fracture shape was symmetrical relative to the plane of the middle fracture, and the deflection direction of the fractures on both sides of the symmetrical plane was opposite.

The final propagation pattern and the first principal stress field of the hydraulic fracture under two types of improved and conventional alternate fracturing supplemented at 25 m perforation cluster space are shown in Figure 8.8. Figure 8.8 (b) shows the results of conventional alternate fracturing I in a homogeneous reservoir, where fractures 1, 3, and 5 were fully extended, fractures 3 and 5 were obviously deflected, and fractures 2 and 4 were affected by the action of the adjacent fractures. The influence of fracture 2 was not obvious, and the propagation of fracture 4 was obviously restrained owing to the comprehensive effect of the first four fractures. Figures 8.8 (d) and (f) show the results of improved alternate fracturing for homogeneous reservoirs. Figure 8.8 (d) is the result of improved alternate fracturing II. First, fracture 2 propagated along the plane, while the other fractures had severe deflection, with the deflection directions being opposite on both sides of fracture 2. Figure 8.8 (f) shows the result of improved alternate fracturing III. The propagation and deflection of fracture 3 were restrained owing to the compression effects of fractures 4 and 2.

In summary, the fracturing sequence had a significant impact on the fracture morphology. In the case of a homogeneous reservoir, simultaneous fracturing led to spatial symmetry deflection due to the symmetrical reservoir structure and symmetrical fracturing mode. However, no symmetry was observed in a heterogeneous reservoir. Improved alternate fracturing can significantly improve the fracture propagation pattern. For homogeneous reservoirs, the gain effect of this improved method was not obvious.

8.4.1.2   Varying perforation cluster spaces

Taking sequential fracturing as an example, the effect of different perforation cluster spaces on the unstable fracture propagation was analysed in detail. Figures 8.5 (a) and (b) show the fracture results for the two reservoirs when the perforation cluster space was 100 m. All fractures propagated along the initial perforation plane, and the final fractures and fracture morphologies were plane without obvious deflection. Figures 8.5 (c) and (d) show the fracture results for the two reservoirs when the perforation cluster space was 75 m. Although the deflection of the first three fractures was minimal, the deflection of fractures 4 and 5 was more obvious because of the cumulative effect of the first three fractures. Figures 8.5 (e) and (f) show the fracture results of the two reservoirs when the perforation cluster space was 50 m. For a homogeneous reservoir, fracture 3 showed initial deflection, and fractures 2 and 1 had greater deflection; in the case of a heterogeneous reservoir, all fractures began to deflect. The influence of reservoir heterogeneity on the final fracture propagation pattern began to appear. Figures 8.5 (g) and (h) show the fracture results of the two reservoirs when the perforation cluster space is 25 m. As shown in Figure 8.5 (h), the fracture begins to deflect from fracture 2. Due to the cumulative effect of the first fracture, the subsequent fracture deflects more toward the x-axis, and the degree of deflection increases gradually; as shown in Figure 8.5 (g), all fractures show obvious deflection. The upper half of the fracture deflected in the positive direction of the x-axis, while the lower half deflected in the negative direction. In alternate and simultaneous fracturing, the fracture deflection was not obvious when the distance between the perforating clusters was large. Generally, when the distance between the perforating clusters was

25 m, the degree of unstable fracture propagation was significant. In summary, with a decrease in the perforation cluster space, the fractures are more prone to deflection, the deflection degree is greater, and the effect of the first on the later fracture can be superimposed.

8.4.1.3 Homogeneous and heterogeneous reservoirs

Figures 8.5 (g) and (h) show the fracture results of the two reservoirs when the perforation cluster space was 25 m. As shown in Figure 8.5 (h), the fracture began to deflect from fracture 2. Due to the cumulative effect of the first fracture, the subsequent fracture deflected more toward the x-axis, and the degree of deflection increased gradually; as shown in Figure 8.5 (g), all fractures showed obvious deflection. The upper half of the fracture deflected in the positive direction, while the lower half deflected in the negative direction of the x-axis. The existence of reservoir heterogeneity had an important influence on the direction of the fracture deflection and the final fracture propagation. When the reservoir was homogeneous, fracture 1 that propagated first hardly deflected; however, in the heterogeneous reservoir, fracture 1 deflected. For simultaneous fracturing with a 25 m perforation cluster space, the homogeneous reservoir result is shown in Figure 8.7 (h), where the compressions of fractures 1, 2, 4, and 5 made fracture 3 propagate along the initial perforation plane. Fractures 1, 5, 2, and 4 were symmetrical to the plane of fracture 3. In the case of the heterogeneous reservoir results (Figure 8.7 (g)), each fracture had an obvious deflection and no obvious symmetry. Based on these results, for a homogeneous reservoir, simultaneous fracturing had a symmetrical reservoir structure and symmetrical fracturing mode, which led to spatial symmetry deflection.

The results of conventional alternate fracturing with a 25 m perforation cluster space in the homogeneous reservoir are shown in Figure 8.6 (h). Fractures 1, 3, and 5 propagated sufficiently, fractures 3 and 5 deflected, and fractures 2 and 4 were affected by the action of the adjacent fractures. The influence of fracture 2 was not obvious, and the propagation of fracture 4 was obviously restrained owing to the comprehensive effect of the first four fractures. In the heterogeneous reservoir (Figure 8.6 (g)), the deflection of fracture 5 was more obvious. Owing to the compression of fractures 3 and 5, fracture 4 hardly deflected. Figures 8.8 (c) and (e) show the results of the improved alternate fracturing considering a heterogeneous reservoir. The first four fractures were fully propagated, and the deflection degrees and directions were identical. Owing to the cumulative effect of the first four fractures, the propagation of fracture 5 was obviously inhibited. Figures 8.8 (d) and (f) show the results of improved alternate fracturing for homogeneous reservoirs. Figure 8.8 (b) shows the result of improved alternate fracturing II. First, fracture 2 propagated along the plane, the other fractures deflected significantly, and the deflection directions on both sides of fracture 2 were opposite. Figure 8.8 (d) is the result of improved alternate fracturing III. The propagation and deflection of fracture 3 were restrained owing to the compression effect of fractures 2 and 4 on fracture 3. In conclusion, the existence of reservoir heterogeneity had an important impact on the fracture deflection direction and final fracture propagation. For homogeneous reservoirs, the first fracture 1 did not deflect, but the deflection of the final fracture shape was very large. For heterogeneous reservoirs, the first fracture 1 deflected, but the deflection degree of all fractures is small.

8.4.2   Disturbance of stress field during dynamic propagation of multiple hydraulic fractures

8.4.2.1   Effect of fracture propagation on *in-situ* stresses

To study the change in reservoir stress caused by fracture, the disturbance of fracture propagation on reservoir *in-situ* stress was analysed. On the premise of generality, the stress difference in the *x*-direction can express the change in behaviour of the entire stress field, as proposed previously (Ju, *et al.*, 2020):

$$\Delta\sigma_x(t) = \sigma_x(t) - S_h, \tag{8.12}$$

where $\sigma_x(t)$ is the effective stress solution at the current time t. The stress difference is a variable used to distinguish between the effects of the initial *in-situ* stress and injection water pressure on hydraulic fracture propagation in reservoir rocks. The stress difference to eliminate the influence of the initial *in-situ* stress reflects the influence of the injection water pressure on the hydraulic fracture propagation. Using the above definition as a criterion, we determined the stress concentration and disturbance areas associated with fractures.

Figure 8.9 shows the stress difference evolution of improved alternate fracturing III for a perforation cluster space of 25 m. For heterogeneous and homogeneous reservoirs, the stress difference far away from the fracture area was 0, which indicated that the location far away from the fracture area was only affected by the initial *in-situ* stress field; that is, the stress field changes caused by fracture propagation can only

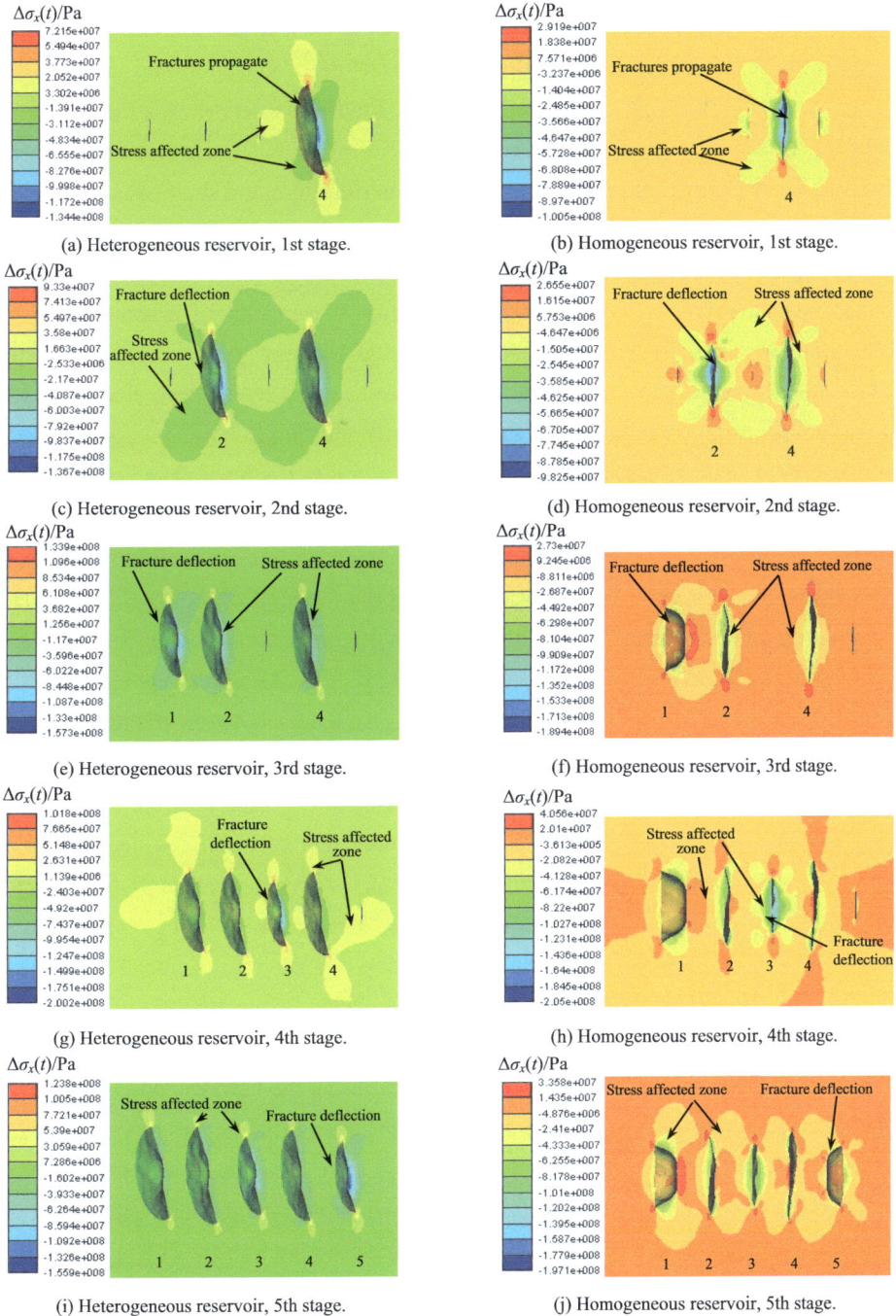

(a) Heterogeneous reservoir, 1st stage.

(b) Homogeneous reservoir, 1st stage.

(c) Heterogeneous reservoir, 2nd stage.

(d) Homogeneous reservoir, 2nd stage.

(e) Heterogeneous reservoir, 3rd stage.

(f) Homogeneous reservoir, 3rd stage.

(g) Heterogeneous reservoir, 4th stage.

(h) Homogeneous reservoir, 4th stage.

(i) Heterogeneous reservoir, 5th stage.

(j) Homogeneous reservoir, 5th stage.

Figure 8.9. Evolution of stress difference $\Delta\sigma_x(t)$ in alternate fracturing III for a perforation cluster space of 25 m.

affect the area around the fracture. The actual stress field in the fracture propagation area was the result of the superposition of the fracture propagation and initial *in-situ* stress caused by the injection water pressure. Figure 8.9 (a) shows the results of fracturing in the heterogeneous reservoir in the first stage. The stress-affected area appeared at the fracture tip and adjacent initial perforation, and there was an obvious stress concentration at the fracture tip. Figure 8.9 (b) shows the results of fracturing in the homogeneous reservoir in the first stage. The stress-affected area was X-shaped, located near the fracture. Figure 8.9 (c) shows the results of fracturing in the heterogeneous reservoirs in the second stage. Stress concentration occurred at the tips of fractures 2 and 4, and the stress-affected areas caused by the two fractures were superimposed. Figure 8.9 (d) shows the results of fracturing in homogeneous reservoir in the second stage. The stress-affected areas of the two fractures were superimposed over a large area, and the stress concentration occurred at the fracture tip. Figure 8.9 (e) shows the results of fracturing the heterogeneous reservoir in the third stage. There was a stress concentration at the fracture tip, and the range of the stress influence area was very small, concentrated near the fracture. Figure 8.9 (f) shows the results of fracturing in the homogeneous reservoir in the third stage. Owing to the superposition of the stress-affected areas of fractures 1 and 2, stress area between fractures 1 and 2 was significantly enhanced. Figure 8.9 (g) shows the results of fracturing in the heterogeneous reservoir in the fourth stage. Owing to the superposition of the stress field, the stress influence area of the four fractures was concentrated near the stress concentration area at the fracture tip. Figure 8.9 (h) shows the results of fracturing in the homogeneous reservoir in the fourth stage. The influence range of the stress shadow area was large, and the stress superposition was obvious. Figure 8.9 (i) shows the results of fracturing in the heterogeneous reservoir in the fifth stage. The stress influence area was negligible, and the stress concentrated at the tips of the five fractures. Figure 8.9 (j) shows the results of fracturing in the homogeneous reservoir in the fifth stage. Owing to the superposition of the stress influence areas from the five fracture, there was a significantly enhanced stress area between fractures. Comparing the stress difference evolution results of the two reservoirs in different stages, the existence of reservoir heterogeneity had a significant impact on the shape and area of the water injection pressure-affected area.

8.4.2.2   Shadow effect analysis of shear stress field

Propagation deformation of multiple fractures may occur in many cases of fracturing processes, owing to the change in the stress field. The superposition of stress fields between fractures causes the interaction of fractures in the process of propagation, particularly the interaction of shear stress fields. To understand the causes of the formation of the 3D fracture network, the evolution of the shear stress field was used to reflect the stress superposition and reduction between fractures, and the stress shadow effect was detected. Owing to the difference in the positive and negative shear stresses on both sides of the fracture, the superposition effect of different fractures easily weakened. These phenomena fully explain the interaction of the stress field between fractures. According to the computed results in Section 4.1 (the initial perforation cluster space of a horizontal well was set to 25 m), obvious fracture deflection appeared under all fracturing scenarios owing to the interaction of the stress field caused by the close perforation cluster space. The following is a comparative analysis of the stress field and fracture evolution process in the hydrofracturing process of the proposed 3D model. Under various fracturing scenarios, the initial perforation cluster space of horizontal wells was set at 25 m, and the evolution of shear stress σ_{xz} for a cluster space of 25 m was analysed. In these figures, the symbol '+' represents positive shear stress and the symbol '−' represents the negative shear stress.

Figure 8.10 shows the shear stress evolution diagram of sequential fracturing. A heterogeneous reservoir is shown regarding the stress results of the first-stage fracturing (Figure 8.10 (a)). The positive shear stress was generated on the left side of the upper fracture tip, while negative shear stress was produced on the right side. At the same time, the left side of the lower fracture tip was under negative shear stress, while the right side was under positive. Figure 8.10 (b) shows that there was a homogeneous reservoir, that negative shear stress was generated on the left side of the upper fracture tip, while positive shear stress was generated on the right side. At the same time, the left side of the lower fracture tip was under the positive shear stress and the right side was under the negative. In addition, the stress around fracture 1 was nearly symmetrical (Figure 8.10 (b)), leading to the propagation of the fracture in the plane direction without deflection. Fracture 1 deflected mainly because of the action of the heterogeneous reservoir (Figure 8.10 (a)). Then, according to the stress results of the second-stage fracturing (Figure 8.10 (c)), a heterogeneous reservoir is shown. Fracture 2 began to propagate, and the left and right sides of the upper and lower fracture tips generated positive and negative shear stresses, respectively, similarly to those of fracture 1. Figure 8.10 (d) shows the stress field in the homogeneous reservoir in the second stage, where fracture 2 began to deflect, and the stress distribution

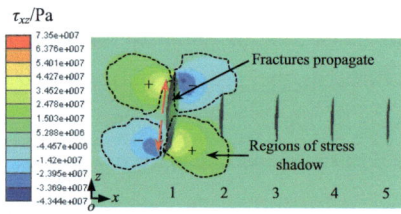

(a) Heterogeneous reservoir, 1st stage.

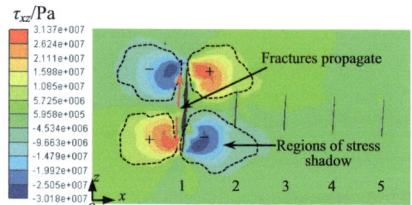

(b) Homogeneous reservoir, 1st stage.

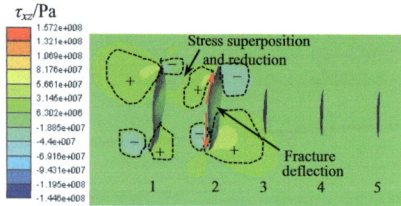

(c) Heterogeneous reservoir, 2nd stage.

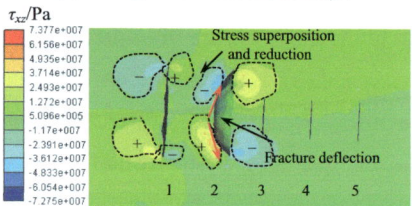

(d) Homogeneous reservoir, 2nd stage.

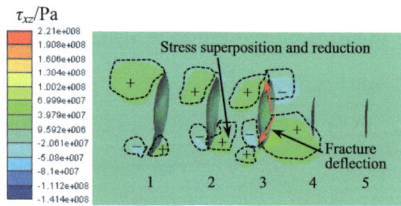

(e) Heterogeneous reservoir, 3rd stage.

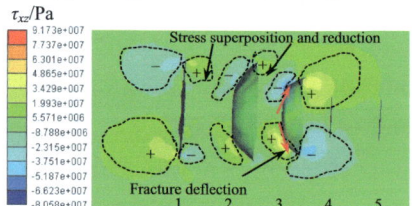

(f) Homogeneous reservoir, 3rd stage.

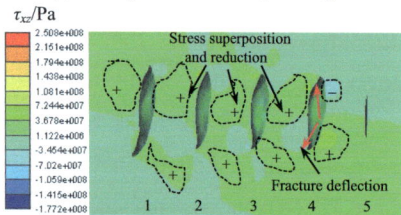

(g) Heterogeneous reservoir, 4th stage.

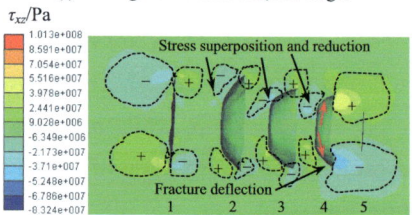

(h) Homogeneous reservoir, 4th stage.

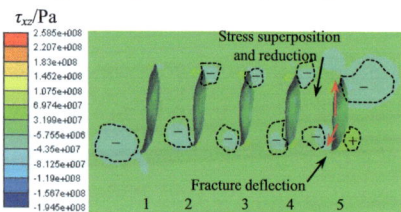

(i) Heterogeneous reservoir, 5th stage.

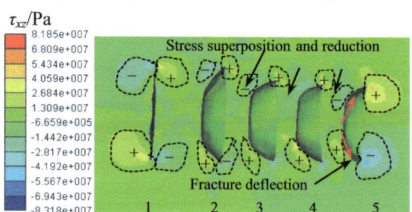

(j) Homogeneous reservoir, 5th stage.

Figure 8.10. Evolution of shear stress τ_{xz} (Pa) in sequential fracturing for a cluster space of 25 m ('+' represents positive shear stress and '−' represents negative shear stress).

pattern was the same as that of fracture 1. The results show that the shear stress field caused by the first two fractures was superimposed and decreased. Figure 8.10 (d) shows that the shear stress field on the left side of fracture 2 obviously weakened, and began to propagate and deflect to the right side of the larger stress area. Figure 8.10 (c) shows that there was no significant change in the deflection direction of fracture 2 because the natural fracture offset the effect of the partial stress shadow effect. In Figures 8.10 (e)–(j), based on the same analysis and reasoning, according to the distribution of the shear stress field of fractures 3, 4, and 5 without fracture reservoir, the shear stress area of each fracture had similar superposition and reduction. The fracture began to propagate on the right side of the larger stress area, showing significant fracture deflection, because the fracture in later fracturing was more likely to form larger stress accumulation and severe deflection. The distribution of the shear stress field in the propagation of fractures 3, 4, and 5 in the heterogeneous reservoir shows that although the shear stress area of each fracture had a similar superposition and reduction, the fracture deflection showed no obvious change. The results show that the fracture propagation interfered with the initial stress during fracturing, and the stress concentration at the fracture tip increased. Stress concentration area affected the interaction of fractures and caused fracture deflection through stress superposition and stress reduction, namely the stress shadow effect, where the existence of heterogeneous properties inhibited the effect of stress shadow.

Figure 8.11 shows the shear stress evolution of alternate fracturing I. For Figures 8.11 (a)–(d), according to the stress results of the first-stage fracturing, almost symmetrical and equivalent positive and negative shear stress regions appeared on both sides of the fracture tips of the two reservoirs. Fracture 3 began to propagate in the second stage of fracturing, and the area of the induced shear stress change was slightly affected by fracture 1. There was no obvious deflection of the fracture in either reservoir. We note that the space between fractures 1 and 3 was 50 m, consistent with the observation that there was no significant fracture deflection during the first two stages of continuous fracturing at the 50 m spacing. In the next stage, fracture 2 began to propagate, as shown in Figures 8.11 (e) and (f), the superposition of stress fields of fractures 1 and 3 on the left and right sides decreasing the induced shear stress variation area of fracture 2, and restraining the deflection of fracture 2 in the two reservoirs. Fracture 5 then propagated in the fourth stage of fracturing. Because the distance between fractures 5 and 3 was 50 m (Figures 8.11 (g) and (h)), the area of induced shear stress change was less affected by fractures 1, 2, and 3, and the fracture deflection in the two reservoirs was smaller than that in sequential fracturing. In the next stage of fracturing, fracture 4 began to propagate. As shown in Figure 8.11 (i),

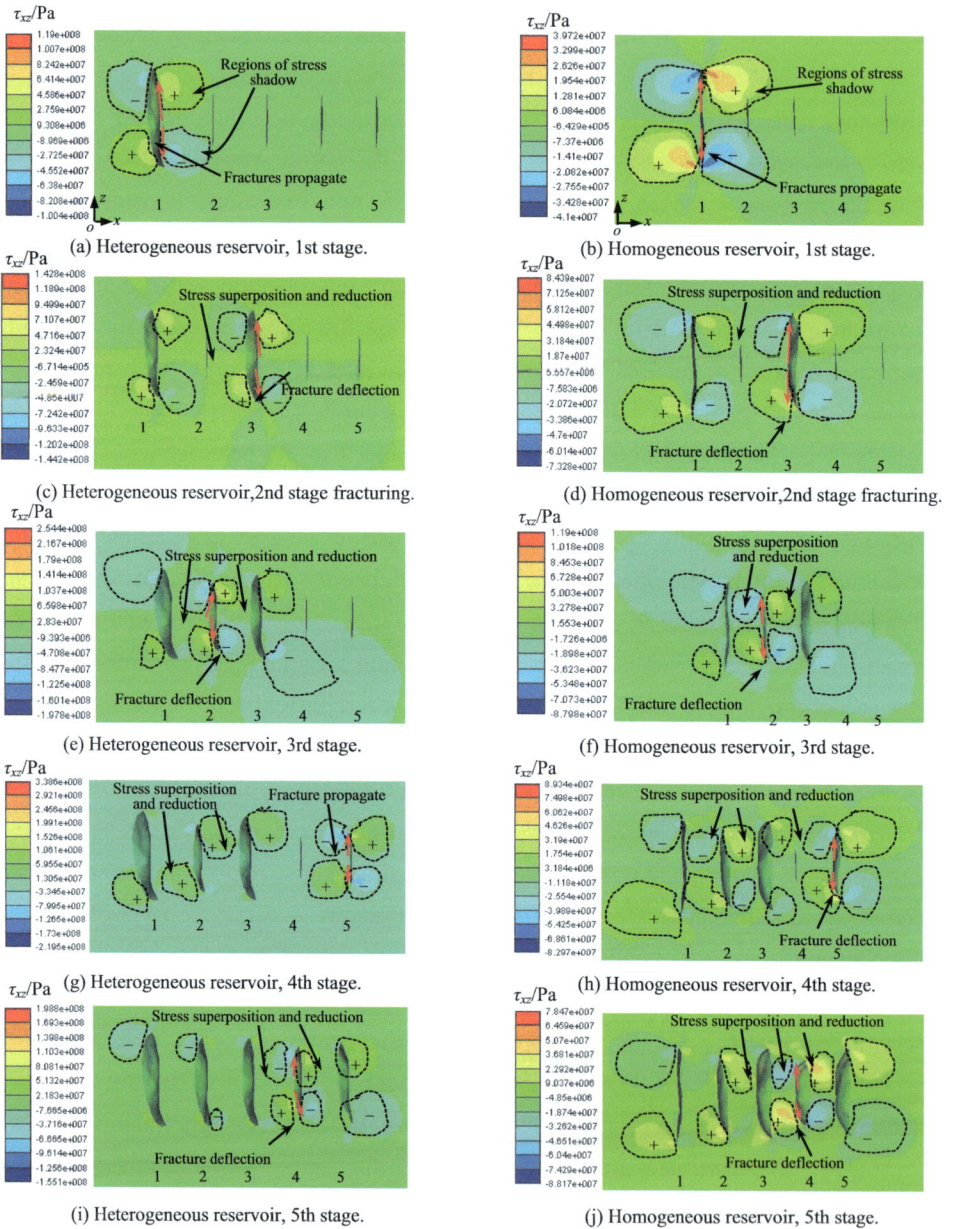

(a) Heterogeneous reservoir, 1st stage.

(b) Homogeneous reservoir, 1st stage.

(c) Heterogeneous reservoir, 2nd stage fracturing.

(d) Homogeneous reservoir, 2nd stage fracturing.

(e) Heterogeneous reservoir, 3rd stage.

(f) Homogeneous reservoir, 3rd stage.

(g) Heterogeneous reservoir, 4th stage.

(h) Homogeneous reservoir, 4th stage.

(i) Heterogeneous reservoir, 5th stage.

(j) Homogeneous reservoir, 5th stage.

Figure 8.11. Evolution of shear stress τ_{xz} (Pa) in alternate fracturing I for a cluster space of 25 m ('+' represents positive shear stress and '−' represents negative shear stress).

under the combined action of heterogeneous properties and stress shadow effect, fractures almost spread along the plane without deflection in a heterogeneous reservoir. As shown in Figure 8.11 (j), there was a homogeneous reservoir. The fracture continued to propagate in the plane mode and formed a short fracture area due to compression. In conclusion, compared with sequential fracturing, alternate fracturing can reduce the interaction between fractures, fracture deformation, and the stress shadow effect, implying an improved fracturing mechanism. The stress shadow effect caused by the post-propagating fracture also affected the shape of the previously propagating fracture.

Figure 8.12 shows the shear stress evolution of the first improved alternate fracturing II. From the stress results of the first stage of fracturing, a heterogeneous reservoir is shown in Figure 8.12 (a), and a homogeneous reservoir is shown in Figure 8.12 (b). Negative shear stress was generated on the left side of the upper fracture tip of fracture 2, while a positive shear stress was generated on the right side; at the same time, the left side of the lower fracture tip exhibited positive shear stress, while the right side exhibited negative. From the stress results of the second-stage fracturing, because the distance between fractures 4 and 2 was 50 m, the stress shadow effect was very weak. Regarding the results of the heterogeneous reservoir (Figure 8.12 (c)), fracture 4 was similar to fracture 2; the fracture 4 deflection of the homogeneous reservoir (Figure 8.12 (d)) was very small. From the stress results of the third stage fracturing, the shear stress field of fracture 1 and the stress field of fracture 2 in the heterogeneous reservoir (Figure 8.12 (e)) superimposed, resulting in the upper fracture tip deflecting more to the left; in the homogeneous reservoir (Figure 8.12 (f)), the shear stress fields of fractures 1 and 2 superposed, the shear stress field on the right side weakened significantly, and began to propagate and deflect to the left side. From the stress results of the fourth stage fracturing, owing to the superposition and compression of the stress fields of fractures 2, 3, and 4, the deflection of fracture 3 was slightly restrained in both heterogeneous (Figure 8.12 (g)) and homogeneous (Figure 8.12 (h)) reservoirs. From the stress results of the fourth stage fracturing, fracture 5 of the heterogeneous reservoir (Figure 8.12 (i)) had a larger deflection degree under the action of the stress field, but a smaller fracture area; in the homogeneous reservoir (Figure 8.12 (j)), the shear stress field of fracture 5 overlapped with that of fractures 4 and 3, the shear stress field on the left side weakened significantly, and began to propagate and deflect to the right side. In conclusion, the improved alternate fracturing II improved the effect on a heterogeneous reservoir more significantly than the conventional alternate fracturing, while that on a homogeneous reservoir was negligible compared to the conventional alternate fracturing.

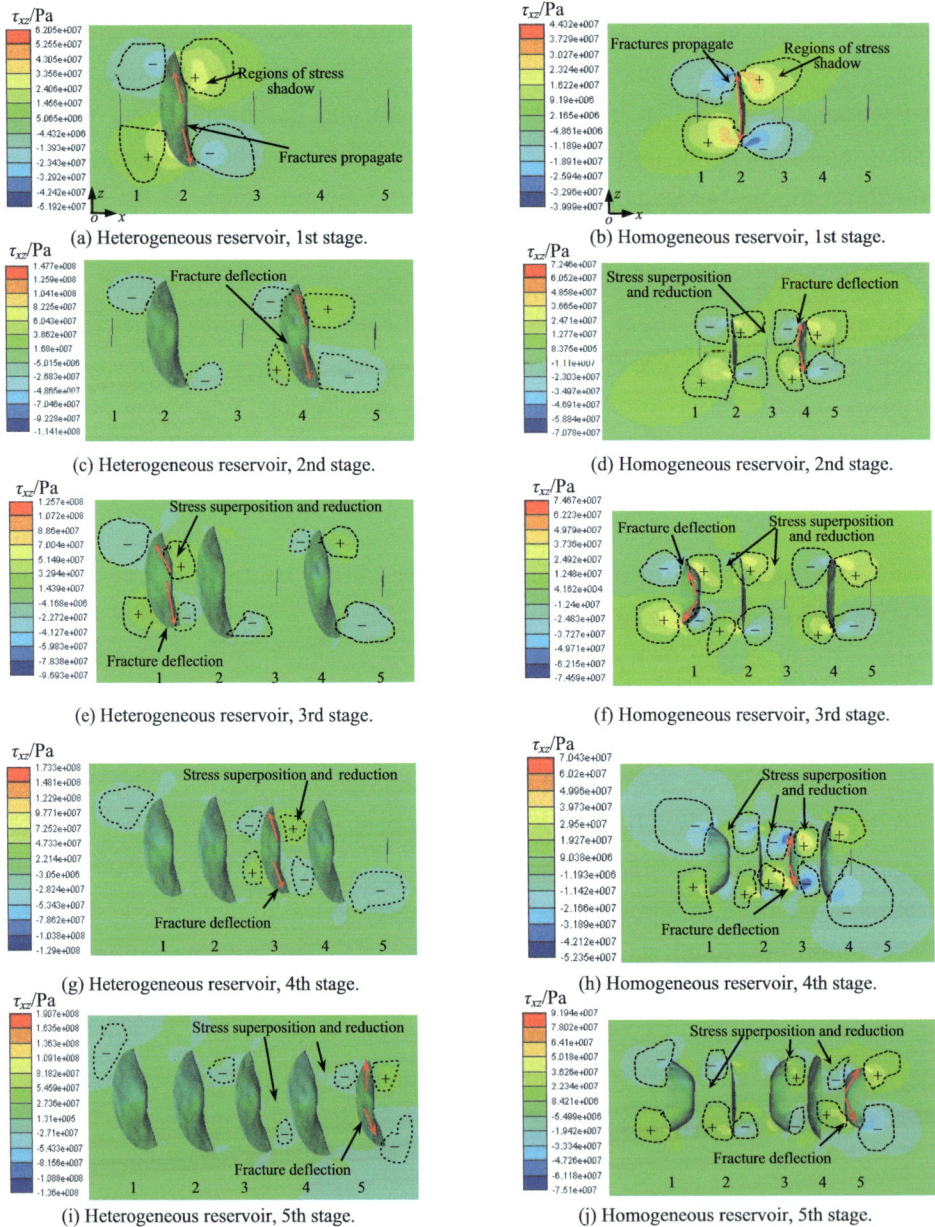

(a) Heterogeneous reservoir, 1st stage.
(b) Homogeneous reservoir, 1st stage.
(c) Heterogeneous reservoir, 2nd stage.
(d) Homogeneous reservoir, 2nd stage.
(e) Heterogeneous reservoir, 3rd stage.
(f) Homogeneous reservoir, 3rd stage.
(g) Heterogeneous reservoir, 4th stage.
(h) Homogeneous reservoir, 4th stage.
(i) Heterogeneous reservoir, 5th stage.
(j) Homogeneous reservoir, 5th stage.

Figure 8.12. Evolution of shear stress τ_{xz} (Pa) in alternate fracturing II for a cluster space of 25 m ('+' represents positive shear stress and '–' represents negative shear stress).

Figure 8.13 shows the shear stress evolution of the second improved alternate fracturing III. In Figures 8.13 (a) and (b), from the stress results of the first stage of fracturing, regardless of whether the heterogeneous or homogeneous reservoir was analysed, negative shear stress occurred on the left side of the upper fracture tip of fracture 2, and positive shear stress occurred on the right side; at the same time, the left side of the lower fracture tip experienced a positive shear stress, while the right side experienced a negative. In Figures 8.13 (c) and (d), from the stress results of the second-stage fracturing, because the distance between fractures 4 and 2 was 50 m, the stress shadow effect was weak; therefore, the fracture 4 propagation shape of the heterogeneous reservoir (Figure 8.13 (c)) was very similar to that of fracture 2. For the homogeneous reservoir (Figure 8.13 (b)), the stress field at the upper tip of fractures 2 and 4 was superimposed, the shear stress field on the right side of fracture 2 weakened, and began to propagate and deflect to the left. For Figures 8.13 (e)–(j), based on the same analysis and reasoning, the stress field evolution process of fractures 1, 3, and 5 in the two reservoirs was consistent with that of the improved alternate fracturing II. In summary, the improved alternate fracturing II improved the effect on the heterogeneous reservoir more significantly than the conventional alternate fracturing, while the improvement of the homogeneous reservoir was negligible compared with the conventional alternate fracturing.

Figure 8.14 shows the shear stress evolution diagram of the heterogeneous and homogeneous reservoirs with simultaneous fracturing. The continuous stages of simultaneous fracturing are divided into 90, 170, 250, 330, and 410 s, as shown in the subfigures. The five perforations started to propagate at the same time, and the induced shear stress change area between the fractures significantly superimposed and reduced. For a heterogeneous reservoir, each fracture propagated along the plane except for the upper and lower tip deflections. For the homogeneous reservoir, fractures 2, 3, and 4 propagated in plane form, while fractures 1 and 5 in the latter three stages deflected outward owing to the stress interference of the middle three fractures. Owing to the stress shadow effect of the middle fracture, the deflections of fractures 1 and 5 at both ends were greater. The results show that, in the case of a narrow initial perforation cluster space, simultaneous fracturing caused a severe stress shadow effect between fractures, hindered the fracture propagation, and had a greater impact on heterogeneous than on homogeneous reservoirs.

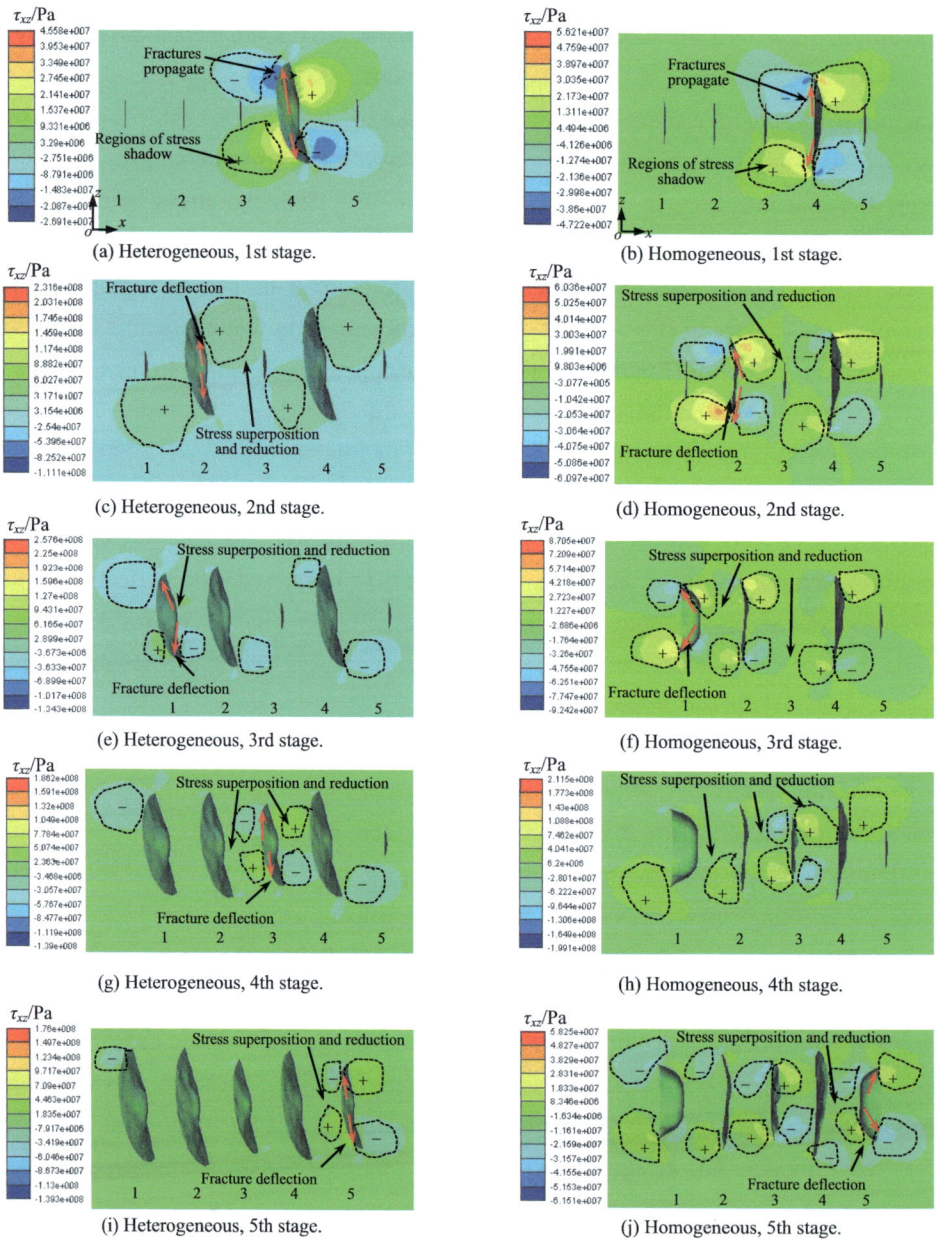

(a) Heterogeneous, 1st stage.

(b) Homogeneous, 1st stage.

(c) Heterogeneous, 2nd stage.

(d) Homogeneous, 2nd stage.

(e) Heterogeneous, 3rd stage.

(f) Homogeneous, 3rd stage.

(g) Heterogeneous, 4th stage.

(h) Homogeneous, 4th stage.

(i) Heterogeneous, 5th stage.

(j) Homogeneous, 5th stage.

Figure 8.13. Evolution of shear stress τ_{xz} (Pa) in alternate fracturing III for a cluster space of 25 m ('+' represents positive shear stress and '−' represents negative shear stress).

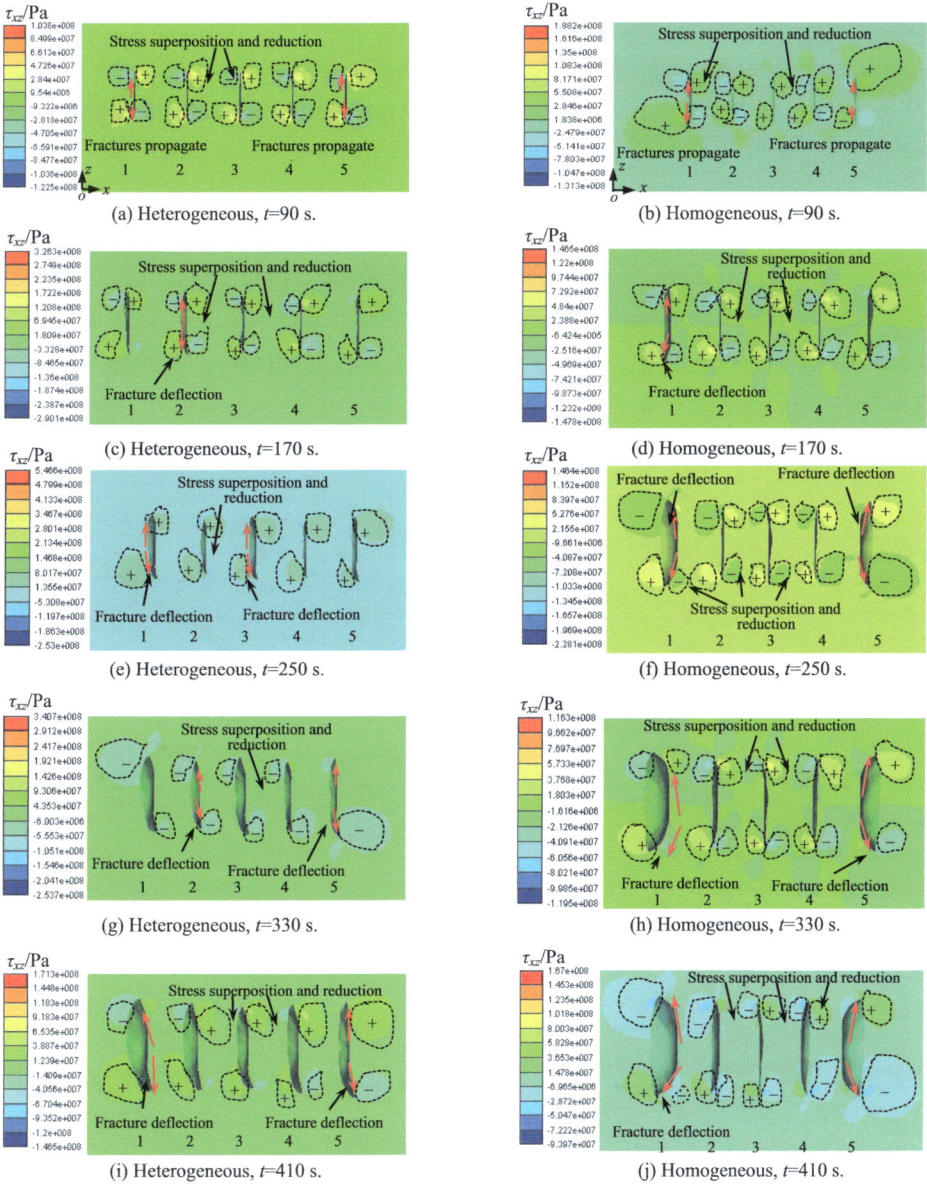

Figure 8.14. Evolution of shear stress τ_{xz} (Pa) in simultaneous fracturing for a cluster space of 25 m ('+' represents positive shear stress and '−' represents negative shear stress).

8.4.3   Quantitative area and volume of three-dimensional fracturing fracture networks

In this section, the effects of heterogeneous properties and the stress shadow effect on the fracture area and volume are compared and analysed. Figure 8.15 shows the evolution of the areas and volumes of the fracture network for sequential, alternate I, and simultaneous fracturing with a perforation cluster space of 25 m in homogeneous and heterogeneous reservoirs. It should be noted that the total time of simultaneous fracturing at 410 s is approximately one-fourth of the total time of continuous or alternate fracturing for 2010 s, so the simultaneous fracturing curve is shorter than that of the sequential or alternate fracturing. Figure 8.16 shows the evolution of the areas and volumes of the fracture network for alternate I, alternate II, and alternate III fracturing with a perforation cluster space of 25 m in homogeneous and heterogeneous reservoirs. Figure 8.17 shows a comparison of the final areas and volumes of the

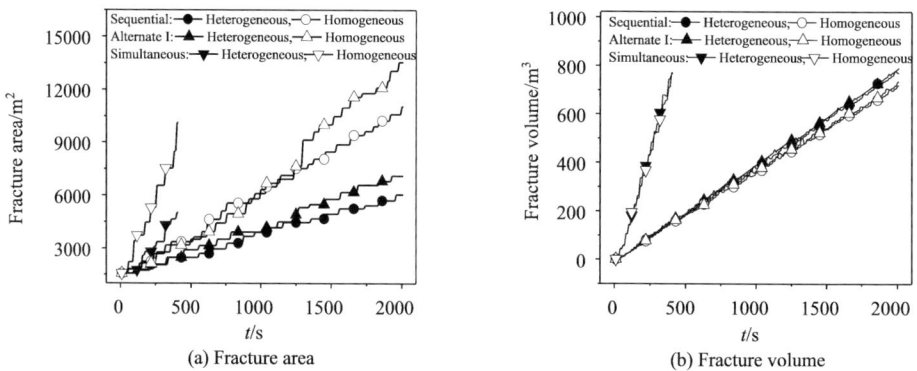

(a) Fracture area   (b) Fracture volume

Figure 8.15. Evolution of areas and volumes of fracture network for sequential, alternate I, and simultaneous fracturing with a perforation cluster space of 25 m in homogeneous and heterogeneous reservoirs.

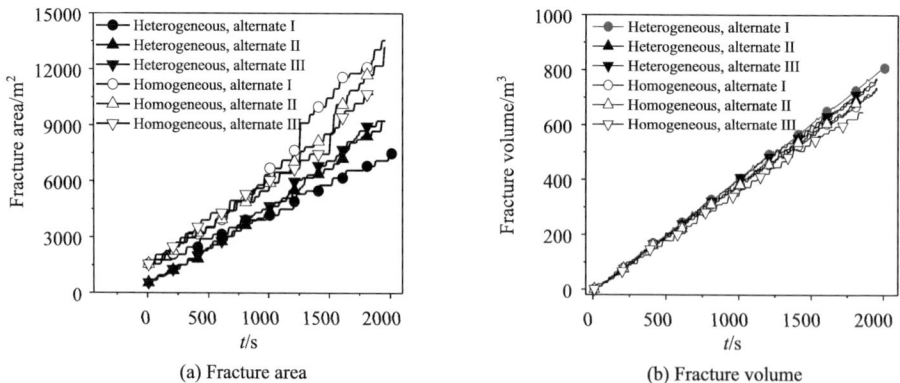

(a) Fracture area   (b) Fracture volume

Figure 8.16. Evolution of areas and volumes of fracture network for alternate I, alternate II, and alternate III fracturing with a perforation cluster space of 25 m in homogeneous and heterogeneous reservoirs.

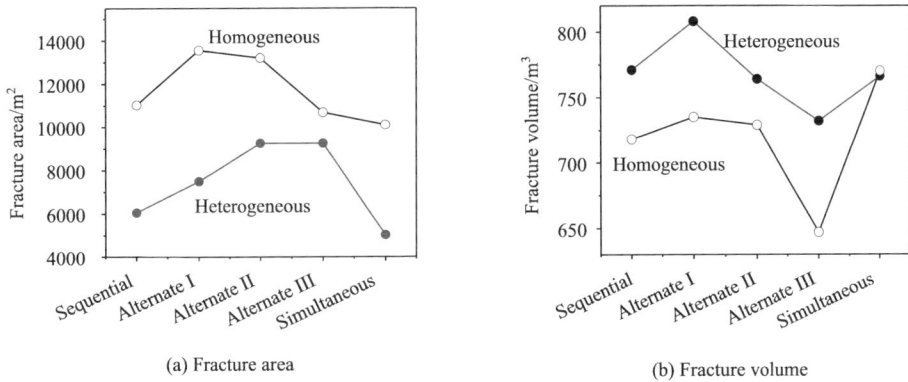

(a) Fracture area

(b) Fracture volume

Figure 8.17. Comparison of final areas and volumes of fracture network for sequential, alternate I, alternate II, alternate III, and simultaneous fracturing with a perforation cluster space of 25 m in homogeneous and heterogeneous reservoirs.

fracture network for sequential, alternate I, alternate II, alternate III, and simultaneous fracturing with a perforation cluster space of 25 m in homogeneous and heterogeneous reservoirs.

8.4.3.1   Fracture area

Figure 8.15 (a) shows the change in the fracture area with time under the three fracturing modes (sequence, alternation, and simultaneous). It can be observed that for the three fracturing modes, the hydraulic fracture area in a homogeneous reservoir was always larger than that in a heterogeneous reservoir. In a homogeneous reservoir, there was little difference in the evolution of the fracture propagation area between sequential and alternate fracturing, while the difference between them starts to be obvious in the later stage, that is, at 1250 s. The fracture area of alternate fracturing is larger than that of sequential fracturing; in heterogeneous reservoirs, the fracture area of alternate fracturing is slightly larger than that of sequential fracturing, and the gap increases gradually after 1250 s. Figure 8.16 (a) shows the change in fracture area with time of the conventional alternate fracturing and the two types of improved alternate fracturing. The initial fracture area of the two types of improved alternate fracturing in a heterogeneous reservoir is smaller because the initial grid was modified in order to get through the two models. As shown in Figure 8.16 (a), for the three alternate fracturing methods, the hydraulic fracture area in a homogeneous reservoir was always larger than that in a heterogeneous reservoir. In a homogeneous reservoir, the fracture area of the improved alternate fracturing was slightly lower than that of the conventional alternate fracturing, and the improved alternate fracturing showed no improvement compared with the conventional alternate fracturing. For a heterogeneous reservoir, although the initial fracture area of the two types of improved alternate fracturing was small, it exceeded the conventional alternate fracturing early in the third stage of fracturing. It can be seen that the effect of improved alternate fracturing on conventional alternate fracturing was significant in heterogeneous reservoirs. Figure 8.17 (a) shows a comparison of the final fracture area of the five fracturing sequences, and that the final fracture area of the homogeneous reservoir was larger than that of the heterogeneous reservoir. In the two reservoirs, the fracture area of simultaneous fracturing was the smallest, because the stress shadow effect among the fractures was the most severe, while the hydraulic fracture propagation was most significantly inhibited, consistent with the evolution result of the shear stress field. For heterogeneous reservoirs, the final fracture areas of the two improved alternate fracturing methods were larger than those of the conventional alternate fracturing method; for homogeneous reservoirs, the final fracture area of the two improved alternate fracturing methods was smaller than that of the conventional alternate fracturing method.

8.4.3.2   Fracture volume

Figure 8.15 (b) shows the change in fracture volume with time under three fracturing modes: sequential, alternate, and simultaneous. It can be observed that for the three fracturing modes, the volume of the hydraulic fracture in a homogeneous reservoir was always smaller than that in a heterogeneous reservoir. This is because the fracture area in the heterogeneous reservoir was smaller, and the pressure of injecting the same volume of fracturing fluid on the fracture was greater, so that there was no larger fracturing volume in the heterogeneous reservoir. Figure 8.16 (b) shows the change in fracture volume of conventional alternate fracturing with time and two improved alternate fracturing. For the three alternate fracturing methods, the volume of the hydraulic fracture in a heterogeneous reservoir was always larger than that in a homogeneous reservoir. Figure 8.17 (b) shows the change in the final fracture volume of the five fracturing sequences. For the three alternate fracturing methods, the hydraulic fracture volume in a heterogeneous reservoir was always larger than that in a homogeneous reservoir, and the fracture volume in simultaneous fracturing was also larger. In a homogeneous reservoir, conventional alternate fracturing had the largest fracture volume, which shows that the alternate fracturing scenario also had advantages in improving the fracture volume.

8.5   Conclusions

In this study, using numerical models, we analysed the unstable propagation of multiple fractures, *in-situ* stress disturbance around fractures, as well as the evolution of the shear stress field, fracture area, and volume. The specific conclusions are summarised as follows:

(1) 3D multistage fracturing models were established to simulate the unstable dynamic propagation and interaction of multiple hydraulic fractures. To study the influence of reservoir heterogeneity, a poroelastic effective medium model was used to simulate a heterogeneous reservoir. This study discusses the effects of sequential, improved alternate, and simultaneous fracturing, as well as different perforation cluster spaces (100 m, 75 m, 50 m, and 25 m) in heterogeneous and homogeneous reservoirs.

(2) For the spatial propagation morphology of the fracturing fracture network, with a decrease in the perforation cluster space, the fracture was more prone to deflection, and the degree of deflection was greater. Alternate fracturing was found to significantly improve fracture deformation and interaction between fractures. The existence of reservoir heterogeneity influenced the fracture deflection direction and final fracture propagation. In case of a homogeneous reservoir, simultaneous fracturing led to spatial symmetry deflection due to the symmetrical reservoir structure and symmetrical fracturing mode, while symmetry did not appear in a heterogeneous reservoir. Improved alternate fracturing significantly improved the fracture propagation pattern. For homogeneous reservoirs, the gain effect of this improved method was negligible.

(3) For the disturbance of stress field during dynamic propagation of multiple hydraulic fractures, in the process of fracturing, the propagation of fracture interfered with the initial *in-situ* stress, and the stress concentration at the fracture tip increased. The stress concentration area affected the interaction of fractures and caused fracture deflection through stress superposition and stress reduction, that is, the stress shadow effect. When the initial perforation cluster space was narrow, simultaneous fracturing led to a severe stress shadow effect between fractures, hindered fracture propagation, and had a greater impact on heterogeneous than homogeneous reservoirs.

(4) Regarding the quantitative area and volume of 3D fracturing fracture networks, alternate fracturing produced a larger fracture area and improved the

fracturing effect better than sequential and simultaneous fracturing. The two improved alternating fracturing sequences proposed in this study effectively reduced the influence of stress shadows between multiple fractures and increased the area of fracturing networks. Because of the existence of some hard heterogeneous planes in heterogeneous reservoirs, it was difficult for fractures to propagate in these places to form an extended fracture surface. Although the fracture area of the heterogeneous reservoir was small, the fracture volume was larger owing to the water pressure in the fracture.

This study provides a method to understand the mechanisms of stress-dependent unstable propagation of hydraulic fractures from the perspective of numerical analysis, and the main influencing factors are discussed through representative numerical cases. Future research should address the control of stress shadow effects between multiple hydraulic fractures, the stable propagation of fracturing fracture networks, and the optimisation of fracturing scenarios. Addressing these issues is challenging and progress will be reported in the future.

References

Abass, H.H., Soliman, M.Y., Tahini, A.M., Surjaatmadja, J., Meadows, D.L., Sierra, L. (2009), "Oriented fracturing: a new technique to hydraulically fracture an open hole horizontal well", SPE Annual Technical Conference and Exhibition, New Orleans, Louisiana, USA, SPE-124483-MS.

Abousleiman, Y.N., Tran, M.H., Hoang, S., Bobko, C.P., Ortega, A., Ulm, G.J. (2007), "Geomechanics field and laboratory characterization of Woodford shale: the next gas play", SPE Annual Technical Conference and Exhibition, Anaheim, California, USA, SPE-110120-MS.

Bai, T., Pollard, D.D., Gao, H. (2007), "Explanation for fracture spacing in layered materials", Nature, Vol. 403 No.6771, pp. 753–756.

Bazant, Z.P., Salviato, M., Chau, V.T., Visnawathan, H., Zubelewicz, A. (2014), "Why fracking works", International Journal of Applied Mechanics, Vol. 81 No. 10, pp. 1–10.

Bunger, A.P., Zhang, X., Jeffrey, R.G. (2012), "Parameters effecting the interaction among closely spaced hydraulic fractures", Society of Petroleum Engineers, Vol. 17 No. 1, pp. 292–306.

Carter, E. (1957), "Optimum fluid characteristics for fracture extension, in: Howard, G, Fast, C. (Eds.), Drilling and Production Practices", Tulsa, Oklahoma: American Petroleum Institute, API-57-261.

Cheng, W., Jin, Y., Chen, M. (2015a), "Experimental study of step-displacement hydraulic fracturing on naturally fractured shale outcrops", Journal of Geophysics and Engineering, Vol. 12 No. 4, pp. 714–723.

Cheng, W., Jin, Y., Chen, M. (2015b), "Reactivation mechanism of natural fractures by hydraulic fracturing in naturally fractured shale reservoirs", Journal of Natural Gas Science and

Engineering, Vol. 27, pp. 1357–1365.

Cipolla, C., Lolon, E., Erdle, J., Rubin, B. (2010), "Reservoir modeling in shale-gas reservoirs", SPE Reservoir Evaluation & Engineering, Vol. 13 No. 4, pp. 638–653.

Ding, Y., Liu, X., Luo, P. (2017), "The analytical model of hydraulic fracture initiation for perforated borehole in fractured formation", Journal of Petroleum Science and Engineering, Vol. 162, pp. 502–512.

Dutton, S., Flanders, W., Barton, M. (2003), "Reservoir characterization of a Permian deep-water sandstone, East Ford field, Delaware basin, Texas", AAPG Bulletin, Vol. 87, pp. 609–627.

Ebigbo, A., Lang, P.S., Paluszny, A., Zimmerman, R.W. (2016), "Inclusion-based effective medium models for the permeability of a 3D fractured rock mass", Transport Porous Media, Vol. 113 No. 1, pp. 137–158.

ELFEN TGR user and theory manual. (2016), Rockfield Software Ltd., United Kingdom, 2016.

Germanovich, L.N., Astakhov, D.K. (2004), "Fracture closure in extension and mechanical interaction of parallel joints", Journal of Geophysical Research, Vol. 109 No. B2, pp. 1–22.

Germanovich, L.N., Ring, L.M., Astakhov, D.K., Shlyopobersky, J., Mayerhofer, M.J. (1997), "Hydraulic fracture with multiple segments II: modeling", International Journal of Rock Mechanics and Mining Sciences, Vol. 34 No. 3–4, pp. 98–112.

Ghaderi, A., Taheri-Shakib, J., Nik, M. (2018), "The distinct element method (DEM) and the extended finite element method (XFEM) application for analysis of interaction between hydraulic and natural fractures", Journal of Petroleum Science and Engineering, Vol. 171, pp. 422–430.

Guo, T., Zhang, S., Qu, Z., Zhou, T., Xiao, Y., Gao, J. (2014), "Experimental study of hydraulic fracturing for shale by stimulated reservoir volume", Fuel, Vol. 128, pp. 373–380.

Gutierrez, R., Sanchez, E., Roehl, D., Romanel, C. (2019), "XFEM modeling of stress shadowing in multiple hydraulic fractures in multi-layered formations", Journal of Natural Gas Science and Engineering, Vol. 70, pp. 1–15.

He, Q., Suorineni, F., Ma, T., Oh, J. (2017), "Effect of discontinuity stress shadows on hydraulic fracture re-orientation", International Journal of Mining Science and Technology, Vol. 91, pp. 179–194.

Hornby, B.E., Schwartz, L.M., Hudson, J.A. (1994), "Anisotropic effective-medium modeling of the elastic properties of shales", Geophysics, Vol. 59 No. 10, pp. 1570–1583.

Hossain, M., Rahman, M. (2008), "Numerical simulation of complex fracture growth during tight reservoir stimulation by hydraulic fracturing", Journal of Petroleum Science and Engineering, Vol. 60 No. 2, pp. 86–104.

Jo, H. (2012), "Optimizing fracture spacing to induce complex fractures in a hydraulically fractured horizontal wellbore", SPE Americas Unconventional Resources Conference. Society of Petroleum Engineers, SPE-154930-MS.

Ju, Y., Li, Y., Wang, Y., Yang, Y. (2020), "Stress shadow effects and microseismic events during hydrofracturing of multiple vertical wells in tight reservoirs: A three-dimensional numerical model", Journal of Natural Gas Science and Engineering, Vol. 84, pp. 103684.

Kresse, O., Weng, X., Gu, H., Wu, R. (2013), "Numerical modeling of hydraulic fractures interaction in complex naturally fractured formations", Rock Mechanics and Rock Engineering, Vol. 46 No. 3, pp. 555–568.

Kumar, D., Ghassemi, A. (2016), "A three-dimensional analysis of simultaneous and sequential fracturing of horizontal wells", Journal of Petroleum Science and Engineering Vol. 146, pp. 1006–1025.

Lecampion, B., Desroches, J. (2015), "Simultaneous initiation and growth of multiple radial hydraulic fractures from a horizontal wellbore", Journal of the Mechanics and Physics of Solids, Vol. 82, pp. 235–258.

Lisjak, A., Mahabadi, O., He, L., Tatone, B., Kaifosh, P., Haque, S., Grasselli, G. (2018), "Acceleration of a 2D/3D finite-discrete element code for geomechanical simulations using general purpose GPU computing", Computers and Geotechnics, Vol. 100, pp. 84–96.

Manríquez, A. (2018), "Stress behavior in the near fracture region between adjacent horizontal wells during multistage fracturing using a coupled stress-displacement to hydraulic diffusivity model", Journal of Petroleum Science and Engineering, Vol. 162, pp. 822–834.

Manríquez, A.L., Sepehrnoori, K., Cortes, A. (2017), "A novel approach to quantify reservoir pressure along the horizontal section and to optimize multistage treatments and spacing between hydraulic fractures", Journal of Petroleum Science and Engineering, Vol. 149, pp. 579–590.

Marco, A., Moraes, S. (1991), "Diagenesis and microscopic heterogeneity of lacustrine deltaic and turbiditic sandstone reservoirs (Lower Cretaceous), Potiguar Basin, Brazil", AAPG Bulletin, Vol. 75, pp. 1758–1771.

Munjiza, A., Owen, D., Bicanic, N. (1995), "A combined finite-discrete element method in transient dynamics of fracturing solids", Engineering Computations, Vol. 12 No. 2, pp. 145–174.

Olson, J.E., Taleghani, A.D. (2009), "Modeling simultaneous growth of multiple hydraulic fractures and their interaction with natural fractures", SPE Hydraulic Fracturing Technology Conference, The Woodlands, Texas, January, SPE-119739-MS.

Paluszny, A., Tang, X.H., Zimmerman, R.W. (2013), "Fracture and impulse based finite-discrete element modeling of fragmentation", Compute Mechanics, Vol. 52 No. 5, pp. 1071–1084.

Profit, M., Dutko, M., Yu, J., Cole, S., Angus, D., Baird, A. (2015), "Complementary hydro-mechanical coupled finite/discrete element and microseismic modelling to predict hydraulic fracture propagation in tight shale reservoirs", Computational Particle Mechanics, Vol. 3 No. 2, pp. 229–248.

Roussel, N., Sharma, M. (2011a), "Strategies to minimize frac spacing and stimulate natural fractures in horizontal completions", SPE Annual Technical Conference and Exhibition, Denver, Colorado, USA, October, SPE-146104-MS.

Roussel, N.P., Sharma, M.M. (2011b), "Optimizing fracture spacing and sequencing in horizontal-well fracturing", SPE Production & Operations, Vol. 26 No. 02, pp. 173–184.

Roussel, N.P., Manchanda, R., Sharma, M.M. (2012), "Implications of fracturing pressure data recorded during a horizontal completion on stage spacing design", Society of Petroleum Engineers, SPE-152631-MS.

Saberhosseini, S.E., Ahangari, K., Mohammadrezaei, H. (2019), "Optimization of the horizontal-well multiple hydraulic fracturing operation in a low-permeability carbonate reservoir using fully coupled XFEM model", International Journal of Mining Science and Technology, Vol. 114, pp. 33–45.

Sarmadivaleh, M., Rasouli, V. (2014), "Test design and sample preparation procedure for experimental investigation of hydraulic fracturing interaction modes", Rock Mechanics and Rock Engineering, Vol. 48 No. 1, pp. 93–105.

Shan, Q., Jin, Y., Chen, M., Hou, B., Zhang, R., Wu, Y. (2017), "A new finite element method to predict the fracture initiation pressure", Journal of Natural Gas Science and Engineering, Vol. 43, pp. 58–68.

Sobhaniaragh, B., Mansur, W.J., Peters, F.C. (2018a), "The role of stress interference in hydraulic fracturing of horizontal wells", International Journal of Mining Science and Technology, Vol. 106, pp. 153–164.

Sobhaniaragh, B., Nguyen, V.P., Mansur, W.J., Peters, F.C. (2018b), "Pore pressure and stress coupling in closely-spaced hydraulic fracturing designs on adjacent horizontal wellbores", European Journal of Mechanics - A/Solids, Vol. 67, pp. 18–33.

Sriram, G., Dewangan, P., Ramprasad, T. (2014), "Modified effective medium model for gas hydrate bearing, clay-dominated sediments in the Krishna–Godavari Basin", Marine and Petroleum Geology, Vol. 58, pp. 321–330.

Steacy, S., Gomberg, J., Cocco, M. (2005), "Introduction to special section: Stress transfer, earthquake triggering, and time-dependent seismic hazard", Journal of Geophysical Research: Solid Earth, Vol. 110, pp. 1–12.

Sylvia, M., Luiz, C.D.A. (2000), "Depositional and diagenetic controls on the reservoir quality of Lower Cretaceous Pendência sandstones, Potiguar rift basin, Brazil", AAPG Bulletin, Vol. 84, pp. 1719–1742.

Taghichian, A., Zaman, M., Devegowda, D. (2014), "Stress shadow size and aperture of hydraulic fractures in unconventional shales", Journal of Petroleum Science and Engineering, Vol. 124, pp. 209–221.

Terry, D.A., Knapp, C.C. (2018), "A unified effective medium model for gas hydrates in sediments", Geophysics, Vol. 83 No. 6, pp. 317–332.

Wang, Y. (2021a), "Adaptive finite element-discrete element analysis for stratal movement and microseismic behaviours induced by multistage propagation of three-dimensional multiple hydraulic fractures", Engineering Computations, Vol. 38 No. 5, pp. 1350–1371.

Wang, Y. (2021b), "An h-version adaptive FEM for eigenproblems in system of second order ODEs: vector Sturm-Liouville problems and free vibration of curved beams", Engineering Computations, Vol. 38 No. 4, pp. 1807–1830.

Wang, Y., Ju, Y., Yang, Y. (2018a), "Adaptive finite element-discrete element analysis for microseismic modelling of hydraulic fracture propagation of perforation in horizontal well considering pre-existing fractures", Shock and Vibration, Vol. 2018, pp. 1–14.

Wang, Y., Ju, Y., Zhuang, Z., Li, C. (2018b), "Adaptive finite element analysis for damage detection of non-uniform Euler-Bernoulli beams with multiple cracks based on natural frequencies", Engineering Computations, Vol. 35 No. 3, pp. 1203–1229.

Wang, Y., Ju, Y., Chen, J., Song, J. (2019), "Adaptive finite element-discrete element analysis for the multistage supercritical CO_2 fracturing and microseismic modelling of horizontal wells in tight reservoirs considering pre-existing fractures and thermal-hydro-mechanical coupling", Journal of Natural Gas Science and Engineering, Vol. 61, pp. 251–269.

Wang, Y., Ju, Y., Zhang, H., Gong, S., Song, J., Li, Y., Chen, J. (2021), "Adaptive finite element-discrete element analysis for the stress shadow effects and fracture interaction behaviours in three-dimensional multistage hydrofracturing considering varying perforation cluster spaces and fracturing scenarios of horizontal wells", Rock Mechanics and Rock Engineering, Vol. 54, pp. 1815–1839.

Williams, B. (1970), "Fluid loss from hydraulically induced fractures", Journal of Petroleum Exploration and Production Technology, Vol. 22 No. 7, pp. 882–888.

Wong, S., Geilikman, M., Xu, G. (2013), "Interaction of multiple hydraulic fractures in horizontal wells", SPE Journal, SPE-163982-MS.

Wu, Y., Cheng, L., Huang, S., Jia, P., Jin, Z., Xiang, L., Huang, H. (2016), "A practical method for production data analysis from multistage fractured horizontal wells in shale gas reservoirs", Fuel, Vol. 186 No. 15, pp. 821–829.

Yan, C., Jiao, Y. (2018), "A 2D fully coupled hydro-mechanical finite-discrete element model with real pore seepage for simulating the deformation and fracture of porous medium driven by fluid", Computers & Structures, Vol. 196, pp. 311–326.

Yoon, J., Zimmermann, G., Zang, A. (2015), "Numerical investigation on stress shadowing in fluid injection-induced fracture propagation in naturally fractured geothermal reservoirs", Rock Mechanics and Rock Engineering, Vol. 48 No. 4, pp. 1439–1454.

Yu, W., Sepehrnoori, K. (2013), "Optimization of multiple hydraulically fractured horizontal wells in unconventional gas reservoirs", Journal of Petroleum Exploration and Production Technology, Vol. 2013, pp. 1–16.

Yu, W., Zhang, T., Du, S., Sepehrnooria, K. (2015), "Numerical study of the effect of uneven proppant distribution between multiple fractures on shale gas well performance", Fuel, Vol. 142, pp. 189–198.

Zhang, X., Jefrey, R. (2012), "Fluid-driven multiple fracture growth from a permeable bedding plane intersected by an ascending hydraulic fracture", Journal of Geophysical Research: Solid Earth, Vol. 117 No. B12, pp. 1–12.

Zhou, J., Chen, M., Jin, Y., Zhang, G. (2008), "Analysis of fracture propagation behavior and fracture geometry using a tri-axial fracturing system in naturally fractured reservoirs", International Journal of Rock Mechanics and Mining Sciences, Vol. 45 No. 7, pp. 1143–1152.

Zhou, J., Jin, Y., Chen, M. (2010), "Experimental investigation of hydraulic fracturing in random naturally fractured blocks", International Journal of Rock Mechanics and Mining Sciences, Vol. 47 No. 7, pp. 1193–1199.

Chapter 9   Unstable propagation of multiple three-dimensional hydraulic fractures and shear stress disturbance considering thermal diffusion

9.1   Introduction

Hydrofracturing is a core technology for the development of deep tight rock reservoirs. The multistage fracturing of multiple perforation clusters in horizontal wells involves thermal diffusion between the fluid in the reservoir and the pores and fractures, fluid flow, and rock matrix deformation. The thermal diffusion effect and multiple physical field couplings are typical characteristics of deep tight rock fracturing. In the process of hydrofracturing, the stress shadow effect and deflection behaviour of multiple fracture disturbances between the three-dimensional (3D) fracture network propagation are important factors affecting the propagation morphology of the spatial fracture network and the fracturing effect (He *et al.*, 2017; Gutierrez *et al.*, 2019; Sobhaniaragh *et al.*, 2018; Yoon *et al.*, 2015). In the process of hydraulic fracture propagation, 3D fractures are accompanied by spatial deflection and compression between fractures, resulting in the unstable propagation of fractures (Wong *et al.*, 2013). Some studies have found that the perforation cluster spacing and initiation sequence of multistage fracturing will result in varying degrees of fracture deflection, and the unstable propagation of fractures will affect the fracture orientation control and fracture network design (Bažant *et al.*, 2014; Zhang and Jeffrey, 2012). The formation stress interference region caused by the fracture propagation shows superposition and overlay behaviours in multiple fractures, and the resulting stress shadow region will disturb fracture propagation (Sobhaniaragh *et al.*, 2018; Lecampion and Desroches, 2015; Manríquez, 2018). The influence of the thermal diffusion effect on the 3D fracture propagation, different perforation cluster spacings, and propagation disturbance behaviour of fracture networks under typical fracturing scenarios (sequential, simultaneous, and alternate fracturing) have become the focus of theoretical and practical research.

Numerical computation has become an important means to study multi-physical field coupling, unstable propagation of multiple fractures, and engineering-scale analysis of the 3D propagation disturbance of hydraulic fracture networks. Using numerical values, models, and complex condition analysis scenarios of fracture propagation in rock mass, we quantitatively analysed the interaction between the fracture network and stress shadow effect, and explored the mechanisms of fracture initiation, propagation disturbance, and deflection (Ghaderi et al., 2018; Hossain and Rahman, 2008; Paluszny et al., 2013). Taghichian et al. (Taghichian et al., 2014) developed a numerical model to analyse the interaction between the stress shadow and fractures by combining the finite element model of fractures with simple geometric shapes and the analytic solution of plane fractured fractures. They studied the stress shadow area and fracture propagation height between fractures in a tight rock mass. Based on the enhanced two-dimensional (2D) displacement discontinuity method, Kresse et al. (Kresse et al., 2013) established stress shadow and fracture propagation height correction models for branch fractures in complex hydraulic fracture networks to study the disturbance deflection behaviour of fractures. Kumar and Ghassemi (Kumar and Ghassemi, 2016) simulated the fracture propagation and the in-situ stress field redistribution behaviour of multiple perforation clusters in horizontal wells based on the finite element and boundary element methods. The adaptive finite element method shows good efficiency in vibration, stability, and damage analysis, and other complex problems (Wang et al., 2018, 2021a; Wang, 2021b; Wang and Wang, 2022). The traditional finite element method cannot reliably and effectively solve fracture problems owing to the limitation of meshing in the fracture tip region (Azadi and Khoei, 2011); however, the adaptive finite element method enables the same. In previous studies by the author of this book, the adaptive mesh refinement method and the coupling technology of finite and discrete elements were introduced to study the dynamic propagation of fractures by considering the thermal diffusion effect (Wang et al., 2019) and the 3D propagation evolution behaviour of hydraulic fractures (Wang, 2021a). Other studies using this model also found that the deflection behaviours of fractures depended on the heterogeneity of rock mass and the distribution of natural fractures. This will change the varying range and area of stress in the reservoir rock, thus affecting the initiation and propagation of fractures (Wang and Liu, 2021; Wang et al., 2021b).

In this study, the thermal-hydro-mechanical coupling effect was considered to further study the stress shadow effect and deflection behaviour of multiple fracture disturbances in the 3D propagation of a hydraulic fracture network. The adaptive finite element-discrete element algorithm is introduced to obtain high-precision stress solutions using adaptive mesh refinement in the local domain around the fracture tip to effectively describe the dynamic propagation of the fracture; thus, the analysis strategy and computation scheme are developed. We established an engineering-scale 3D numerical model of horizontal well multistage fracturing. Typical cases were used to compute and analyse the thermal diffusion effect of 3D fracture propagation, the spatial propagation disturbance behaviour of the fracture network under different perforation cluster spacings, and different fracturing scenarios (sequential, simultaneous, and alternate fracturing).

9.2   Governing equation for hydrofracturing by considering thermal-hydro-mechanical coupling

In this study, the physical fields involved in the fracturing process of a rock mass include the temperature, fluid, and solid fields. We further introduce differential control equations for the solid deformation of the rock mass matrix, fluid flow in pores and fractures, and thermal diffusion.

9.2.1   Solid deformation of rock matrix

The matrix deformation control equation of rock mass is given by (Wang *et al.*, 2019; Wang, 2021a):

$$L^{\mathrm{T}}(\boldsymbol{\sigma}^{\mathrm{e}} - \alpha \boldsymbol{m} p_{\mathrm{s}}) + \rho_{\mathrm{b}} \boldsymbol{g} = \boldsymbol{0}, \tag{9.1}$$

where L denotes the differential operator, $\boldsymbol{\sigma}^{\mathrm{e}}$ denotes the effective stress tensor, α denotes the Biot's coefficient, $\boldsymbol{m}$ denotes the identity tensor, p_{s} denotes the pore fluid pressure in the rock formation, ρ_{b} denotes the wet bulk density, and $\boldsymbol{g}$ denotes the gravity vector.

9.2.2   Fluid flow in the rock matrix and fracture networks

The governing equation of seepage in a rock matrix is given by:

$$\mathrm{div}\left[\frac{k}{\mu_{\mathrm{l}}}(\nabla p_{\mathrm{l}} - \rho_{\mathrm{l}} g)\right] = \left(\frac{\phi}{K_{\mathrm{l}}} + \frac{\alpha - \phi}{K_{\mathrm{s}}}\right)\frac{\partial p_{\mathrm{l}}}{\partial t} + \alpha \frac{\partial \varepsilon_{v}}{\partial t}, \tag{9.2}$$

where k is the inherent permeability of the porous rock, μ_{l} is the viscosity of the pore liquid, K_{l} represents the bulk stiffness of the pore liquid, K_{s} represents the bulk stiffness of the solid grains, ε_v represents the volumetric strain, and t is the current moment.

The fluid flow control equation in the fracture is given by:

$$\frac{\partial}{\partial x}\left[\frac{k^{fr}}{\mu_n}\left(\nabla p_n - \rho_{fn} g\right)\right] = S^{fr}\frac{d p_n}{d t} + \alpha\left(\Delta\dot{e}_{\varepsilon}\right), \tag{9.3}$$

where k^{fr} is the intrinsic permeability, μ_n is the viscosity of the fracturing fluid, p_n is the fluid pressure in the fracture, ρ_{fn} is the fluid density in the fracture, S^{fr} describes the storage coefficient, and $\Delta\dot{e}_{\varepsilon}$ represents the aperture strain rate. According to the fluid plate theory, the fluid permeability in the fracture is given by:

$$k^{fr} = \frac{e^2}{12}, \tag{9.4}$$

where e denotes the fracture width. The compressibility parameters are expressed as follows:

$$S^{fr} = \left(\frac{1}{e}\right)\left[\left(\frac{1}{K_n^{fr}}\right)+\left(\frac{e}{K_f^{fr}}\right)\right], \tag{9.5}$$

where K_n^{fr} is the fracture normal stiffness and K_f^{fr} is the bulk modulus of the fracturing fluid.

Fluid leak-off occurs when the injected fluid enters the fracture, which is sometimes as high as 50–80% in the engineering application of hydrofracturing. Fluid leak-off is vital in accurately describing the fracture propagation, and the following computation method for fluid leak-off velocity q_1 is often adopted:

$$t - t_{exp} < t_{sp}, \quad q_1 = \frac{V_{sp}}{t_{sp}}, \tag{9.6a}$$

$$t - t_{exp} \geqslant t_{sp}, \quad q_1 = \frac{C}{\sqrt{t - t_{sp}}}, \tag{9.6b}$$

with the two-term parameters given by:

$$C = \frac{2C_I C_{II}}{C_I + \sqrt{C_I^2 - 4C_{II}^2}}, \quad C_I = \left[\frac{k_f \phi_f \Delta p}{2\mu_f}\right]^{0.5}, \quad C_{II} = \left[\frac{k_r \phi_r c_T}{\mu_r}\right]^{0.5}\Delta p,$$

where t_{sp} is the period from fluid flow to the section to the outflow to the surrounding rock, t_{exp} is the moment of fluid flow to the section, t is the moment of fluid flow through the section, V_{sp} is the fluid leak-off volume, and C, C_I, and C_{II} are the leak-off parameters.

9.2.3   Thermal diffusion

The governing equation of thermal diffusion between the rock mass matrix and fluid in the pores and fractures is given by (Wang *et al.*, 2019):

$$\text{div}\left[k_b \nabla T_f \right] = \rho_b c_b \frac{\partial T_f}{\partial t} + \rho_f c_f \boldsymbol{q}_f \nabla T_f , \tag{9.7}$$

where k_b is the thermal conductivity, T_f is the fluid temperature, ρ_b is the bulk density, c_b is the specific heat coefficient, ρ_f is the fluid density, c_f is the specific heat coefficient of the fluid, and $\boldsymbol{q}_f$ is the Darcy fluid flux.

In the process of thermal-hydro-mechanical coupling and parameter transfer, the temperature transfers between the fluid and solid is achieved by thermal diffusion, and the change in temperature causes a change in the stress field of the solid. The effect of the pore fluid on the solid is realised by the effective stress principle, wherein the fluid transfers the pore fluid pressure field to the solid. The effect of the fracture fluid on the solid is realised by the water pressure of the fracture wall, and the fluid transfers the fluid pressure field to the solid. The effect of solids on the pore fluid is realised by the volumetric strain, which transfers the strain field to the fluid.

9.3   Heat transfer and mesh adaptive analysis of the fracture propagation process

9.3.1   Heat transfer between the finite element nodes

The differential governing Eqs. (9.1)–(9.3) of the solid deformation of the rock mass matrix, fluid flow in pores and fractures, and thermal diffusion are discretized using the conventional finite element method   (Profit *et al.*, 2016). After discretization by finite element elements, the rock matrix and fracture regions use the fracture plane nodes for heat transfer. The form of heat transfer between the element nodes is shown in Figure 9.1, and the temperature and heat flux at the nodes are (Wang *et al.*, 2019):

$$q_c^1 = \alpha_c \left(T_N \right)\left(T_N - T_f^1 \right), \tag{9.8a}$$

$$q_c^2 = \alpha_c \left(T_N \right)\left(T_N - T_f^2 \right), \tag{9.8b}$$

where q_c^1 and q_c^2 are the contact heat fluxes between the network and formation nodes, respectively; T_f^1 and T_f^2 are the corresponding formation nodal temperatures, respectively; T_N is the temperature value of the node within the fracture; and α_c is the contact thermal conductivity. Temperature changes in the rock masses cause volume change.

$$\frac{\Delta V}{V} = \alpha_T \Delta T, \tag{9.9}$$

where ΔT is the temperature change in the rock, ΔV is the volume change, V is the initial volume, and α_T is the linear thermal propagation coefficient of the rock matrix.

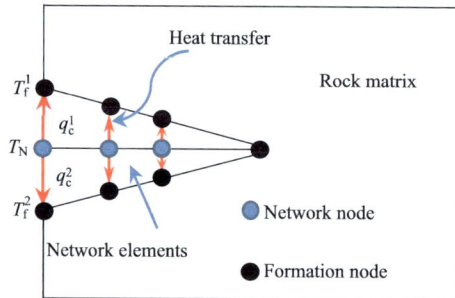

Figure 9.1. Heat transfer between the finite element nodes in the formation and network.

9.3.2 Fracture criteria

The tensile strength and fracture energy of the element were σ_t and G_f, respectively. The stress-strain relationship of the element is shown in Figure 9.2; when the maximum principal tensile stress reaches σ_t, the element begins to undergo damage ($d = 0$); when the principal tensile stress is 0, the damage of the element reaches its maximum ($d = 1$) and results in fracturing. The area enclosed by the stress-strain curve and the x-axis is the fracture energy G_f; the slope of the stress-strain curve at the damage stage is H; the elastic modulus at the damage stage is $\tilde{E}$; and E is the elastic modulus at the elastic stage. Implementing the mesh refinement in fracture propagation: in the process of fluid-driven fracture propagation, stress concentration appears at the fracture tip and forms a damage zone ($d = 1$). The damage zone indicates that the element was damaged, and the fracture propagated vertically along the direction of the maximum principal stress. The fracture length is the straight-line distance in the damage zone. When the predicted fracture length reached the propagation length given in advance, the conventional discrete element method was used to deal with the fracture between elements for the fracture propagation. The propagation length was the predicted fracture length. The mesh was refined in the specified region of the fracture tip, and the new damage region was computed.

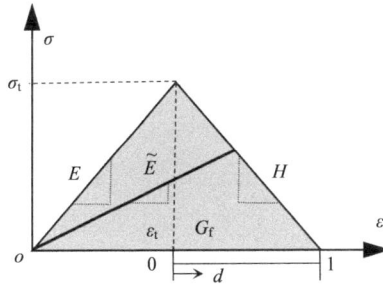

Figure 9.2. Relationship of stress-stain and damage analysis.

9.3.3   Local mesh refinement

In this study, the convergence of the high-precision stress solution computed using the superconvergent patch method (Zienkiewicz and Zhu, 1992) is higher by at least one order of magnitude compared to that of the original finite element solution. This can be used to estimate and control the error of the original finite element solution. The error estimation parameters are defined for each element K (Wang et al., 2021a, 2017; Wang and Wang, 2022):

$$\xi_K = \frac{\|e\|_K}{\bar{e}_m} \leqslant 1, \tag{9.10}$$

where $\bar{e}_m$ represents the average error of all elements in the domain, m represents the number of patched elements, and $\|e\|$ represents the total error of all elements in the form of energy modules. If Equation (9.10) is not satisfied, the error is too large, and the mesh size needs to be adjusted. The new mesh size was estimated as follows:

$$h_{\text{new}} = \xi_K^{-1/m} h_K, \tag{9.11}$$

where h_K is the current mesh size of element K.

   To generate meshes more reasonably and efficiently, the mesh can only be refined in the fracture tip region. Error estimation and mesh refinement of the elements at the fracture tip can significantly reduce the computing time of the adaptive solution of the 3D problem, and provide a reliable description of the fracture propagation path.

9.4   Three-dimensional numerical model with multiple perforations for parallel hydraulic fractures

In this study, we established an engineering-scale initial geometric model of multiple perforation cluster multistage fracturing of a horizontal well in a deep tight rock mass, as shown in Figure 9.3. The model sizes were a=600 m and b=400 m. The basic physical parameters of the model are presented in Table 9.1. Five perforation cluster locations were chosen and numbered from 1 to 5. According to the different fracturing sequences of the perforation clusters, the sequence of sequential fracturing scenarios is 1→2→3→4→5, that of alternate fracturing scenarios is 1→3→2→5→4, and that of simultaneous fracturing scenarios is 1→2→3→4→5 (i.e. five perforation clusters simultaneously undergo fracturing and propagation). Table 9.2 presents the duration and total time of each stage of sequential and simultaneous fracturing. Five perforation clusters were evenly distributed along the horizontal well section, and the coordinate position was y_i (i is the perforation cluster numbering value, which can range from 1 to 5):

$$y_i = (300 - 2a) + a(i-1), \ i = 1,2, \cdots,5, \tag{9.12}$$

where a is the cluster spacing.

To study the influence of perforation cluster spacing on the fracture propagation, this study computed and analysed the fracturing conditions of perforation cluster spacings of 100, 75, 50, and 25 m, as presented in Table 9.3. To study the influence of the reservoir temperature gradient on the fracture propagation, the temperature gradient conditions of the different fracturing fluids and rock matrices are listed in Table 9.4. Notably, the computation and comparative analysis were mainly conducted for two cases with the temperature effect (thermal-hydro-mechanical coupling) and those without the temperature effect (hydro-mechanical coupling only). This study focused on the fracture propagation behaviour with or without the temperature effect and with different temperature gradients, and did not study the fracture propagation behaviour with different temperature gradients. To reliably describe fracture initiation at the initial stage, initial mesh refinement was conducted in the perforation area using 150 m $\leqslant x \leqslant$ 250 m, y_1–50 m $\leqslant y \leqslant y_5$+50 m, and 150 m $\leqslant z \leqslant$ 250 m. In this study, linear tetrahedral elements were used to discretize the model. Figure 9.4 shows the cross-section of the model meshing; a finer initial mesh was used in each perforation area. The computation of the aforementioned 3D numerical model of multistage fracturing in a horizontal well was conducted using the rock mechanics computation and analysis program package ELFEN (Rockfield Software Ltd., 2016).

Table9.1. Basic physical parameters of the model.

Parameter	Value
Vertical *in-situ* stress (z direction) S_v /MPa	40
Horizontal minimum *in-situ* stress (x direction) S_h /MPa	46
Horizontal maximum *in-situ* stress (y direction) S_H /MPa	60
Fluid injection rate Q/ (m^3/s)	0.5
Leak-off coefficient C_{I} /(m^3/ s$^{1/2}$)	1×10^{-16}
Leak-off coefficient C_{II} /(m^3/ s$^{1/2}$)	1×10^{-16}
Pore pressure p_s /MPa	10
Biot's coefficient α	0.75
Elastic modulus E /GPa	31
Poisson's ratio v	0.22
Permeability k /nD	50
Porosity ϕ	0.05
Dynamic viscosity coefficient of the fracturing fluid μ_n /(Pa · s)	1.67×10^{-3}
Bulk modulus of the fracturing fluid K_f^{fr} /MPa	2000
Tensile strength σ_t /MPa	5.26
Fracture energy G_f /(N · m)	165

Table 9.2. Duration and total time of the multiple fracturing stages for sequential and alternate fracturing.

Fracturing stages	Duration Δt /s	Total time t /s
Initial equilibrium stage	10	10
First stage	400	410
Second stage	400	810
Third stage	400	1210
Fourth stage	400	1610
Fifth stage	400	2010

Table 9.3. Perforation cluster spacing for fracturing scenarios.

Fracturing scenarios	Perforation cluster spacing a /m			
Sequential	100	75	50	25
Simultaneous	100	75	–	–
Alternate	100	75	–	–

Table 9.4. Temperature gradients for fracturing scenarios.

Category	Temperature T /℃	
Fracturing fluid	20	35
Rock matrix	60	60

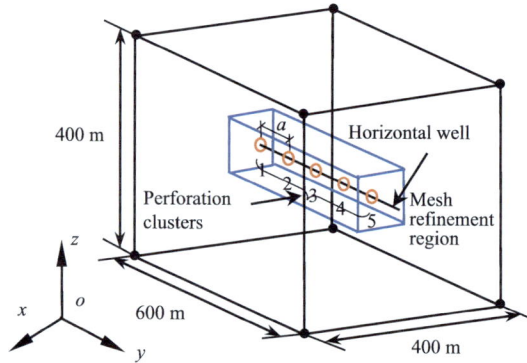

Figure 9.3. Initial geometric engineering-scale 3D model of multistage hydrofracturing of a horizontal well with multiple perforation clusters.

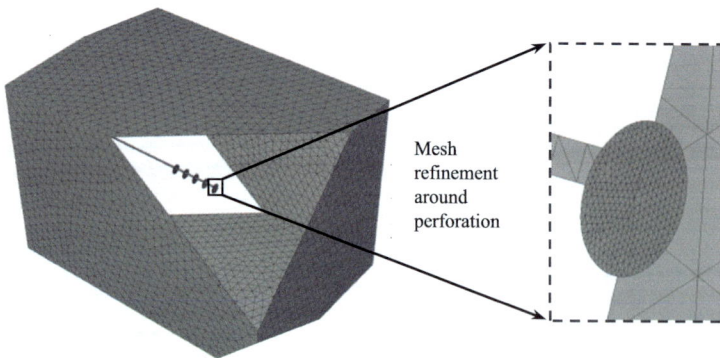

Figure 9.4. Finite element model and initial mesh.

9.5   Results and analysis of unstable dynamic propagation of the three-dimensional parallel hydraulic fractures

9.5.1   Perforation cluster spacing

The mesh adaptive optimisation refinement for the analysis of the numerical model fracturing process is described below, and the disturbance behaviour of the fracture network propagation under different adjacent perforation cluster spacings is studied using each optimisation mesh.

Figure 9.5 shows the local mesh refinement in the fracture area under sequential fracturing and the results of fracture propagation and local mesh refinement in the first, third, and fifth stages, respectively. An external rough mesh was used outside the reservoir model, and a finer mesh was used in the area around the fracture. With the dynamic propagation of the fracture, the mesh around the fracture was adaptively refined and enriched, and the resulting high-quality mesh ensured the accuracy of the stress field solution in the fracture tip region and the reliability of the fracture propagation path.

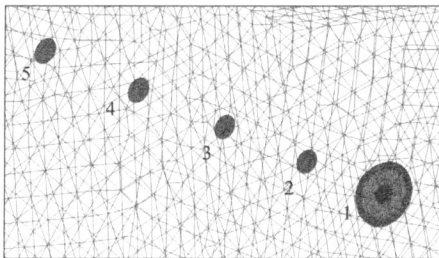

(a)First stage, t=410 s

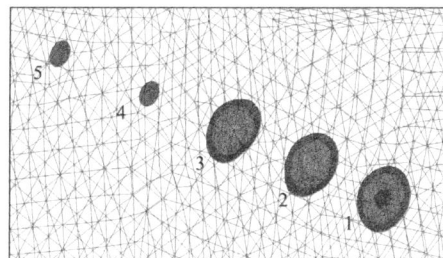

(b) Third stage, t=1210 s

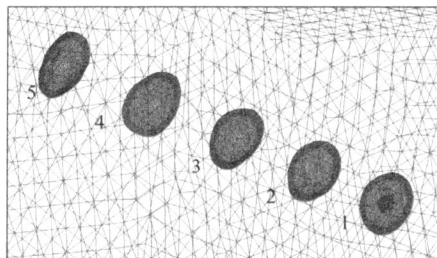

(c) Fifth stage, t=2010 s

Figure 9.5. Local mesh refinement around the fracture domains in sequential fracturing.

Figure 9.6 shows the final fracture morphology and stress field results for different adjacent perforation cluster spacings during sequential fracturing. It also provides the first principal stress in the propagation process of hydrofracturing tensile fracture. The fractures had different degrees of mutual disturbance and deflection. Figure 9.6 (a) shows the fracture propagation morphology with a perforation cluster spacing of $a=100$ m. Owing to the large spacing of each perforation cluster, the fracture is close to parallel and has stable straight propagation. Figure 9.6 (b) shows the fracture propagation morphology with a perforation cluster spacing of $a=75$ m. Owing to the reduction in the perforation cluster spacing, the disturbance stress field between the fractures interfered with each other and there was a slight deflection from the second fracture to the left (away from other fractures), and the cumulative disturbance and deflection of subsequent fracture propagation increased. Figure 9.6 (c) shows the fracture propagation morphology for a perforation cluster spacing of $a=50$ m. As the spacing of the perforation clusters continued to shrink, the disturbance stress field between the fractures appeared as superposition and interference, and the deflection intensified from the second fracture. Figure 9.6 (d) shows the fracture propagation morphology with a perforation cluster spacing of $a=25$ m. Owing to the sudden decrease in the distance of each perforation cluster, the disturbance stress field between the fractures shows the severe interference, and the deviation starts from the second fracture. Perforation cluster spacing is an important factor affecting the unsteady propagation of 3D fractures.

In order to reveal the deflection behaviour of fractures between adjacent perforation clusters, Figure 9.7 shows the evolution process of shear stress field in the process of fracture propagation. It can be seen that under the sequential fracturing with the perforation cluster spacing of 75m, the in-situ stresses in the local area at the tip of adjacent fractures have changed greatly. The positive and negative shear stresses change appears on both sides of the first fracture tips, and the shear stress at the second fracture tip offsets the stress area (stress shadow area) on the left side of the first fracture, resulting in large shear stress value and deflection on the left side of the second fracture. If the spacing between adjacent perforation clusters continues to decrease, the superposition and disturbance of shear stress (stress shadow effect) will be intensified, resulting in the deflection of the adjacent fractures.

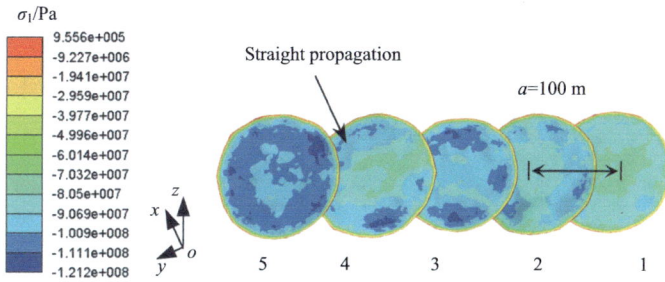

(a) Perforation cluster spacing of a=100 m.

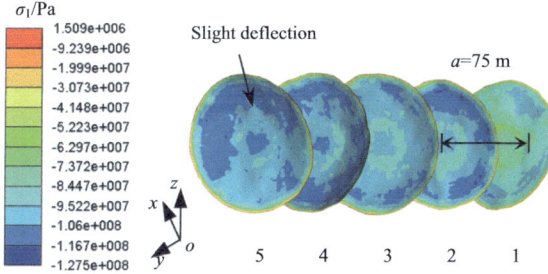

(b) Perforation cluster spacing of a=75 m.

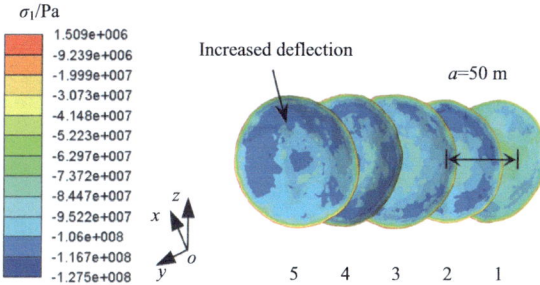

(c) Perforation cluster spacing of a=50 m.

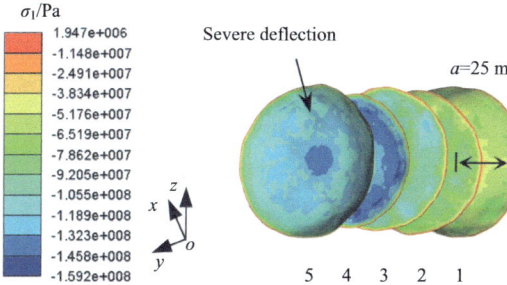

(d) Perforation cluster spacing of a=25 m.

Figure 9.6. Final morphology of the fracture network and stress results (first principal stress, Pa) in sequential fracturing.

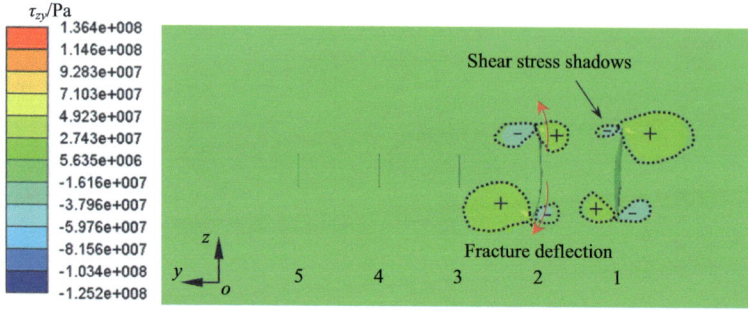

Figure 9.7. Disturbance of shear stress τ_{zy} (Pa) in sequential fracturing for a cluster spacing of 75m ("+" represents positive shear stress, and "−" represents negative shear stress).

9.5.2   Sequential, simultaneous, and alternate fracturing

Under the physical and mechanical parameters and cases of the model in this study, when the spacing between perforation clusters is a=75 m, the disturbance phenomenon of the fracture space propagation occurs. Based on this cluster spacing, the disturbance behaviour of the fracture network propagation under sequential, simultaneous, and alternate fracturing is discussed below.

Figure 9.8 shows the results of the fracture propagation and stress field evolution at each stage of sequential fracturing. Figure 9.8 (a) shows the fracture distribution during the first stage. The first fracture straightly propagated in a space close to the plane because of the lack of interference from other fractures. Figure 9.8 (b) shows the fracture morphology in the second stage. The second fracture exhibited a slight deflection in space owing to the disturbance of the first fracture to the *in-situ* stress field. Figure 9.8 (c) shows the fracture morphology in the third stage. The disturbance of the first and second fractures in the *in-situ* stress field intensified the deflection of the third fracture. Figures 9.8 (d) and 8 (e) show the fracture morphology of the fourth

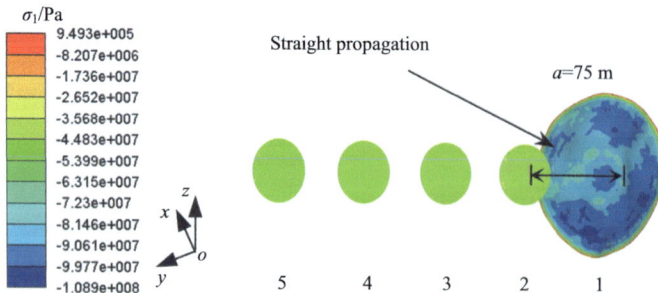

(a) First stage, t=410 s

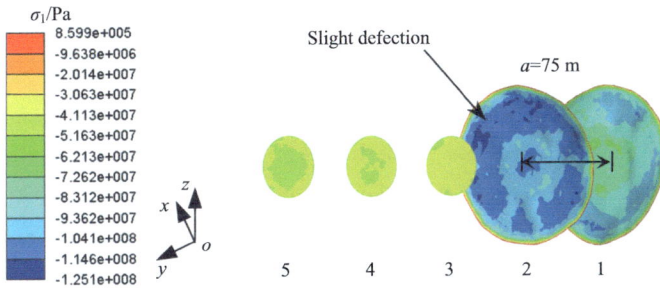

(b) Second stage, t=810 s

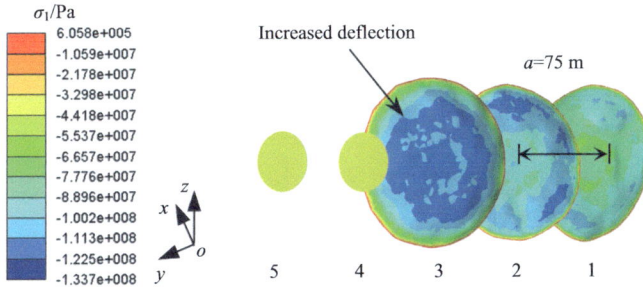

(c) Third stage, t=1210 s

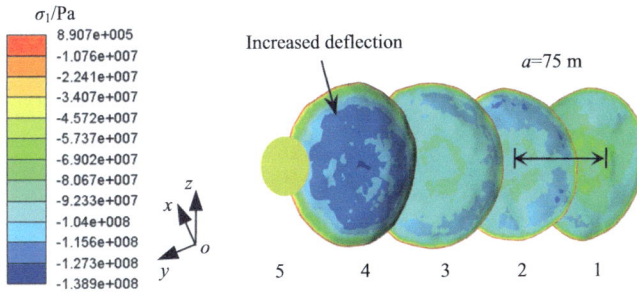

(d) Fourth stage, t=1610 s

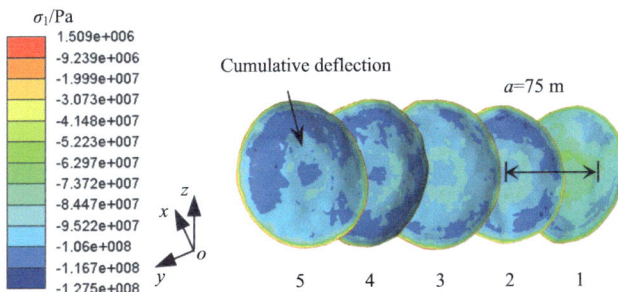

(e) Fifth stage, t=2010 s

Figure 9.8. Dynamic propagation of the fracture network and evolution of stress results in sequential fracturing.

and fifth stages. Cumulative deflection occurs in the fracturing process of the remaining perforation clusters and the fractures are all tilted in the same direction, which is typical fracture propagation behaviour of sequential fracturing.

Figure 9.9 shows the results of the fracture propagation and stress field evolution at each stage of simultaneous fracturing. Figure 9.9 (a) shows the distribution of fractures in the intermediate stage (t=1210 s). All fractures were close to the plane and propagated stably. Figure 9.9 (b) shows the fracture morphology at the final moment. The first and fifth fractures, located on both sides of the perforation cluster, deflect outwards, which is typical fracture propagation behaviour for simultaneous fracturing.

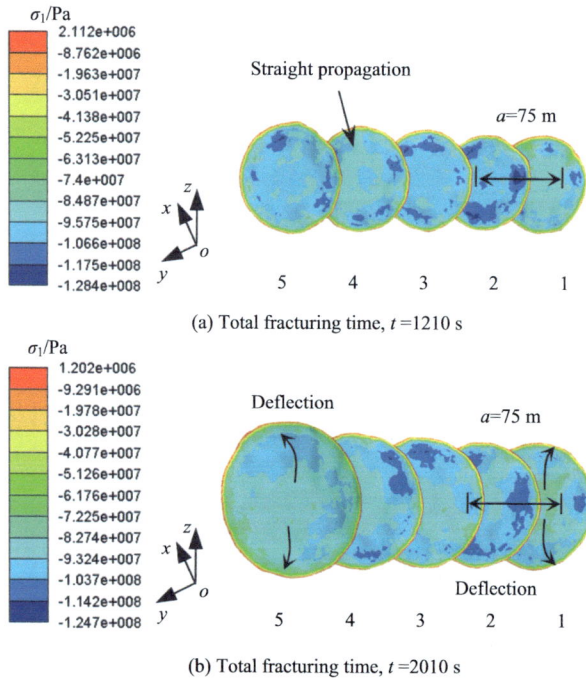

(a) Total fracturing time, t =1210 s

(b) Total fracturing time, t =2010 s

Figure 9.9. Dynamic propagation of the fracture network and evolution of stress results in simultaneous fracturing.

In order to reveal the fracture deflection behaviour caused by different initiation sequences of adjacent perforation clusters, Figure 9.10 shows the shear stress distribution on both sides of the fracture tip during simultaneous fracturing. As the fractures propagate at the same time, the second, third and fourth fractures in the middle are affected by the shear stress field of the cracks on both sides simultaneously, which makes the fractures not easy to deflect. For the first and fifth fractures located in the initial and last stages of horizontal wells, they are only affected by the shear stress field of one side of the fracture, which makes these fractures prone to deflection.

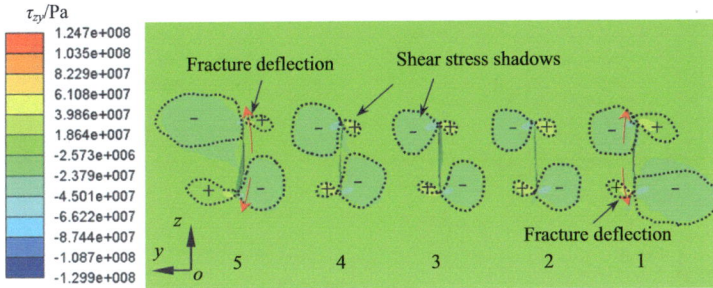

Figure 9.10. Disturbance of shear stress τ_{zy} (Pa) in simultaneous fracturing for a cluster spacing of 75m ("+" represents positive shear stress, and "–" represents negative shear stress).

Figure 9.11 shows the results of fracture propagation and stress field evolution at each stage under alternate fracturing. Figure 9.11 (a) shows the fracture distribution of the first stage, which achieved the same fracturing results under the same conditions as those of the first stage, as shown in Figure 9.8 (a). In particular, because there is no interference from other fractures, the first fracture propagates close to the plane in space. From the stress field analysis process shown in Figure 9.7, it can be seen that due to the implementation of alternate fracture initiation, the spacing of perforation clusters with successive fracture initiation actually becomes larger, which results in low stress shadow effect and weakens the deflection of fractures. Figure 9.11 (b) shows the fracture morphology in the second stage. The third fracture begins to propagate alternately. Currently, the spacing between the first and third fractures is twice that of the original perforation cluster, which reduces the mutual disturbance between the fractures and causes the third fracture to propagate in the plane. Figure 9.11 (c) shows the fracture morphology during the third stage. The second fracture, which lies between the first and third fractures, begins to propagate. The second fracture finally spreads in a plane under the influence of fractures on both sides. Similarly, the fourth and fifth fractures propagated alternately. As shown in Figures 9.11 (d) and (e), only a slight deflection occurred during the fracture propagation. In conclusion, multiple fractures can be stably propagated by controlling the initiation sequence without changing the perforation cluster spacing.

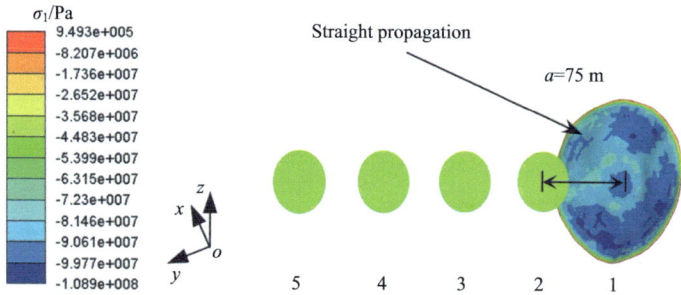

(a) First stage, $t = 410$ s

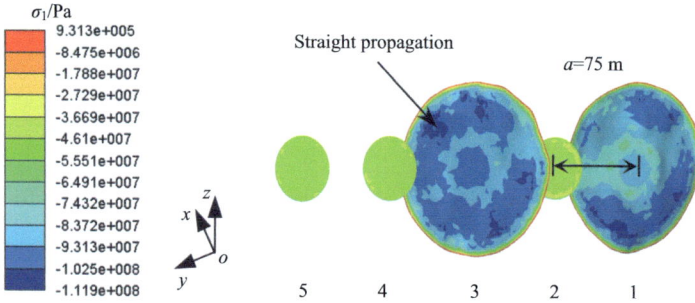

(b) Second stage, $t = 810$ s

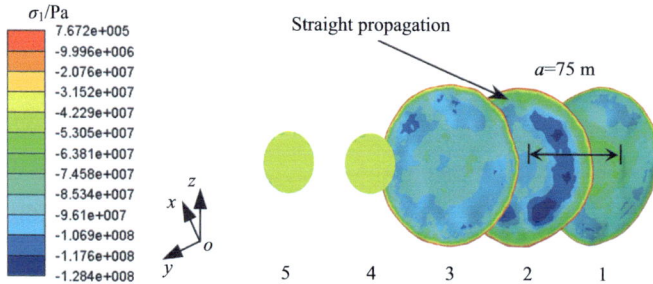

(c) Third stage, $t = 1210$ s

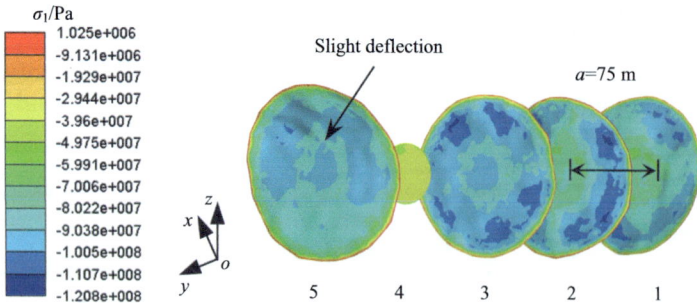

(d) Fourth stage, $t = 1610$ s

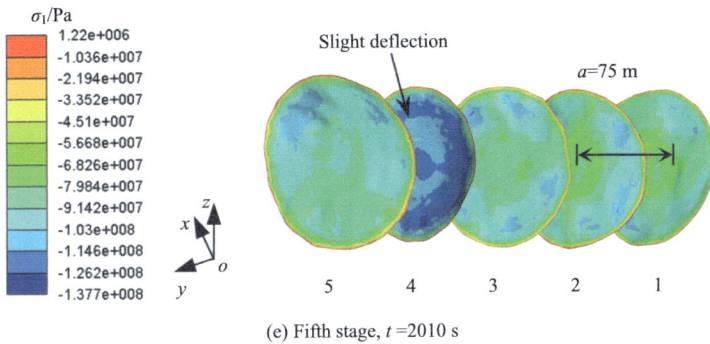

(e) Fifth stage, t =2010 s

Figure 9.11. Dynamic propagation of the fracture network and evolution of stress results in alternate fracturing.

To quantitatively analyse the fracture propagation behaviour, this study analysed the fracture area and volume results under various fracturing scenarios, as presented in Table 9.5, and presented the results for the perforation cluster spacings of 100 and 75 m. In different fracturing scenarios, the final total area and volume of the fractures decreased with decreasing perforation cluster spacing, and the spacing of the perforation clusters plays a controlling role. By comparing various fracturing scenarios, simultaneous fracturing resulted in a smaller fracture area and volume than sequential fracturing, whereas alternate fracturing resulted in a larger fracture area and volume. The initiation sequence of perforation clusters plays a controlling role. When studying the significant influence and stress interference effect between fractures, the author and the research group continued to reduce the spacing of perforation clusters to a=50, 25, and 12.5 m. Consequently, significant differences were observed in the fracture area and volume, and the influences of fracture deflection and stress field interference on the spacing of perforation clusters were obtained (Wang *et al.*, 2021a). In

Table 9.5. Results of the fracture areas and volumes for different fracturing scenarios.

Fracturing scenarios	Category	a =100 m	a =75 m
Sequential	Fracture area S /m^2	100.66	99.346
	Fracture volume V /m^3	846.58	845.46
Simultaneous	Fracture area S /m^2	100.29	98.523
	Fracture volume V /m^3	844.76	842.15
Alternate	Fracture area S /m^2	103.23	99.467
	Fracture volume V /m^3	849.99	847.13

conclusion, under the same perforation cluster spacing, alternate fracturing becomes an effective scheme to alleviate the disturbance of 3D propagation of the fracture network and optimise the spatial fracture network morphology.

9.5.3   Thermal diffusion effects

Thermal diffusion occurs within the fluid in the rock mass matrix, pores, and fractures. Figure 9.12 shows the thermal diffusion behaviour in the fracture region. The temperature gradient between the fracturing fluid and rock matrix causes thermal diffusion in the annular region of the three-dimensional fracture plane.

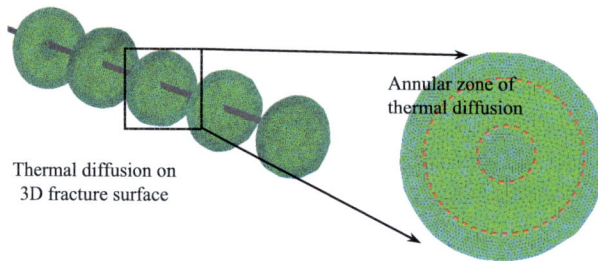

Figure 9.12. Thermal diffusion around fracture domains.

The depth of the reservoir studied in this study was approximately 2000 m. Multistage fracturing of horizontal wells was conducted on this reservoir. The reservoir temperature at this depth was approximately 60 °C, with a temperature gradient of 30 °C/km. To analyse the heat transfer between the fracturing fluid and rock matrix in deep tight rock fractures, we quantitatively analysed different temperature gradients (20 °C and 60 °C, 35 °C and 60 °C), fracture areas, and volumes. To analyse the influence of the thermal diffusion effect, this study computes the fracturing process by considering the thermal-hydro-mechanical coupling effect of the temperature field, but does not consider the hydro-mechanical coupling effect of the temperature field. Figure 9.13 shows a comparison of the fracture areas under the different fracturing scenarios. The final fracture area when considering the effect of the temperature field was larger than that without the effect of the temperature field. Ignoring the thermal diffusion effect underestimates the fracture network propagation. For the conditions with weak (a=100 m) and enhanced (a=75 m) fracture propagation disturbances, as shown in Figures 9.13 (a) and (b), respectively, the fracture area shows the same result: the larger the temperature gradient, the stronger the thermal diffusion effect. Compared with sequential and simultaneous fracturing, the thermal diffusion effect can significantly improve the fracture propagation in the alternative fracturing scheme, which is favourable for stable fracture propagation.

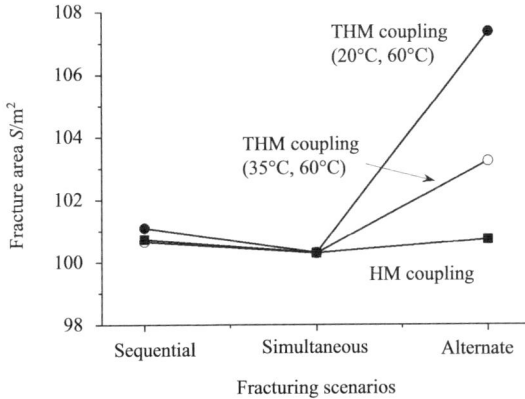

(a) Perforation cluster spacing of a=100 m.

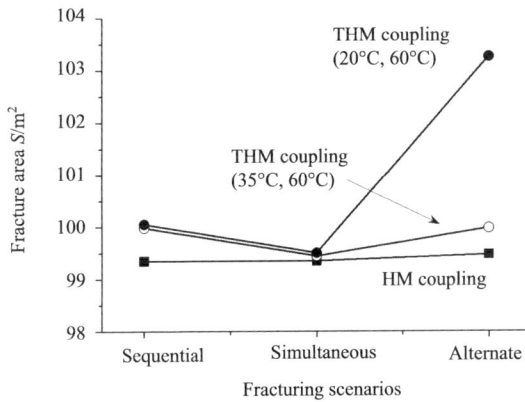

(b) Perforation cluster spacing of a=75 m.

Figure 9.13. Comparison of the fracture areas in different fracturing scenarios.

Figure 9.14 shows a comparison of the fracture volumes under different fracturing scenarios. The final fracture volumes with the effect of the temperature field were all larger than those without it. Ignoring the thermal diffusion effect underestimates the fracture network propagation. For the conditions with weak and enhanced propagation disturbances, as shown in Figures 9.14 (a) and 14 (b), respectively, the results of the fracture area are consistent: the larger the temperature gradient, the stronger the thermal diffusion effect. The thermal diffusion effect can significantly improve the fracture propagation in sequential, simultaneous, and alternate fracturing.

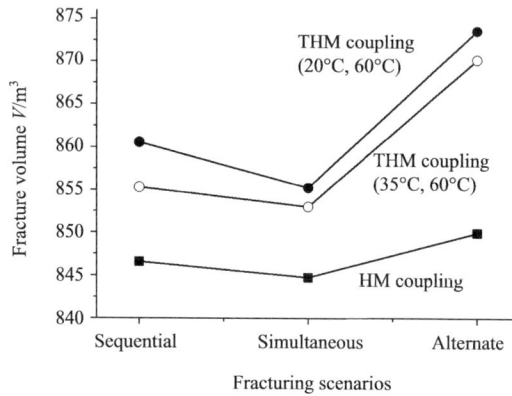

(a) Perforation cluster spacing of a =100 m.

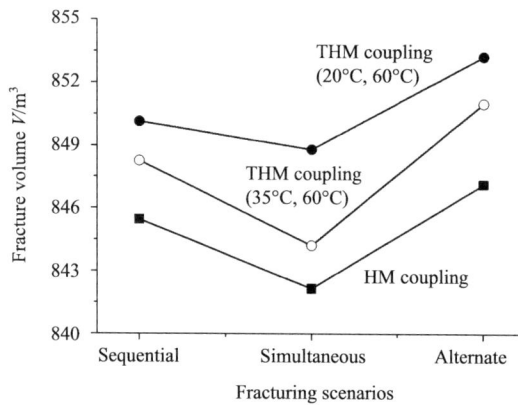

(b) Perforation cluster spacing of a =75 m.

Figure 9.14. Comparison of the fracture volumes in different fracturing scenarios.

9.6   Conclusions

The multistage fracturing of multiple perforation clusters in a horizontal well in a deep tight rock mass involves thermal diffusion, fluid flow, and rock matrix deformation between the fluid in the reservoir and the pores and fractures. In this study, the thermal-hydro-mechanical coupling effect is comprehensively considered to study the stress shadow effect and multiple fracture disturbance deflection behaviour in the 3D propagation of a hydraulic fracture network. The main conclusions are as follows.

   (1) A finite element-discrete element method was introduced to obtain a

high-precision stress solution and effectively describe the dynamic propagation of the fracture through the adaptive mesh refinement of the local region of the fracture tip, thereby providing an analysis strategy and solution scheme. We established an engineering-scale 3D numerical model for the multistage fracturing of a horizontal well. We computed and analysed the thermal diffusion effect of the 3D fracture propagation and disturbance behaviour of the fracture network propagation under different adjacent perforation cluster spacings (100 and 75 m) and different fracturing scenarios (sequential, simultaneous, and alternate fracturing) using typical cases.

(2) The stress disturbance region caused by the dynamic propagation of fractures exhibits superposition and overlay behaviours in multiple fractures, resulting in stress shadow effects and fracture deflection. The reduction in spacing between multiple perforation clusters in horizontal wells increases the stress shadow area and aggravates the mutual interference between fractures.

(3) Compared to sequential fracturing with multiple perforation clusters, simultaneous fracturing increases the stress shadow area, whereas alternate fracturing can reduce it. Alternate fracturing is an effective scheme for alleviating the disturbance of the 3D dynamic propagation of the fracture network and optimising the spatial fracture network morphology.

(4) Heat transfer between the fracturing fluid in deep tight rock fractures and the rock matrix was evaluated, and different temperature gradients (20 °C and 60 °C, 35 °C and 60 °C) and behaviours of thermal diffusion promoting fracture propagation were quantitatively analysed. The larger the temperature gradient between the fracturing fluid and rock matrix, the stronger the thermal diffusion effect. In sequential, simultaneous, and alternate fracturing, the thermal diffusion effect plays a significant role in improving the fracture propagation, and the fracture area and volume are larger.

This study investigated the influence of the stress and temperature fields on fracture propagation in the process of hydrofracturing; in particular, the varying area and maximum value of shear stress can reflect the mechanisms of fracture propagation to a certain extent. However, how the temperature field affects the fracture deflection requires further research, which will become the focus of future research.

References

Azadi, H., Khoei, A.R. (2011), "Numerical simulation of multiple crack growth in brittle materials with adaptive remeshing", International Journal for Numerical Methods in Engineering, Vol. 85 No. 8, pp. 1017–1048.

Bažant, Z. P., Salviato M., Chau V. T., Visnawathan H, Zubelewicz, A. (2014), "Why fracking works", Journal of Applied Mechanics, Vol. 81 No. 10, pp. 1–10.

ELFEN TGR user and theory manual. (2016), Rockfield Software Ltd., United Kingdom, 2016.

Ghaderi, A., Taheri-Shakib, J., Nik, M. (2018), "The distinct element method (DEM) and the extended finite element method (XFEM) application for analysis of interaction between hydraulic and natural fractures", Journal of Petroleum Science and Engineering, Vol. 171, pp. 422–430

Gutierrez, R., Sanchez, E., Roehl, D., Romanel, C. (2019), "XFEM modeling of stress shadowing in multiple hydraulic fractures in multi-layered formations", Journal of Natural Gas Science and Engineering, Vol. 70, pp. 1–15.

He, Q., Suorineni, F.T., Ma, T., Oh, J. (2017), "Effect of discontinuity stress shadows on hydraulic fracture re-orientation", International Journal of Rock Mechanics and Mining Sciences, Vol. 91, pp. 179–194.

Hossain, M.M., Rahman, M.K., (2008), "Numerical simulation of complex fracture growth during tight reservoir stimulation by hydraulic fracturing", Journal of Petroleum Science and Engineering, Vol. 60 No. 2, pp. 86–104.

Kresse, O., Weng, X., Gu, H., Wu, R. (2013), "Numerical modeling of hydraulic fractures interaction in complex naturally fractured formations", Rock Mechanics and Rock Engineering, Vol. 46 No. 3, pp. 555–568.

Kumar, D., Ghassemi, A.A. (2016), "Three-dimensional analysis of simultaneous and sequential fracturing of horizontal wells", Journal of Petroleum Science and Engineering, Vol. 146, pp. 1006–1025.

Lecampion, B., Desroches, J. (2015), "Simultaneous initiation and growth of multiple radial hydraulic fractures from a horizontal wellbore", Journal of the Mechanics and Physics of Solids, Vol. 82, pp. 235–258.

Manríquez, A.L. (2018), "Stress behavior in the near fracture region between adjacent horizontal wells during multistage fracturing using a coupled stress-displacement to hydraulic diffusivity model", Journal of Petroleum Science and Engineering, Vol. 162, pp. 822–834.

Manríquez, A.L., Sepehrnoori, K., Cortes, A. (2017), "A novel approach to quantify reservoir pressure along the horizontal section and to optimize multistage treatments and spacing between hydraulic fractures", Journal of Petroleum Science and Engineering, Vol. 149, pp. 579–590.

Paluszny, A., Tang, X.H., Zimmerman, R.W. (2013), "Fracture and impulse based finite-discrete element modeling of fragmentation", Computational Mechanics, Vol. 52 No. 5, pp. 1071–1084.

Profit, M., Dutko, M., Yu, J., Cole, S., Angus, D., Baird, A. (2016), "Complementary hydro-mechanical coupled finite/discrete element and microseismic modelling to predict hydraulic fracture propagation in tight shale reservoirs", Computational Particle Mechanics, Vol. 3 No. 2, pp. 229–248.

Sobhaniaragh, B., Mansur, W., Peters, F. (2018), "The role of stress interference in hydraulic fracturing of horizontal wells", International Journal of Rock Mechanics and Mining Sciences, Vol. 106, pp. 153–164.

Taghichian, A., Zaman, M., Devegowda, D. (2014), "Stress shadow size and aperture of hydraulic fractures in unconventional shales", Journal of Petroleum Science and Engineering, Vol. 124, pp. 209–221.

Wang, Y. (2021a), "Adaptive finite element–discrete element analysis for stratal movement and microseismic behaviours induced by multistage propagation of three-dimensional multiple hydraulic fractures", Engineering Computations, Vol. 38 No. 6, pp. 2781–2809.

Wang, Y. (2021b), "An h-version adaptive FEM for eigenproblems in system of second order ODEs: vector Sturm-Liouville problems and free vibration of curved beams", Engineering Computations, Vol. 38 No. 4, pp. 1807–1830.

Wang, Y., Liu, X. (2021), "Stress-dependent unstable dynamic propagation of three-dimensional multiple hydraulic fractures with improved fracturing sequences in heterogeneous reservoirs: Numerical cases study via poroelastic effective medium model", Energy & Fuels. Vol. 35 No. 22, pp. 18543–18562.

Wang, Y., Wang, J. (2022), "An hp-version adaptive finite element algorithm for eigensolutions of free vibration of moderately thick circular cylindrical shells via error homogenization and higher-order interpolation", Engineering Computations, Vol. 39 No. 5, pp. 1874–1901.

Wang, Y., Ju, Y., Zhuang, Z., Li, C. (2018), "Adaptive finite element analysis for damage detection of non–uniform Euler–Bernoulli beams with multiple cracks based on natural frequencies", Engineering Computations, Vol. 35 No. 3, pp. 1203–1229.

Wang, Y., Ju, Y., Chen, J., Song, J. (2019). "Adaptive finite element–discrete element analysis for the multistage supercritical CO_2 fracturing of horizontal wells in tight reservoirs considering pre-existing fractures and thermal-hydro-mechanical coupling", Journal of Natural Gas Science and Engineering, Vol. 61, pp. 251–269.

Wang, Y., Ju, Y., Zhang, H., Gong, S., Song, J., Li, Y., Chen, J. (2021a), "Adaptive finite element-discrete element analysis for the stress shadow effects and fracture interaction behaviours in three-dimensional multistage hydrofracturing considering varying perforation cluster spaces and fracturing scenarios of horizontal wells", Rock Mechanics and Rock Engineering, Vol. 54 No. 4, pp. 1815–1839.

Wang, Y., Duan, Y., Liu, X., Huang, J., Hao, N. (2021b), "Dynamic propagation and intersection of hydraulic fractures and pre-existing natural fractures involving the sensitivity factors: Orientation, spacing, length, and persistence", Energy & Fuels. Vol. 35 No. 19, pp. 15728–15741.

Wang, Y., Hu, J., Wang, J., Wu, J. (2022), "Adaptive mesh refinement for finite element analysis of the free vibration disturbance of cylindrical shells due to circumferential micro-crack damage", Engineering Computations, Vol. 39 No. 9, pp. 3271-3295.

Wong, S.W., Geilikman, M., Xu, G. (2013), "Interaction of multiple hydraulic fractures in horizontal wells", SPE Unconventional Gas Conference and Exhibition, Muscat, Oman, SPE-163982-MS.

Yoon, J.S., Zimmermann, G., Zang, A. (2015), "Numerical investigation on stress shadowing in fluid injection-induced fracture propagation in naturally fractured geothermal reservoirs", Rock Mechanics and Rock Engineering, Vol. 48 No. 4, pp. 1439–1454.

Zhang, X., Jeffrey, R. (2012), "Fluid-driven multiple fracture growth from a permeable bedding plane intersected by an ascending hydraulic fracture", Journal of Geophysical Research: Solid Earth, Vol. 117, pp. 1–12.

Zienkiewicz, O.C., Zhu, J.Z. (1992), "The superconvergent patch recovery (SPR) and adaptive finite element refinement", Computer Methods in Applied Mechanics and Engineering, Vol. 101 No. 1, pp. 207–224.

Chapter 10   Summary and prospect

10.1   Summary

The chapters of the book can be summarized as follows:

(1) In Chapter 1, the research background and significances of stress shadow effects and continuum-discontinuum methods for stress-dependent unstable dynamic propagation of multiple hydraulic fractures are well summarized and analysed. This section can provide a state-of-art review for the research of unstable dynamic propagation of multiple hydraulic fractures, and have a comprehensive grasp of the research in this field.

(2) In Chapter 2, a dual bilinear cohesive zone model based on energy evolution was introduced to detect the initiation and propagation of fluid-driven tensile and shear fractures. The model overcomes the limitations of classical linear fracture mechanics, such as the stress singularity at the fracture tip, and considers the important role of fracture surface behaviour in the shear activation. The bilinear cohesive criterion based on the energy evolution criterion can reflect the formation mechanism of complex fracture networks objectively and accurately. Considering the hydro-mechanical coupling and leak-off effects, the combined finite element-discrete element-finite volume approach was introduced and implemented successfully, and the results showed that the models considering hydro-mechanical coupling and leak-off effects could form a more complex fracture network. The musicale (laboratory- and engineering-scale) Mode I/II fractures can be simulated in hydrofracturing process. Based on the proposed method, the accuracy and applicability of the algorithm were verified by comparing the analytical solution of KGD and PKN models. The effects of different *in-situ* stresses and flow rates on the dynamic propagation of hydraulic fractures at laboratory and engineering scales were investigated. When the ratio of *in-situ* stress is small, the fracture propagation direction is not affected, and the fracture morphology is a cross-type fracture. When the ratio of *in-situ* stress is relatively large, the propagation direction of the fracture is affected by the maximum *in-situ* stress, and it is more inclined to propagate along the direction of the maximum *in-situ* stress,

forming double wing-type fractures. Hydrofracturing tensile and shear fractures were identified, and the distribution and number of each type were obtained. There are fewer hydraulic shear fractures than tensile fractures, and shear fractures appear in the initial stage of fracture propagation and then propagate and distribute around the perforation. The proposed dual bilinear cohesive zone model is effective for simulating the types of Mode I/II fractures and seizing the fluid-driven propagation of multiscale tensile and shear fractures. Practical fracturing process involves the multi-type and multiscale fluid-driven fracture propagation. This study introduces general fluid-driven fracture propagation, which can be extended to the fracture propagation analysis of potential fluid fracturing, such as other liquids or supercritical gases.

(3) In Chapter 3, a multi-thread parallel computation scheme for solid and fluid analysis was developed to improve computational efficiency in large-scale models and multi-physical field coupling (hydro-mechanical) problems, and fracture criteria via a dual bilinear cohesive zone model was introduced. Considering the hydro-mechanical coupling and leak-off effects, the combined finite element-discrete element-finite volume approach was introduced and implemented successfully, and a multi-thread parallel computation method and global procedure was proposed. This study provides valuable results for parallel computation of dynamic fluid-driven propagation of hydraulic fractures in porous elastic rock mass and lays the foundation for further development of efficient and reliable simulations for dynamic multiscale complex fractures under multiphysical field coupling.

(4) In Chapter 4, based on the development of combined finite element-discrete element-finite volume method, a novel heterogeneous continuum-discontinuum computation method and models are proposed to investigate the effects of deviation angle, distribution, and geomaterial property of bedding and granule on diversion and penetration behaviours of hydraulic fractures. The representative numerical cases and results are derived, and the results show that, once the hydraulic fracture encountered bedding, the closer the angle between the fracture tip and bedding is to 90°, the greater the possibility of hydraulic fracture penetrating bedding was; thus, smaller angles lead to a greater degree of inhibition of hydraulic fracture propagation by bedding. When the geomaterial properties of bedding are extremely strong or weak, the hydraulic fracture could not easily penetrate the bedding: strong bedding properties will hinder the penetration of hydraulic fractures; whereas when the bedding properties was weaker than that of the rock matrix, the hydraulic fracture could easily propagate along the weak bedding plane. Different granule distribution patterns not only affected the

fractures in contact with granules but also significantly impacted the overall number of fractures and the direction of fracture propagation. Once the granule property was strong, the hydraulic fracture did not easily penetrate the granule; when the granule property was weaker than that of the rock matrix, the hydraulic fracture easily penetrated the granule.

(5) In Chapter 5, the combined finite element-discrete element method and discrete fracture network model were used to simulate the hydrofracturing process of naturally fractured reservoirs by controlling different sensitive factors (orientation, spacing, length, and persistence) of natural fractures. To investigate the sensitive factors, typical numerical cases were carefully designed and established, and the results of the dynamic propagation and intersection of hydraulic fractures and pre-existing natural fractures were derived. Two typical types of fracture network morphologies are detected when hydraulic fractures intersect the natural fractures: centre-type (intersection of hydraulic fractures and crossed cluster of the natural fractures) and edge-type (intersection of hydraulic fractures and the edge of the natural fractures). Quantitative results of the length and volume of the fracture networks and gas production in enhanced permeability fractured reservoirs were analysed. The mechanisms by which the sensitivities of natural fractures affect and control the optimal fracturing behaviours are well understood; the conditions of small fracture spacing, large fracture length, or small fracture persistence of natural fractures are conducive for hydraulic fractures to intersect with the natural fractures and form a complex centre-type fracture network and improve gas production.

(6) In Chapter 6, the numerical models of the tight multilayered reservoirs containing multiple interlayers were established to study hydrofracturing of multiple perforation clusters and its influencing factors on unstable propagation and deflection of hydraulic fractures. Brittle and plastic multilayered reservoirs fully considering the influences of different *in-situ* stress ratio and physical attributes for reservoir and interlayer strata on propagations of hydraulic fractures were investigated. The combined finite element-discrete element method and mesh refinement strategy were adopted to guarantee the accuracy of stress solutions and reliability of fracture path in computation. Results show that the shear stress fields between adjacent multiple hydraulic fractures are superposed to cause fractures deflection. Stress shadows induce the shielding effects of hydraulic fractures and inhibit fractures growth to emerge unstable propagation behaviours, and a main single fracture and several minor fractures develop. As the *in-situ* stress ratio increases, hydraulic fractures more easily

deflect towards the direction of maximum *in-situ* stress, and stress shadow and mutual interaction effects between them are intensified. Compared to brittle reservoir, plastic-enhanced reservoir may limit fracture growth and cannot form long fracture length; nevertheless, plastic properties of reservoir are prone to induce more microseismic events with larger magnitude. The obtained fracturing behaviours and mechanisms based on engineering-scale multilayered reservoir may provide effective schemes for controlling and estimating the unstable propagation of multiple hydraulic fractures.

(7) In Chapter 7, the adaptive finite element–discrete element method is introduced to overcome the shortcomings of the traditional finite element method for simulate 3D fracture propagation, in which the local remeshing and coarsening strategy are used to guarantee the accuracy of solutions, reliability of fracture propagation path, and computation efficiency. The 3D engineering-scale numerical models involved the crucial hydro-mechanical coupling and of fracturing fluid leak-off are proposed to simulate the 3D multistage hydrofracturing, by which the evolution and disturbance in stress filed, and fracture interaction behaviours are computed. The fracture network and evolution of shear stress induced by multistage hydrofracturing considering varying perforation cluster spaces (i.e., 100 m, 75 m, 50 m, 25 m, and 12.5 m) and fracturing scenarios (i.e., sequential, alternate and simultaneous fracturing) are compared. Numerical results show that the stress shadow effects and fracture interaction behaviours will become intense and obvious once the propagation of fractures fractures get closer, causing by the superposition and reduction of fracturing-induced shear stress variation areas. When the perforation cluster spaces are very narrow, e.g. the values less than 25 m under the working conditions of the numerical cases investigated in this research, the alternate fracturing can bring more fracturing fracture areas and improve the fracturing effects than sequential and simultaneous fracturing.

(8) In Chapter 8, a 3D finite element-discrete element model considering fluid-solid coupling and fracturing fluid leak-off effect is proposed to simulate hydrofracturing, and an effective medium model to describe heterogeneous reservoir rock mass is established. Several typical fracturing sequences (sequential, alternate, simultaneous, and two improved alternate scenarios), perforation cluster spaces (100 m, 75 m, 50 m, and 25 m), and heterogeneous reservoir conditions were analysed. The spatial unstable propagation, stress field interference, and evolution behaviour of the pressure fracture network were simulated, and the quantitative results of the dynamic

propagation shape, fracture propagation area, and spatial volume of the pressure fracture network were obtained. The numerical results show that when the perforation cluster space decreases, the stress shadow effect between hydraulic fractures intensifies, and the 3D fracture exhibits unstable propagation and spatial deflection. Compared with sequential and simultaneous fracturing, all types of improved alternate fracturing can effectively reduce the stress interference between fractures, fracture deflection, and form the developed fracture network area. Compared with the homogeneous reservoir model, the heterogeneity of the reservoir rock mass becomes an important factor that limits the propagation of the fracture network. These results can provide a basis for understanding the mechanisms of stress-dependent unstable dynamic propagation of 3D multiple hydraulic fractures of improved fracturing sequences in heterogeneous reservoirs.

(9) In Chapter 9, a 3D engineering-scale numerical model is established under different fracturing scenarios (sequential, simultaneous, and alternate fracturing) and different perforation cluster spacings while considering the thermal-hydro-mechanical coupling effect. Stress disturbance region caused by fracture propagation in a deep tight rock mass is superimposed and overlaid with multiple fractures, resulting in a stress shadow effect and fracture deflection. The results show that the size of the stress shadow areas and the interaction between fractures increase with decreasing multiple perforation cluster spacing in horizontal wells. Alternate fracturing can produce more fracture areas and improve the fracturing effect compared with those of sequential and simultaneous fracturing. The larger the temperature gradient between the fracturing fluid and rock matrix, the stronger the thermal diffusion effect, and the effect of thermal diffusion on the fracture propagation is significant.

(10) In Chapter 10, all chapters in the book are summarized, and the prospect for future work is introduced.

10.2   Prospect

Based on the research work presented in this book, further work can be carried out:

(1) **Computational theory for hydrofracturing-induced multiscale damage and fracture evolution analysis under multiphases and multiphysical fields coupling conditions.** In the deep reservoirs, rock is influenced by the coupling effects of multiphysical fields, such as solids, fluids, temperature, and chemistry, and under complex conditions owing to the existence of multiple phases, such as solids, liquids,

and gaseous. Therefore, it is necessary to establish a refined computational theory for simulating these working conditions. To comprehend the real mechanical behaviours of deep reservoir rocks, it is necessary to characterize the continuous evolution of hydrofracturing-induced multiscale damages and fractures in the rocks and form the refined computational theories.

(2) **Three-dimensional adaptive finite element-discrete element algorithm for accurate and efficient continuous-discontinuous analysis of rocks.** To characterize the evolution of continuous stress fields and discrete fracture fields in deep reservoir rocks, it is necessary to establish numerical methods of continuous-discontinuous analysis and develop adaptive finite element–discrete element algorithms. The high-precision solutions of this algorithm should satisfy a pre-set tolerance and ensure an efficient computational process via mesh refinement. To characterize the three-dimensional occurrence state and the non-planar damage evolution and crack propagation behaviours of deep reservoir rock, a three-dimensional model to simulate rock analysis is needed and the need to establish a three-dimensional adaptive finite element-discrete element algorithm has become urgent.

(3) **High-performance parallel computing software for large-scale rock engineering problem computation and analysis.** To determine the damage evolution and fracture propagation behaviours of rocks under multiphysical fields, multiphases, and multiscale conditions, using the three-dimensional adaptive algorithm, a large amount of computation is needed. Thus, higher efficiency in computing is required. An important way to achieve this is to introduce parallel computing methods to enable multicore and multi-threaded parallel computing. To develop this proposed program into a computing platform that can be used by all kinds of users, it is necessary to develop a user-friendly computing software that is suitable for analysing the large-scale rock engineering problems.

Abstract

Unconventional new mining energy resources (i.e. coalbed methane, shale oil and gas, tight oil and gas) are exploited through hydraulic fracture network to enhance the permeability of reservoir rock mass. The recovery effects depend on the morphology and complexity of fracture network. The unstable propagation in the deflection, intersection and penetration of hydraulic fractures has become an important factor affecting the distribution form of fracture network, including many kinds of crucial influencing factors, which are not only affected by the heterogeneous attributes of bedding, granules and natural fractures in rock mass, but also disturbed by the adjacent multiple hydraulic fractures. Through the relationship between fracture propagation and induced stress field evolution, this book proposed the numerical methods, models and simulation investigations on unstable hydraulic fracture and stress disturbance in heterogeneous rock. The book covers the following main contents: (1) dual bilinear cohesive zone model-based fluid-driven propagation of multiscale tensile and shear fractures, (2) multi-thread parallel computation method for dynamic propagation of hydraulic fracture networks, (3) heterogeneous continuum-discontinuum computation method for dynamic diversion and penetration of hydraulic fractures contacting multi-layers and granules, (4) dynamic propagation and intersection of hydraulic fractures and pre-existing natural fractures involving the sensitivity factors: orientation, spacing, length, and persistence, (5) unstable propagation of multiple hydraulic fractures and stress shadow effects in multilayered reservoirs, (6) unstable propagation of multiple hydraulic fractures and shear stress disturbance in multi-well hydrofracturing, (7) unstable propagation of multiple three-dimensional hydraulic fractures and shear stress disturbance in heterogeneous reservoirs, and (8) unstable propagation of multiple three-dimensional hydraulic fractures and shear stress disturbance considering thermal diffusion.

Given its scope, the book offers a valuable reference guide for researchers, postgraduates and undergraduates majoring in engineering mechanics, civil engineering, mining engineering, petroleum engineering, and geological engineering.

编　后　记

　　"博士后文库"是汇集自然科学领域博士后研究人员优秀学术成果的系列丛书。"博士后文库"致力于打造专属于博士后学术创新的旗舰品牌,营造博士后百花齐放的学术氛围,提升博士后优秀成果的学术影响力和社会影响力。

　　"博士后文库"出版资助工作开展以来,得到了全国博士后管委会办公室、中国博士后科学基金会、中国科学院、科学出版社等有关单位领导的大力支持,众多热心博士后事业的专家学者给予积极的建议,工作人员做了大量艰苦细致的工作。在此,我们一并表示感谢!

"博士后文库"编委会